Invertebrate Receptors

SYMPOSIA OF THE ZOOLOGICAL SOCIETY OF LONDON

NUMBER 23

Invertebrate Receptors

(The Proceedings of a Symposium held at The Zoological Society of London on 30 and 31 May, 1967)

Edited by

J. D. CARTHY

Field Studies Council, London, England

and

G. E. NEWELL

Queen Mary College, London, England

Published for

THE ZOOLOGICAL SOCIETY OF LONDON

BY

ACADEMIC PRESS

1968

ACADEMIC PRESS INC. (LONDON) LTD
Berkeley Square House
Berkeley Square
London, W.1

U.S. Edition published by

ACADEMIC PRESS INC.
111 Fifth Avenue
New York, New York 10003

Library of Congress Catalog Card Number: 68–9105

PRINTED IN GREAT BRITAIN BY
R. MACLEHOSE AND CO. LTD,
THE UNIVERSITY PRESS, GLASGOW

CONTRIBUTORS

BAILEY, D. F., *Department of Pharmacy, University of Aston in Birmingham, Birmingham, England* (p. 263)

BARBER, V. C., *Department of Zoology, University of Bristol, Bristol, England* (p. 37)

BENJAMIN, P. R., *Department of Zoology, University of Durham, Durham, England* (p. 263)

CARTHY, J. D., *Field Studies Council, London, England* (p. 251)

FINLAYSON, L. H., *Department of Zoology and Comparative Physiology, University of Birmingham, Birmingham, England* (p. 217)

HODGSON, E. S., *Department of Biological Sciences, Columbia University, New York, U.S.A.* (p. 269)

HOWSE, P. E., *Department of Zoology, The University, Southampton, England* (p. 167)

KIRSCHFELD, KUNO, *Max-Planck-Institut für Biologie, Tübingen, Germany* (p. 165)

LAND, M. F., *Department of Physiology, University of California, Berkeley, California, U.S.A.* (p. 75)

LAVERACK, M. S., *Gatty Marine Laboratory, The University, St. Andrews, Scotland* (p. 299)

MILLOTT, NORMAN, *Department of Zoology, Bedford College, University of London, England* (p. 1)

NEWELL, G. E., *Queen Mary College, University of London, England* (p. 97)

NEWELL, P. F., *Westfield College, University of London, England* (p. 97)

SCHNEIDER, D., *Max-Planck-Institut für Verhaltensphysiologie, Seewiesen über Starnberg (Obb), Germany* (p. 279)

SHAW, S. R., *Gatty Marine Laboratory, and Department of Natural History, University of St. Andrews, Scotland* (p. 135)

SHAW, T. I., *Queen Mary College, University of London, England* (p. 63)

STEINBRECHT, R. A., *Max-Planck-Institut für Verhaltensphysiologie, Seewiesen über Starnberg (Obb), Germany* (p. 279)

THURM, ULRICH, *Max-Planck-Institut für Biologie, Tübingen, Germany* (p. 199)

WOLKEN, JEROME J., *Biophysical Research Laboratory, Carnegie-Mellon University, Pittsburgh, Pennsylvania, U.S.A.* (p. 113)

ORGANIZERS AND CHAIRMEN

ORGANIZERS

G. E. NEWELL and J. D. CARTHY, *on behalf of The Zoological Society of London*

CHAIRMEN OF SESSIONS

J. D. CARTHY, *Field Studies Council, London, England*
O. E. LOWENSTEIN, *Department of Zoology and Comparative Physiology, The University, Birmingham, England*
G. E. NEWELL, *Department of Zoology, Queen Mary College, London, England*

FOREWORD

The numerous phyla into which invertebrates are classified reflect the great range of body form and structure to be found among them, a range far greater than is to be found in the single phylum of the vertebrates. Thus to the comparative physiologist the invertebrates present an array of different mechanisms for the carrying out of the functions of life. Their variety of sense organs is an excellent example of this.

With advances in electrophysiological techniques and with new facts of fine structure from electron microscopy, fresh evidence is now continually forthcoming on the ways in which the essential process of transduction of environmental stimuli is carried out. In this symposium reception of the various major stimuli is reviewed and a range of sense organs throughout the invertebrates is described. This approach was chosen in preference to concentration on one sense organ or sense to the exclusion of the others, and in an endeavour to demonstrate the state of knowledge among invertebrates, other than insects, arthropod sense organs are not represented by an amount of material that is commensurate with the huge volume of work upon them.

June 1968

J. D. CARTHY
G. E. NEWELL

CONTENTS

Functional Aspects of the Optical and Retinal Organization of the Mollusc Eye

M. F. LAND

The Eye of the Slug *Agriolimax reticulatus* (Müll.)

P. F. NEWELL and G. E. NEWELL

The Photoreceptors of Arthropod Eyes

JEROME J. WOLKEN

Organization of the Locust Retina

S. R. SHAW

The *Musca* Compound Eye

KUNO KIRSCHFELD

The Fine Structure and Functional Organization of Chordotonal Organs

P. E. HOWSE

Steps in the Transducer Process of Mechanoreceptors

ULRICH THURM

Proprioceptors in the Invertebrates

L. H. FINLAYSON

The Pectines of Scorpions

J. D. CARTHY

Anatomical and Electrophysiological Studies on the Gastropod Osphradium

D. F. BAILEY and P. R. BENJAMIN

Taste Receptors of Arthropods

EDWARD S. HODGSON

Checklist of Insect Olfactory Sensilla

D. SCHNEIDER and R. A. STEINBRECHT

On Superficial Receptors

M. S. LAVERACK

Symp. zool. Soc. Lond. (1968) No. 23, 1–36.

THE DERMAL LIGHT SENSE

NORMAN MILLOTT

Department of Zoology, Bedford College, University of London, England

SYNOPSIS

This is arbitrarily defined as a widespread photic sense that is not mediated by eyes or eye-spots and in which light does not act directly on an effector. Its existence is revealed by indicator reactions and in a few instances by electrophysiological evidence.

Knowledge of the intimate mechanism has recently been advanced by attention to molluscs such as *Spisula* and *Aplysia*, the echinoid *Diadema* and to several species of crayfish. The study of *Diadema* in particular has revealed in the dermatoptic sense a pattern of nervous interaction which recalls that found in other sensory systems.

Though regions of high sensitivity have been circumscribed, skin receptors have proved exceedingly elusive, but it is suggestive that photosensitivity accompanies nerve concentrations within, or just below, translucent skin. The existence of photosensitive nerves has been shown by behavioural or electrophysiological means in a variety of animals not all of which possess the dermatoptic sense. The properties of such nerves have been extensively studied in *Aplysia*.

Aside from the possession of photoreceptor pigment, the neurons concerned exhibit little structural specialization at the microscopic level and much of the evidence of lamellate sub-microscopic structure in crayfish and *Diadema* is of dubious significance in photosensitivity. Clearer evidence of structural specialization appears in the accessory photoreceptive structures of cephalopods.

The photoreceptive surface of *Diadema* is elaborated by the presence of pigmentary effectors and iridophores, which may play accessory roles.

The identity of the photoreceptive pigments involved has not yet been established, though haem-protein and carotenoid-protein are implicated in the photosensitive neurons of *Aplysia*. Participation of carotenoids, or their derivatives, is indicated in the pallial nerve of *Spisula* and in the epistellar bodies and parolfactory vesicles of cephalopods. In *Spisula* different photoreceptive pigments mediate excitation and inhibition, the latter being primary.

The spectral sensitivity of the radial nerve of *Diadema* is maximal between 455 and 460 nm. This approximates to the absorption maximum of echinochrome when extracted from the radial nerves. Recent microspectrophotometric examination, however, shows that the pigment *in situ* absorbs maximally between 530 and 550 nm, so there is no evidence that it mediates neural photoreception.

Studying the spine reactions of *Diadema* has revealed the importance of photic inhibition and nervous interaction at both central and peripheral sites; and the electrical concomitants of both "on" and "off" reactions have been shown. Recently, evidence of slowly propagated potentials in the radial nerve, which inhibit the shadow response, has been obtained.

The dermal light sense may have a manifold significance, but it is most widely and characteristically involved in defensive shadow reactions, the ubiquity of which adds a new dimension to the significance of "off-responses". As developed in echinoids and molluscs, the sense is highly successful, which suggests it is no mere evolutionary relic.

INTRODUCTION

Sensitivity of the general body surface to light has been recorded in almost all phyla of the animal kingdom. It has been widely assumed that the receptors concerned are superficial and diffuse, but the terms "dermal" or "dermatoptic" are loosely used, for in most cases the location of the receptors is uncertain, and in some they are more deeply situated and localized. For the purpose at hand we shall regard it as sensitivity to electromagnetic radiation of wave lengths between 390 and 760 nm that is not mediated by eyes or eye-spots and in which light does not act directly on the effector.

In general the existence of the dermal light sense is revealed by clear and sometimes striking reactions to light in eyeless species. In other instances it has been claimed that blinding does not abolish or even impair some of the reactions to light. Again, it has been claimed that eyeless bivalves are no less sensitive to changes in light intensity than similar animals with eyes, and effects of light on the speed or orientation of locomotion in a variety of eyeless or blinded forms such as ammocoetes, fish and amphibians have been described. Similarly light affects the locomotion of planarians, and *Daphnia*, but not only through their eyes, though accurate orientation may depend on the eyes (Viaud, 1948).

The responses mediated through the dermal light sense may be widespread or localized in isolated or small groups of organs. They may be integrated as in the withdrawal reactions of coelenterates, tubicolous worms etc., kineses of cyclostomes, kineses or taxes of blind fish, and in tropisms. Coordination is sometimes impressive as in the covering reactions of sea urchins by which opaque covering is strategically placed to shade the photosensitive skin from intense light.

Responses are made to uniform, or directional, steady illumination or to changes in intensity, especially shadows.

In most cases such reactions are the only indications of dermal photosensitivity available and for this reason they have been termed indicator reactions (Millott, 1957a).

Manifestations of the sense have long been known, but many accounts of it are old, purely descriptive or based upon inadequate experimental analyses; thus as Médioni (1961) has indicated, responses to light have not always been distinguished from those to heat. They have been reviewed several times: by Willem (1891), Nagel (1896), Millott (1957a) and Steven (1963). Here we shall be concerned not so much with reviewing manifestations and their distribution as with what has been revealed in more recent work concerning the intimate physiological mechanisms involved. Unfortunately, apart from the isolated instances reported below, analysis refined by electrophysiological methods has not been undertaken. It is therefore salutary to reiterate a previous reminder (Millott, 1957a) that until nerve discharges have been shown to arise from photostimulation of dermal receptors, demonstration of the dermal light sense cannot be regarded as complete. Happily some progress has been made in the animals to which recent attention has been pre-eminently directed, namely, crayfish, echinoids and molluscs. In all these cases, evidence of the electrical correlates of behaviour is now to hand. In the echinoids much has been learned from dismembering more elaborate behaviour into simple reflexes by studying the responses of isolated fragments of the living sea urchin, concentrating attention on particular effectors, notably the spines. The dangers implicit in such convenient methods of analysis should be constantly in mind, for the responses are mere fragments of an integrated behaviour which is not always the simple sum of the reactions of isolated parts—despite von Uexküll's concept of these animals as a *Reflexrepublik* (Millott and Takahashi, 1963). Moreover, in the quest for simplification the receptor input has often been reduced to minute light spots, and the registering effectors to one or a very few spines. Under such conditions a misleadingly simple answer may be the outcome!

The importance of studying dermal photosensitivity has already been urged (Millott, 1957a) on the grounds that photoreception is likely to depend on a pattern of activity that is fundamentally similar over a wide range of animal organization. The quest for a common denominator may therefore yield important information. This statement, originally made with molecular events of photic excitation in mind, remains valid, but by a somewhat unexpected twist, recent work has so far served to emphasize its truth more particularly in relation to the nervous activity that underlies dermal photoreception.

THE INTIMATE MECHANISM

Photoreception

Identification of the receptors concerned is a primary requirement in the study of any sensory system, but those involved in the dermal light sense have proved singularly difficult to trace and in most cases they have eluded detection. This is all the more surprising when it is borne in mind that although the photosensitivity is characteristically diffuse, specially sensitive areas have frequently been identified. Thus the tail of ammocoetes is said to be more sensitive than the remaining body surface (Steven, 1951); the most sensitive region of *Ascidia* is located 1·0–1·5 cm from the tip of the oral siphon (Hecht, 1918); the internal surface of the siphon mid-way along its length is said to be most sensitive in *Mya* (Light, 1930). Similarly the branchiae of dorids (Crozier and Arey, 1919) and the tentacles of tubicolous polychaetes have been circumscribed as photosensitive areas (see Nicol, 1950). In two species of *Diadema* (Millott, 1954; Millott and Yoshida, 1960b; Yoshida, 1966) it has been possible to localize maximal sensitivity at the ambulacral margins. In this precise location is a structure of uncertain significance, briefly described by Yoshida (1966) in *Diadema*

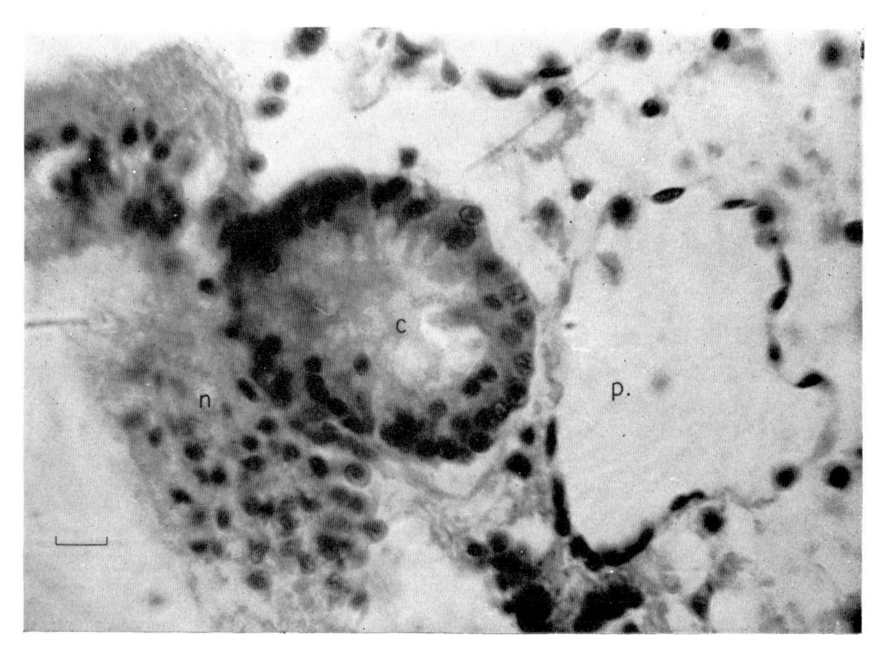

FIG. 1. The podial ganglion and associated structures in *Diadema antillarum*. *c*, cavity of pit, lined by truncated cells and containing secretion; *n*, concentration of nerve elements; *p*, one member of a pore-pair. Fixed: Bouin's fluid. Stained: Masson. Scale 8 μm.

setosum as an assemblage of nerve cells and fibres on the oral aspect of the pore-pairs that give rise to the podia. N. Millott and H. Okumura (unpublished) have recently re-examined the structure in *Diadema antillarum* and find that the nerve elements are associated with a pit open to the exterior and lined by a shallow epithelium of somewhat truncated cells, the processes of which appear to be embedded in secretion (Fig. 1).

Repeated attempts have been made to match the observed distribution of sensitivity with the occurrence of cell types suspected of being photoreceptors. Thus Light (1930) described numerous pear-shaped cells situated just beneath the inner epithelial layer of the siphons of *Mya*. He claimed that such cells are directly connected with siphonal nerves and moreover possess supposedly lens-like bodies and even a "retinella". But such cases fail to convince. Others have been more cautiously advanced. Thus Steven (1951) suspected that pigmented cutaneous cells supplying nerves to the lateral line might be the photoreceptors in ammocoetes. Again, where marked photosensitivity has been pinpointed, as in the podial ganglia of *Diadema setosum*, investigations with the electron microscope (Kawaguti and Ikemoto, unpublished) have shown the existence of lamellated or tubular structures in the ganglion cells, though with commendable caution, Yoshida (1966) has emphasized that it would be premature to attribute a photoreceptive function to them.

At this point it is appropriate to emphasize that in some instances where the dermal light sense has been revealed, well defined and functional eyes also exist. Thus in *Daphnia* according to Viaud (1948) both eyes and dermatoptic sense play their parts (p. 2). In other instances of co-existence their respective functions are not so clearly defined, but we should not be misled into believing that the apparent eyes are necessarily such and the complement of the dermal light sense. Thus in an unspecified species of *Asterias* as revealed by Hartline, Wagner and MacNichol (1952), there can be no doubt that the optic cushion is responsive to light, indeed until Takahashi's (1964) more recent demonstration (p. 9), the slow electrical potential generated by illumination of this organ was the only electrophysiological evidence of photosensitivity in echinoderms. Again, its fine structure, mode of development and affiliation with ciliated cells shows a fundamental specialization that is common to many photoreceptors (Eakin, 1963, 1965). Yet despite all this, the general situation revealed in starfish shows a disconcerting lack of clarity. In species such as *Asterias forbesi*, *Asterina gibbosa*, *Astropecten polyacanthus*, *Pentagonaster placenta*, *Crassispina*, and *Echinaster*, the optic cushions are said to play no important role, whereas in *Asterias rubens*, *Asterina gibbosa* and in *Astropecten irregularis*, other authorities have claimed that they mediate a directive effect of light (see Yoshida, 1966). Recently some clarification has been achieved by Yoshida and Ohtsuki (1966), who showed that removal of the optic cushions of *Asterias amurensis* does not abolish photosensitivity but diminishes it by a factor of ten. Moreover the action spectra of reflexes, obtained from intact arms and from those which have been deprived of their optic cushion, indicate the existence of dual photosensory systems (Fig. 8).

In *Ciona* and *Diadema* well defined cutaneous organs exist which superficially resemble eyes, and their structure has been so interpreted, but they do not function as such (Hecht, 1918; Millott, 1953), indeed those of the latter animal constitute, somewhat ironically, the least photosensitive areas of the body surface and no wonder, for they are not light absorbing structures, but light scattering iridophores. Their structure (see below) is elaborate (Millott and Manly, 1961) but bears no pertinent resemblance to that of a photoreceptor, moreover, it bears almost no resemblance to the structure of the same organs as described by Sarasin and Sarasin (1887) and interpreted as that of a compound eye!

In the absence of any clear sign of structurally differentiated cutaneous photoreceptors some workers have pointed to a direct responsiveness of effectors to light. Although such instances lie outside the arbitrary confines of this account, which is concerned with a sentient system, where an element of uncertainty exists, they merit brief attention. Thus North and Pantin (1958) who were unable to identify the receptors involved in the light-induced bending of *Metridium*, tentatively concluded that they were embodied in the effectors themselves, namely the endodermal musculo-epithelial cells. Again Passano and McCullough (1962) have shown a marked effect of light on the pacemaker systems of *Hydra*, that control the rhythmic body contractions. Light inhibits pacemaker activity, and changes its rhythm and location. The nature of the pacemakers and the photoreceptors is not defined, but it is hinted that light may affect them directly (McCullough, 1962).

Neural photosensitivity

Happily the foregoing disparate or inconclusive information is somewhat offset by recent findings particularly in echinoids and molluscs. It is at least suggestive that dermal photosensitivity often accompanies a significant concentration of nerve elements within, or just below, exposed translucent skin. Thus in *Spisula*, the pallial nerve near its emergence from the visceral ganglion is accessible to light entering the siphonal aperture (Kennedy, 1960). In *Diadema*, as in echinoids generally, a felt of fine, densely packed nerve elements lies at the base of the covering epithelium. Moreover, in *Diadema*, the felt is thickened at the particularly photosensitive ganglia (p. 4) where the branches of the radial nerves emerge to join the superficial system (Fig. 2). The fact that the distribution of photosensitivity and indeed the relative degree of such sensitivity, corresponds with the disposition and extent of superficial light accessible nerve, makes it only reasonable

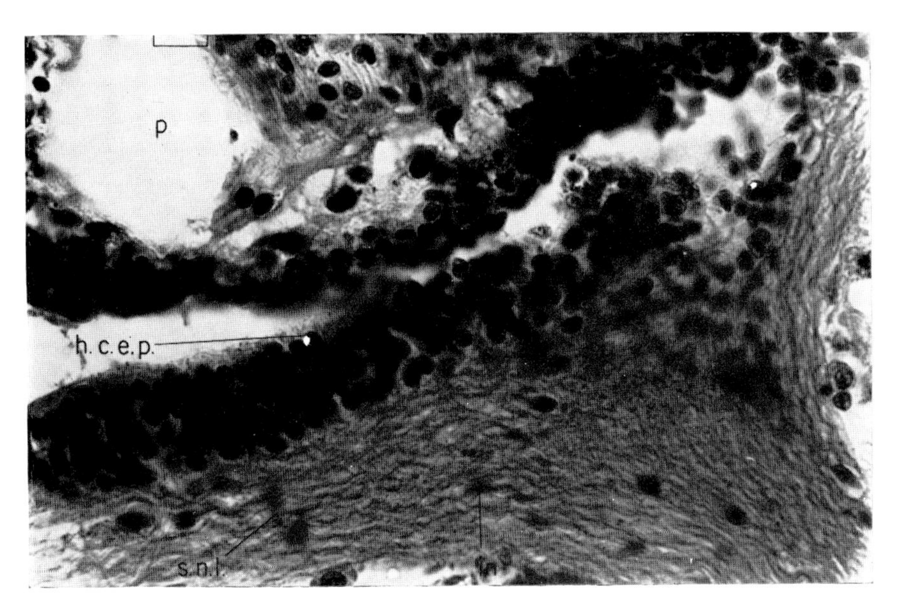

FIG. 2. Portion of a transverse section through the ambulacrum of *Diadema antillarum* showing the thickening (*h.c.e.p.*) of the superficial nerve layer (*s.n.*l.) where it is joined by the integumentary nerve (*i.n.*) at the ambulacral margin. Fixed: Bouin's fluid. Stained: Masson's argentaffine reaction, counterstained Mallory's triple stain. Scale: 12 μm. *p.*, podium. Reproduced with permission from Millott (1954).

to suspect that the nerve elements themselves are light sensitive. There is now abundant evidence that this is so, but the property is not always clearly related to the dermal light sense. Indications of such a property have been reported spasmodically over a long span of years, for example in minnows (von Frisch, 1911), in the sixth abdominal ganglion of crayfish (Prosser, 1934; Welsh, 1934), in ammocoetes (Young, 1935) and in ducks (Benoit and Ott, 1944). In fish belonging to sixteen different families (Motte, 1964) and in two species of *Rana* (Dodt, 1963) photosensitivity of the diencephalon, particularly in the pineal region has been shown, but in fish the central nervous system appears to be photosensitive in the trunk and tail regions also. Sensitivity may be surprisingly high, the threshold in blinded minnows being only some 10^3 times higher than that of human scotopic vision. The first substantial direct demonstration was provided by Arvanitaki and Chalazonitis (1949), who examined the photic responses of the visceral ganglion of *Aplysia*. They showed both excitatory and inhibitory effects of light in an extended electrophysiological study that is surpassingly elegant. The photosensitive neurons appear, however, to be too deeply situated to function normally as photoreceptors.

It would be surprising indeed if such neural photosensitivity had not somewhere been pressed into service for photoreception, by natural selection. A clear instance of this has been demonstrated by Kennedy (1960) by means of single fibre activity in the pallial nerve of *Spisula*. Such activity appears to arise as a spontaneous discharge in a single neuron located within 1 cm from the visceral ganglion, and the discharge is inhibited by light. In the echinoid *Diadema*, the early experiments indicating spatial coincidence of photosensitivity and nerve elements in the skin were extended and their implications strengthened by the demonstration of photosensitivity in the radial nerves (Millott, 1954). In this case the characteristic response to shadows provided the index to sensitivity, spine responses being elicited by illuminating small areas of the nerve *in situ* and then extinguishing the light to release the overt response. Subsequently a more rigorous experimental demon-

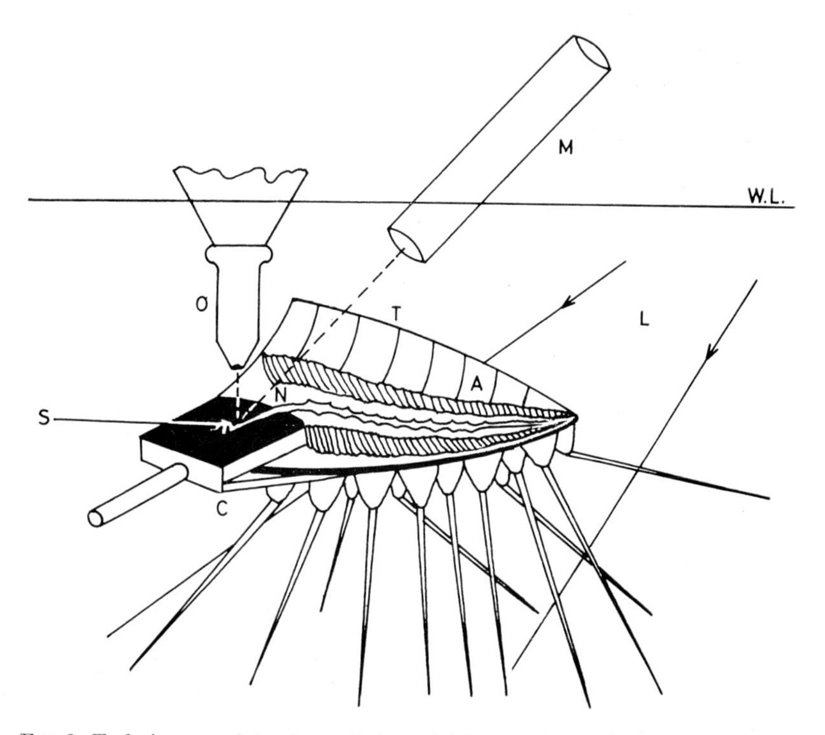

FIG. 3. Technique used to show photosensitivity in the radial nerve of *Diadema antillarum*. *A*, ampullae; *C*, clamp; *L*, light passing below preparation and casting a shadow of the spines, the movement of which is recorded in Fig. 4; *M*, microscope for viewing the stimulating light spot; *N*, radial nerve; *O*, objective lens of compound microscope which projects the light spot, *S*; *T*, piece of test; *W.L.*, water level. Reproduced with permission from Yoshida, M. and Millott, N. (1959).

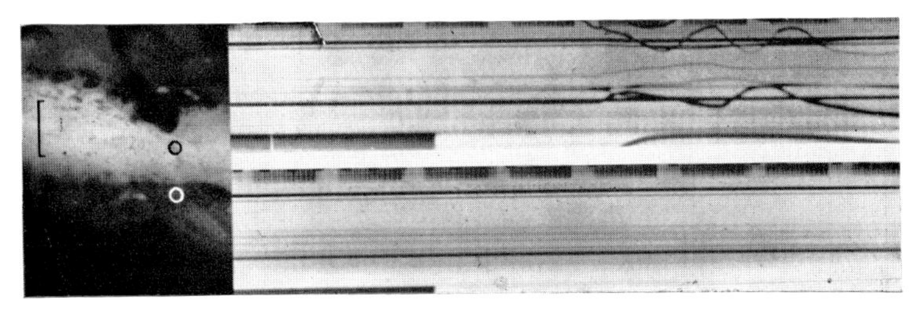

Fig. 4. Behavioural demonstration of photosensitive nerve in *Diadema antillarum*. Left. A light spot of the relative size shown by the black circle was projected on to the radial nerve (appearing as a pale band). Upper right. Spine response which follows cutting off the light at the moment indicated by the disappearance of the black band below the record. Lower right. Absence of the response when the experiment was repeated after the light spot had been moved to a position just outside the edge of the radial nerve (white circle on left). Time scale (above each tracing) in seconds. Scale (on left): 0·5 mm. Reproduced with permission from Yoshide, M. and Millott, N. (1959).

stration was provided by Yoshida and Millott (1959) using partially isolated regions of the nerve as shown in Fig. 3. The light spots employed were minute and the experimental conditions were such as to ensure that only the radial nerve was stimulated (Fig. 4). The implications of this behavioural index were subsequently fully vindicated by

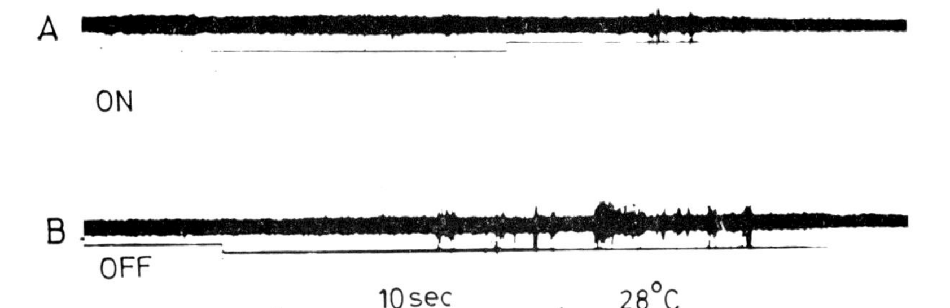

Fig. 5. Electrophysiological demonstration of photosensitive nerve in *Diadema setosum*. The photic responses of the isolated radial nerve. A, "on-response". Note brief discharge following illumination at rise of signal marker. B, "off-response". Note vigorous and prolonged discharge following the cessation of illumination at the fall of signal marker. Reproduced with permission from Takahashi (1964).

Takahashi's (1964) impressively skilful demonstration of its electrical concomitant (Fig. 5) in the isolated radial nerve.

There can therefore be little doubt that neural photosensitivity exists and the correspondence revealed in *Diadema* between the distribution of abundant superficial nerve and that of dermal photosensitivity (p. 6) inevitably implicates such nerve in photoreception. No stronger assertion is warranted for direct proof of photosensitivity in the cutaneous nerve is not yet forthcoming. That the radial nerve is photosensitive there can be scarcely any question, for Takahashi (1964) took the precaution of stripping away all adhering structures such as the coelomic canals from the radial nerve before conducting his experiments. Nevertheless this photosensitivity of the radial nerves is puzzling for it is most unlikely that any significant amount of light will normally reach them, being situated in the perivisceral coelom, screened by the test and its overlying densely pigmented skin on one side and by the deeply pigmented gut and mesenteries on the other.

In view of these findings it is not surprising that several workers have sought for the structural correlates of photosensitivity in the nervous system.

The earliest successful attempts are those of Arvanitaki and Chalazonitis whose findings in connexion with the giant neurons in the visceral ganglion of *Aplysia* have been summarized (Arvanitaki and Chalazonitis, 1961). Here the photoreceptive pigments (see below) are located in cytoplasmic organelles which they termed "grains" or lipochondria, arranged in layers at the periphery of the soma or in islets. They have isolated these bodies and estimate there are about 10^6 per neuron.

A high degree of specialization is not always to be expected. Thus in the case of *Spisula*, Kennedy (1960) did not discover any structural specialization in the neurons suspected of being photosensitive, indeed it was not definitely established whether they were first or second order. Nevertheless physiological specialization at least is implied by their complement of photoreceptive pigment and their characteristic behaviour (see below). Again, in the central nervous system of *Cambarus*, the photosensitive neurons discharge a dual function, being also secondary integrating units for tactile stimuli (Kennedy, 1963a).

Because there have been many clear demonstrations of an ordered, lamellate, sub-microscopic structure in photoreceptors, several workers have sought for corresponding appearances in photosensitive nerve. The quest so far has met with little success and the significance of the findings is not yet clear. Hama (1961) described lamellate bodies in the sheath cells surrounding the giant fibres in the ventral nerve cord of *Cambarus*. The lamellae, which appear to consist of unit membranes, are concentrically arranged forming a cortical zone to the sheath cell as well as separate systems of 10 to 50 parallel lamellae scattered through the cell. He was at pains to point out, however, that such appearances are not confined to the photosensitive regions of the nerve cord. Subsequently, Hermann and Stark (1962) have shown that these structures cannot be related to photoreception because the nerve discharges following photic stimulation of the sixth abdominal ganglion are carried in a few B-fibres, and not in giant fibres.

S. Kawaguti and Ikemoto (private communication) have found lamellated structures in what they have presumed to be sensory cells in the podial ganglion of *Diadema setosum*. Again, N. Millott and H. Okumura (unpublished) have found considerable numbers of lamellated bodies of several types in the radial nerves of *Diadema antillarum*. Because both radial nerves and podial ganglia are clearly photosensitive, the findings are, to a degree, suggestive, but some of the lamellated bodies in *Diadema antillarum* appear suspiciously like myelin figures, and the significance of those in *Diadema setosum* is open to question.

The situation remains puzzling in other respects. Thus Kawaguti, Kamishima and Kobashi (1965) do not mention the occurrence of lamellated bodies in their account of the fine structure of the radial nerves of echinoids. Though their description is primarily concerned with these nerves in *Hemicentrotus*, they state that the photosensitive radial nerves of *Diadema setosum* were also examined and show essentially the same structure. Further, "lamellate bodies", resembling in some respects those we have found in the radial nerve of *Diadema antillarum*, have been noted by Cobb and Laverack (1966) in close association with the lantern muscles of *Echinus*. Though these muscles or structures intimately associated with them are excitable by light in *Parechinus* (Boltt and Ewer, 1963), this property has not yet been demonstrated in *Echinus*.

In a recent communication Chalazonitis, Chagneux-Costa and Chagneux (1966), have extended the description of the pigment-bearing "grains" in the photosensitive neurons of *Aplysia*, by a brief report on their fine structure. The "grains" are described as bounded by a double membrane and as containing stratified lamellar systems formed in places by the alignment of granules 50 Å in diameter.

Although it is still too early to assess the significance of the various lamellated structures revealed in photosensitive nervous systems, the results of studying such systems have wider implications even if only of a salutary character. Thus at the moment, it would be unwise to insist that all photoreceptive elements have their origin in ciliated cells or structures like them, which, of course, is not to deny that many may be so derived. Again, it cannot be said that the study of neural photosensitivity is sufficiently ripe to add its quota of support to Wolken's (1962) hypothesis concerning the necessity for a crystal-like ultrastructure in photoreception. Nevertheless ideas as to photophysical mechanisms possibly involved in the photic excitation of *Aplysia* neurons have already been advanced (Chalazonitis, 1964).

It is now appropriate to mention briefly the accessory photoreceptive structures, namely the epistellar bodies and parolfactory vesicles, of cephalopod molluscs. Though these structures are intimately associated with the nervous system rather than part of it, arising independently in development (see Nishioka, Hagadorn and Bern, 1962), they perhaps qualify for inclusion in this account. This is because of their importance in the extra-ocular photoreception that is widespread among cephalopods (Messenger, 1967) and their significance as representing a measure of specialization intermediate between the structurally unspecialized neural photoreceptors of *Spisula* and the highly specialized photoreceptors of complex eyes. The accessory photoreceptors are deeply pigmented, neuron-like and produced into processes carrying densely and regularly packed microvilli (Nishioka *et al.*, 1962).

These bodies are accessible to light passing through the body wall directly or after it has entered the mantle cavity. Further, the stellate ganglion has the optical properties of a lens, which means that it is potentially capable of concentrating light on the epistellar body which it bears. It is suggested that the bodies may serve as light recorders and in photoperiodic regulation (Nishioka, Yasumasu and Bern; Nishioka, Yasumasu, Packard, Berne and Young, 1966).

Cells with structural features supposedly indicating a receptive capacity in connexion with photoperiodism, have also been described in intimate association with the brain of *Nereis pelagica* (Dhainaut–Courtois, 1965).

Elaboration of the photoreceptive surface

That the extensive and diffuse photoreceptive body surface of a creature such as an echinoid is structurally elaborate is obvious, but in *Diadema* two unusual features are present and both may be significant as accessories to the dermal light sense. They are the pigmentary effector system and the iridophores.

The former is of a type hitherto undescribed, in which pigment (melanin, lipofuscin and echinochrome) is moved within a network of cutaneous channels, apparently by the activities of their walls (Millott, 1964). In young individuals dark- and light-adapted phases occur, which are respectively pale and dark in colour. The pigment which is strongly light absorbing, accumulates with age, to form a permanent light screen, but the interambulacral areas often retain the capacity to disperse and concentrate their pigment.

Since the pigment channels lie outside much of the superficial nerve felt, the pigment and the degree of its dispersion should affect the intensity and spectral quality of the light which reaches the nerve and should affect the level of photosensitivity. Experiments, especially those with young forms, confirm this, darkly coloured phases being less sensitive than the pale (Millott, 1954). However, photochemical changes of this character are unlikely to be wholly responsible for the change in sensitivity and sensory adaptation is more than likely to be a significant factor. Yoshida (1966) has recently revealed some new aspects of the situation in *Diadema setosum*. Thus sensory recovery in darkness occurs

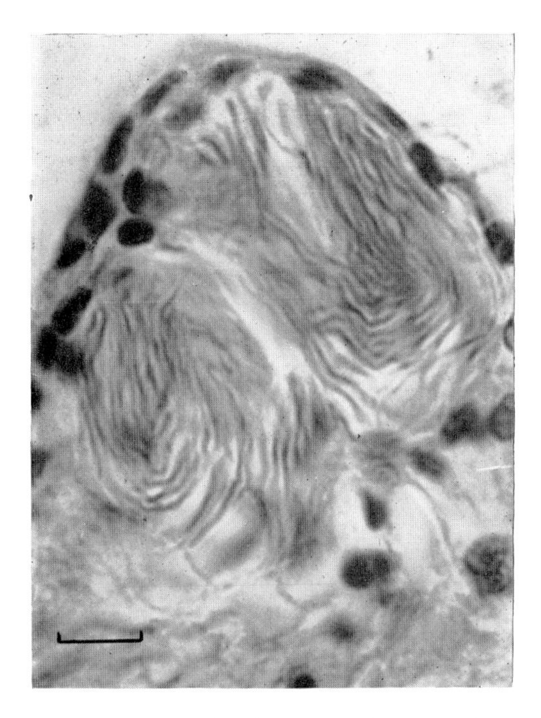

FIG. 6. Transverse section through a band of iridophores of *Diadema antillarum*, showing the systems of gelatinous plates, somewhat distorted by fixation, which appear like "finger-prints". Fixed: Bouin's fluid. Stained: Masson's argentaffine reaction and Mallory's triple stain. Scale: 10 μm.

much more quickly than pigment concentration and changes in sensory threshold may be great when those in pigment distribution are small.

The iridophores form remarkable light scattering structures extending over the skin as slightly elevated ridges running down the inter-ambulacral areas and around the periproct and spine bases. They consist of encapsulated systems of sheathed gelatinous plates (Fig. 6). Their microscopic structure has been described by Millott and Manly (1961) and their sub-microscopic structure by Kawaguti and Kamishima

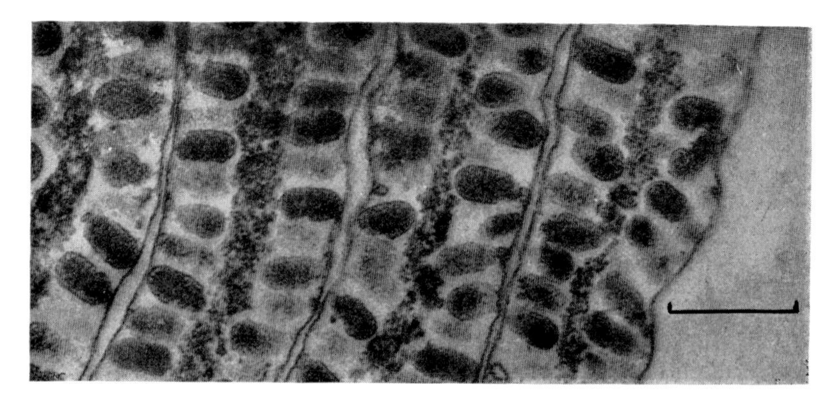

FIG. 7. Electronphotomicrograph of a portion of a section through four plates of an iridophore of *Diadema setosum*, showing the regular palisade of ellipsoids composing the sheath of each plate. Scale: 0·5 μm. Supplied by courtesy of Professor Kawaguti.

(1964), who showed the sheaths to be composed of a regular palisade of ellipsoids (Fig. 7). They reflect an intense blue colour, often visible in the deep recesses between the spine bases, which is probably produced by Rayleigh scattering (Millott and Manly, 1961) in the colloidal plates or by their sheath of regular ellipsoids (Millott, 1966). This could be significant, for the animal is most sensitive to blue light (see below).

Visual pigments

Since light destined to create a physiological effect must first be absorbed, the location and identification of the light-absorbing pigment is necessarily a cardinal problem in the study of any photosensory system. In the case of the dermal light sense, the quest for such information has met with scant success, which is not surprising where the photoreceptors are so diffuse and elusive. Much of the thinking behind the more recent approach to this problem appears to have been conditioned by spectacular researches on the visual pigments of eyes, due notably to Wald and his associates, and by an underlying belief in economy of mechanism. There is good reason for this when we reflect on the suitability of the conjugated double-bond structure of carotenoid molecules for charge transfer, as well as on the fact that visual pigments in the eyes of representatives of such diverse phyla as Arthropoda, Mollusca and Chordata have transpired to be haplo-carotenoid proteins. Nevertheless there appear to be no *a priori* grounds for thinking this way, since the only basic requirement in primary photoreception is a suitably placed light absorber appropriately linked with an electron transport system. For this, the intra-molecular gyrations so elegantly revealed by Wald in the photochemistry of derivatives of the vitamins A, are not a *sine qua non*, indeed haem-proteins can mediate a photic response in *Aplysia* neurons (see below). But even with an approach of this width, nothing corresponding to a well authenticated pigment responsible for primary photoreception in the dermal light sense has yet been forthcoming, though a lively interest in the quest for one is sustained (Fox and Hopkins, 1966).

Carotenoids, sometimes linked to protein, abound in the skin of some of the animal types, notably starfish, in which dermal photosensitivity has been described (Vevers, 1966). Moreover, the same carotenoids (β-carotene and esterified astaxanthin) are present in the skin of *Marthasterias* (Vevers and Millott, 1957) as are present in the optic cushions (Millott and Vevers, 1955), but there is as yet no evidence of their participation in photoreception. Rockstein and collaborators (1957, 1958, 1960) claim to have found a photosensitive pigment in both optic cushions and integument of *Asterias forbesi*. It was not identified, but it absorbed maximally at 525 nm and was obtained by extraction with digitonin. It is stated to regenerate during short exposures to light of wave-lengths netween 600 and 700 nm and to be associated with protein. Yoshida and Ohtsuki (1966), however, could find no evidence of a similar pigment in either podia or optic cushions of *Asterias amurensis*, though they found several carotenoids in the latter. Evidence as to the presence of vitamins A in starfish is contradictory (Vevers, 1966).

In view of the evidence of neural photosensitivity already presented, it is logical to seek evidence of corresponding neural pigmentation. Unfortunately the two best authenticated instances, the deeply pigmented visceral ganglion of *Aplysia* and the accessory photoreceptive vesicles of cephalopods, do not wholly qualify for inclusion in this account, for reasons already stated. Nevertheless it is of some interest to record that the pigments responsible for the photic responses of the visceral ganglion of *Aplysia* are a haem-protein and at least two carotenoid-proteins. The former has three main absorption bands at 418, 542 and 579 nm, the latter show maxima at 463 and 490 nm (Arvanitaki and Chalazonitis, 1949; 1961). The two mediate changes of the cell membrane potential that are opposite in sign. Photic excitation of the haem-protein generally depolarizes the membrane, whereas excitation of the carotenoid-protein usually hyperpolarizes it. In the instance of the epistellar bodies and parolfactory vesicles of cephalopods, vitamin A_1 and considerable quantities of a pigment with the absorption spectrum of rhodopsin can be extracted (Nishioka, Yasumasu and Bern, 1966).

Although Kennedy (1960) did not succeed in identifying the photoreceptive pigments in the nerves of *Spisula*, he advanced some useful information. Thus microspectrophotometric examination of living nerve yielded an absorption spectrum indicating a haem-protein. The absorption spectrum of the extracted pigment substantiated this view.

In other cases indications of the type of pigment involved in the skin or associated nerve have been sought by determining the action spectra of behavioural responses, or of nerve discharges. Unfortunately many of these determinations are of little value because investigators have failed to take into account the differing energy content of the various wavelengths they used. Some of the more reliable and useful results are reviewed by Steven (1963) who drew attention to some similarity in the shape of the curves obtained by various investigators for action spectra in such diverse organisms as *Cerianthus*, *Pholas*, *Diadema* and *Myxine* and to a range of maxima (450–550 nm) which is commensurate with that recorded for the absorption of visual pigments in a variety of organisms. Subsequent studies in *Hydra* extended the range in one direction to 400 nm (Singer, Rushforth and Burnett, 1963).

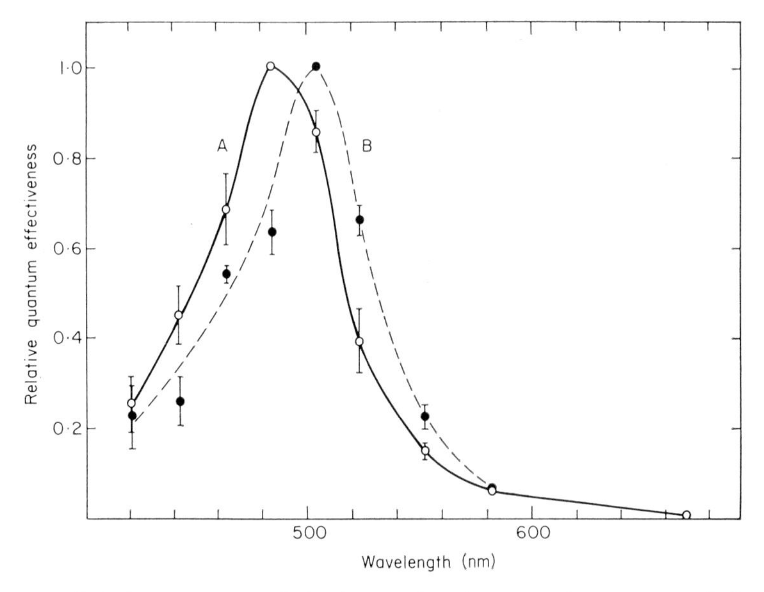

FIG. 8. Relative quantum efficiencies of the photoreceptive systems of *Asterias amurensis*: obtained from arms with (A) and without (B) optic cushions. Vertical bars show ranges of standard error. Reproduced with permission from Yoshida, M. and Ohtsuki, H. (1966). Copyright 1966 by the American Association for the Advancement of Science.

Experiments with blinded minnows (Motte, 1964) show that photosensitivity of the head region is maximal at about 530 nm, the curve of the action spectrum suggesting participation of a pigment of the porphyropsin type. Similarly the inhibition of sustained action potentials in the exposed diencephalon of frogs is maximal with light of 560 nm (Dodt, 1963). The more recent experiments of Yoshida and Ohtsuki (1966), referred to on p. 5, show the dermatoptic sensitivity of *Asterias amurensis,* revealed in terms of relative quantum efficiency, to be maximal at 504 nm. When the optic cushions are present maximal sensitivity is shifted to 485 nm (Fig. 8). What may justifiably be read into these various findings is a matter of argument, but it would not be surprising if, in the end, there proved to be a common photochemical basis vested in haplocarotenoid-proteins.

The results obtained by Kennedy (1960) are particularly interesting. He determined the spectral sensitivity function of the pallial nerve photoreceptor in *Spisula,* using its discharges, which yield a curve for photic inhibition with a bell shape and a maximum at 540 nm. Again Bruno and Kennedy (1962) showed that the photoreceptor neuron in each lateral half of the nerve cord of *Cambarus* has a similarly shaped spectral sensitivity curve with a peak at 500 nm, indicating the presence in these cells of a rhodopsin differing from that in the eyes. Larimer, Trevino and Ashby (1966), have shown that the caudal photoreceptors of the epigeal crayfish *Procambarus simulans, Procambarus clarkii* and *Orconectes virulis* have identical spectral sensitivity functions with maxima at 502 nm, whereas that of the caudal photoreceptor in the blind cavernicolous *Orconectes pellucidus australis* is shifted slightly toward the blue (λ_{max} 497 nm). In this respect therefore, the situation differs from that in the compound eyes, where different species of crayfish have evolved difierent photoreceptive pigments.

The findings in *Spisula* have an added significance. The response studied was the spontaneous discharge in the dark of a single neuron which was abruptly inhibited by illumination. Cessation of illumination is followed by a more vigorous off-discharge. When the maximum impulse frequency of the off-discharge and the duration of the inhibitory interval produced by monochromatic light, are related to the intensity used, the relationship shows marked differences depending on wavelength. Thus when the effects of the intensity of light of 500 nm and 600 nm on the inhibitory interval are compared, it is seen that above a certain intensity, they exert opposite effects. This suggests that more than one photoreceptive pigment is involved. The idea was substantiated by selective adaptation. Thus when exposure to monochromatic light at 475 nm is followed by stimulation at 630 nm an on-discharge as well as an off-discharge is revealed. The response therefore contains both excitatory and inhibitory components which appear to be mediated by different photoreceptive pigments because the response parameters are wavelength specific and a response to blue light cannot be matched by a red stimulus at any intensity. Kennedy therefore suggests that the inhibitory response has a lower threshold and dominates while the light is on, but when stimulation ceases, the excitatory event is revealed because of its longer time course. It is proposed that the two events

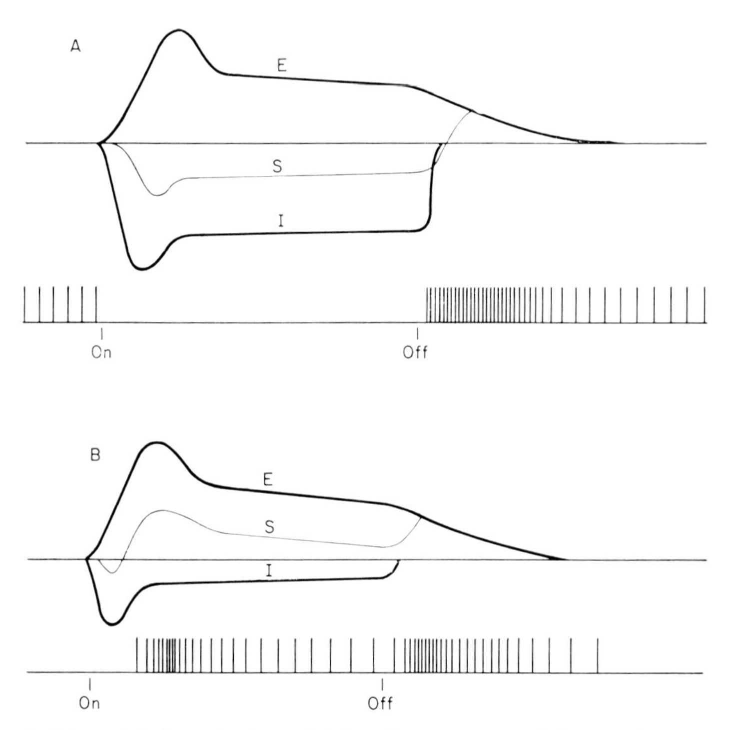

Fig. 9. Kennedy's hypothesis explaining the responses of the receptor neurons in *Spisula* to white (*A*) and red (*B*) light.

In *A* the action of white light generates both inhibitory (curve *I*) and excitatory (curve *E*) potentials, the former of which dominates, so that the sum of the potentials (curve *S*) falls below the firing threshold (base line) for impulse generation. The steady ("dark") discharge of impulses (shown below the figure), is inhibited after the light is admitted at "ON". When the light is extinguished (at "OFF") the excitatory effect (*E*) is revealed because of its slower decay, raising the net potential above the firing threshold, and so generating a burst of impulses at high frequency.

In *B*, the nerve is illuminated with red light after prolonged adaptation to blue. Depletion of a photosensitive pigment by blue light reduces the inhibitory response (*I*) so that the net potential (*S*) quickly rises above the firing threshold, generating impulses at both "ON" and "OFF". From *Invertebrate photoreception* by Kennedy, D.

(presumably potentials) interact and sum algebraically as shown in Fig. 9. The spectral sensitivity function of inhibition suggests that this component may be mediated by carotenoid-protein as in *Aplysia*, though not the same ones for both the sensitivity maxima and the absorption maxima in *Aplysia* are different. Whether the haem-protein present in the nerve (p. 16) of *Spisula* mediates excitation, is open to question.

The inhibitory response in *Spisula* appears to be a first-order effect, because only one neuron may be present in the photosensitive region of the pallial nerve, whereas if the observed discharges were assumed to arise secondarily by synaptic action and to occur in a post-synaptic element, at least three neurons would be necessary.

Successive determinations of the spectral sensitivity in echinoids have been made with increasing accuracy. Hess (1914) claimed that *Centrostephanus* was most responsive to blue-green light. Millott and

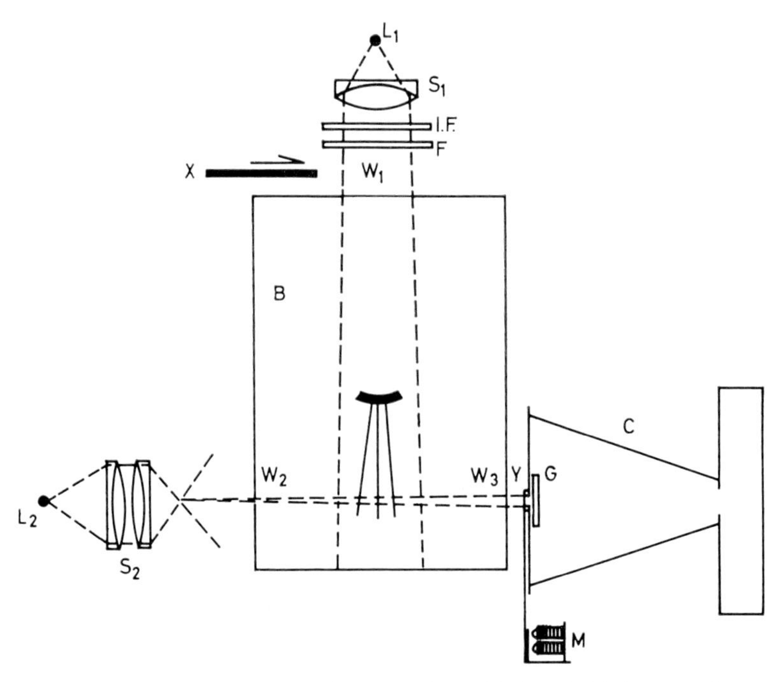

FIG. 10. Method of determining the spectral sensitivity of *Diadema antillarum*, by the relative minimum amounts of energy at various wavelengths necessary to elicit a response to a standard shadow. *B*, tank; *C*, camera; *F*, neutral density filter; *I.F.*, interference filter; *G*, ground glass screen; L_1, L_2, light sources; *M*, time and stimulus signalling device; S_1, S_2, lens systems; W_1, W_2, W_3, windows in opaque sides of tank; *X*, shutter; *Y*, slit. Reproduced with permission from Millott, N. and Yoshida, M. (1957).

Yoshida (1956) showed by means of the withdrawal of podia to shading that *Psammechinus* was most sensitive between 440 and 560 nm. The spectral sensitivity of *Diadema* was investigated by two methods; in both the radial nerve was used. In the earlier method (Fig. 10) sensitivity was determined on a basis of the relative minimum amounts of energy supplied at various wavelengths, that were necessary to elicit the reaction of a small group of spines to a standard shadow (Millott and Yoshida, 1957). The later more accurate method (Fig. 11) was based on determining the relative effectiveness of instantaneous changes from white light to that of various colours, in eliciting the shadow response of a single spine (Yoshida and Millott, 1960), changes to wavelengths at

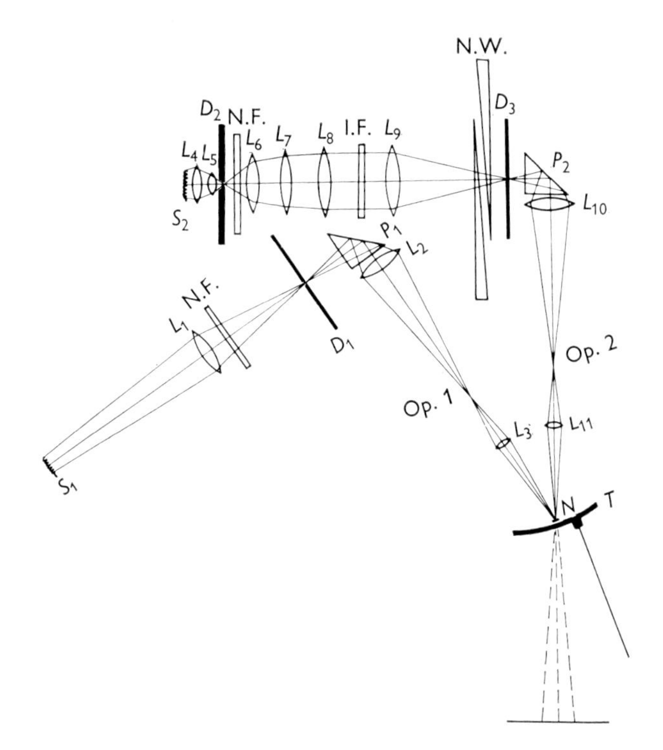

FIG. 11. Method of determining the spectral sensitivity of *Diadema antillarum* by comparing the effectiveness in eliciting a shadow response, of changing from white light to that of various wavelengths. The lower optical path (Op. 1) is that of white light, of constant intensity. The upper optical path (Op. 2) is that of coloured light produced by the interference filter (*I.F.*). The intensity of this beam is controlled by neutral wedges (*N.W.*), so as to be less than that of the white light by an amount just adequate to elicit a shadow response on an instantaneous change from the white to the coloured light. D_1, D_2, D_3, diaphragms; L_1–L_{11}, lenses; *N*, radial nerve; *N.F.*, neutral filter; P_1, P_2, prisms; S_1, S_2, tungsten filament lamps; *T*, piece of test with spine which gave the shadow reaction. Reproduced with permission from Yoshida, M. and Millott, N. (1960).

which the animal is more sensitive being correspondingly less effective in eliciting a shadow response. The results from the two methods agreed reasonably well, the earlier method showing the maximum sensitivity to be at 465 nm and the later showed it to be between 455 and 460 nm (Fig. 12).

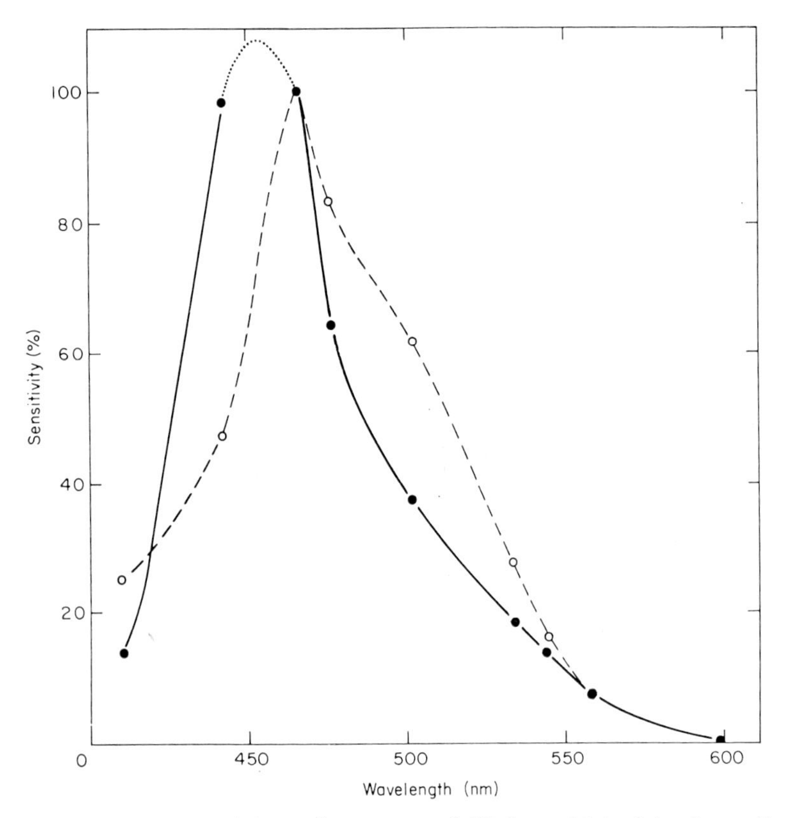

FIG. 12. Comparison of the action spectra of *Diadema* obtained by the methods shown in Figs. 10 (broken line) and 11 (solid line). Reproduced with permission from Yoshida, M. and Millott, N. (1960).

Two points worthy of comment emerge from these determinations. First, the spectral region of maximum sensitivity of the radial nerve corresponds fairly closely with that of the pigmentary effector units in the skin, which respond most effectively to light of 468 nm (Yoshida, 1957). Second, it corresponds with the region of maximum absorption (465–470 nm) in the visible range, shown by the ubiquitous echinochrome pigment when extracted from the skin and radial nerves by acidified diethyl ether, (Millott 1957b). Due caution should be exercised in assessing the meaning of this and it is doubtful whether the correspondence has the significance that could so easily be read into it. Thus not only were the action spectra of the pigmentary effector system and the spine response determined in different species of *Diadema*, but the absorption spectrum was determined with extracted pigment. The colour of echinochrome varies with pH (Ball and Cooper, 1949) and with its state of combination (Wallenfels, 1941). It is therefore significant that the spectral absorption of the radial nerve pigment when *in situ*, as recently determined microspectrophotometrically by N. Millott and H. Okumura (unpublished), is maximal between 530 and 550 nm. Moreover most of the echinochrome found in the radial nerve is due to amoebocytes or to their disorganized remains and there is as yet no clear evidence of its presence in neurons (N. Millott and H. Okumura, unpublished).* Although the results of the elegant analysis of pigmentary effector action in *Diadema setosum* provided by Yoshida (1956, 1957, 1960) are not called into question, following the work of Millott (1964, 1966) on the allied species, they may require some re-interpretation, for in *Diadema antillarum* the precise nature of the effectors and their co-ordination has become a wide open question.

THE IMPORTANCE OF NERVOUS INTERACTION

The importance of nervous interplay in sensory mechanisms has become increasingly evident (Granit, 1947, 1955; Barlow, 1961; Hartline, Ratliff and Miller, 1961; Rushton, 1963). The work of von Buddenbrock (1930) on the shadow reactions of barnacles signalled its possible importance in the dermal light sense. In echinoids this importance has now been clearly revealed by the work of Millott, Yoshida and Takahashi referred to below. In the context of modern neurophysiological concepts this development was inevitable, for in the pioneering work (see Hecht, 1934) and in much of that which has followed from other investigators, the intimate mechanism of the dermatoptic sense was interpreted largely in terms of photochemical models. But Granit has reminded us that in vision at least, there is much more than photochemistry! However, caution is necessary in comparing the findings in one group of animals with those in another. Many of Hecht's impressively stimulating ideas were founded on his studies of the behaviour of bivalve molluscs such as *Mya* and *Pholas* following photic

* Since the above was written, reddish cells resembling neurons have been found in the radial nerve by N. Millott and H. Okumura (unpublished). These cells absorb maximally in the visible range between 540 and 560 nm so that they do not appear to be concerned in the shadow response.

stimulation of their siphons, and judging from Kennedy's findings in *Spisula*, secondary nervous effects may be unimportant in generating or moulding the response that arises from photic stimulation of these animals. The same cannot be said of echinoids.

In this context it is profitable to consider the reflexes executed in response to changes in light intensity by *Diadema*. Responses to shading are characteristic of animals with a dermatoptic sense and those of *Diadema* appear in effectors such as podia and pedicellariae, but they are especially striking in the spines in which the shadow reaction appears as a sharp jerk, normally followed by a more or less protracted oscillation. They were originally examined by von Uexküll (1900) who incorporated his findings into an elusive blend of mechanism and vitalism for which it is hard to find modern physiological equivalents. Since then they have been re-examined by Millott and collaborators and by Yoshida.

Because spine reactions follow both increases and decreases in light intensity they recall "on" and "off" responses (Millott, 1954), but there was a difficulty in making such comparisons because of a sharp distinction drawn by von Uexküll (1900), who claimed that the reactions were different reflexes, those to shading involving the radial nerves, those to increased lighting only the superficial centres. The distinction was unfortunately misleading and threw investigators off the scent of the trail until it was invalidated by Millott and Yoshida (1959), who showed that both reflexes involve the radial nerves.

In a shadow reaction two primary factors must be considered, light as an energy-bearing stimulus and the effect of cutting it off. In the past reactions have sometimes been described as dependent on the parameters of lighting or shading without circumspection in defining what constitutes the stimulus. Here one must be careful not to play the words or to try to apply the empirical approach of Hecht's earlier work to something which is intrinsically more complex than can be represented adequately by simple thermodynamic equations. Had it proved possible to attack the problem by electrophysiological techniques it might have been solved more quickly and completely, but echinoids do not yield readily to such attack. Accordingly Millott and Yoshida (1960a) tackled the problem by studying the responses of individual spines to standardized lighting and shading of the radial nerve. Happily such responses prove reasonably constant in easily comparable features such as latency, duration, amplitude and frequency of oscillation. It is thus possible to determine the effects on each of them, of varying separately light and shade.

In the case of lighting, its intensity and duration exert similar and significant effects on all of these response parameters, whereas in the case of shading they do not. Thus although the proportionate decrease in intensity of illumination (i.e., shading intensity) affects the *whole* reaction, the duration of diminished lighting does not affect the early part of the reaction. Thus the amplitude of the first contraction is unaffected and the latency is affected only near the threshold and then but slightly and erratically (Fig. 13).

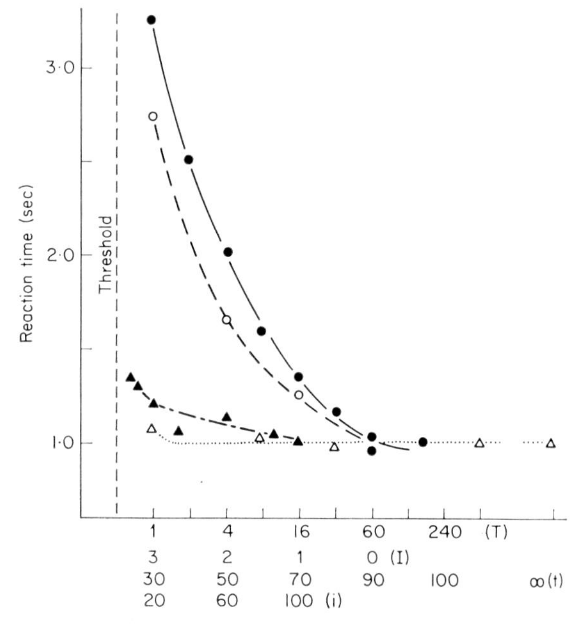

Fig. 13. Comparison of the affects of lighting and shading on the latency of the shadow response of *Diadema antillarum*.
Curve ● —— ●, abscissae (T), duration of lighting in seconds.
Curve △ . . . . . . △, abscissae (t), duration of shading in milliseconds.
Curve o – – – o, abscissae (I), intensity of lighting in arbitrary logarithmic units.
Curve ▲ — · — ▲, abscissae (i), intensity of shading (percentage decrease in intensity of light projected on to the radial nerve).
Reproduced with permission from Millott, N. and Yoshida, M. (1960a).

Since light supplies energy, the observed relationship between its intensity and duration is only to be expected, but what happens in the case of shading? Here two features are suggestive. First, the fact that only the later part of a reaction is susceptible to the effect of varying the duration of shading, suggests that once the reaction has begun, the continued absence of light has some special significance. Second, it is much more difficult to elicit an overt response by admitting light than by cutting it off. The two are compatible if it be supposed that light exerts an inhibitory as well as an excitatory effect and that inhibition, usually over-riding, persists until the light is cut off, when a reaction is released which is in proportion to the preceding illumination. So long as shading is not total the reaction will be inhibited to a degree that depends on the intensity of light remaining (Fig. 14). The shadow therefore interrupts the inhibiting effect of light, precipitating something analogous to a post-inhibitory rebound. When light is re-admitted its over-riding inhibitory effect is again manifest.

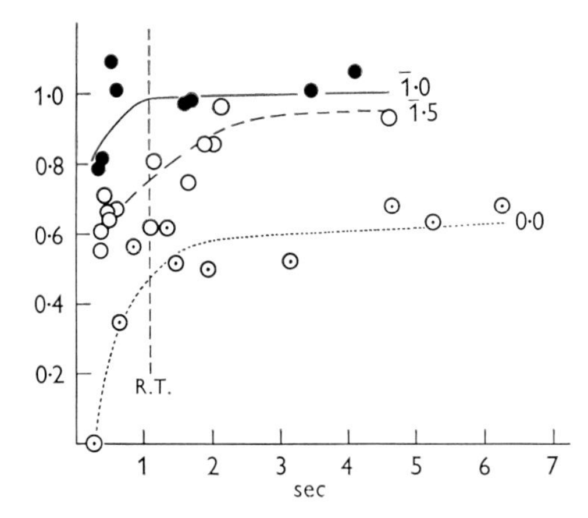

Fig. 14. The effect on the duration of the shadow response of *Diadema antillarum*, of re-admitting light at various intensities. Abscissae, duration of shading in seconds. Ordinates, ratio of the duration of the response to that of a control response during which no light was re-admitted. The vertical dotted line shows the latency (*R.T.*). The figures alongside each curve show the relative intensity of light re-admitted, expressed in arbitrary logarithmic units. Reproduced with permission from Millott, N. and Yoshida, M. (1960b).

Initially such a hypothesis demands substantiation by a direct demonstration of an inhibitory effect of light. This was achieved, so far as the radial nerve is concerned, by Millott and Yoshida (1960b) who obliterated light spots 1·0 mm in diameter projected on to it so as to elicit a shadow response and then, after a finite interval (usually 38 msec), re-admitting light to show a suppressive effect in proportion to its intensity (Fig. 15). Further, since the two interacting light spots could be separated spatially by distances up to 6·0 mm, nervous interaction is clearly indicated. Experiments in which the light spots were made to interact after transection of the radial nerve or its side branches, showed that interaction could occur both within and outside the nerve. Interaction of a similar kind occurring in the skin was shown

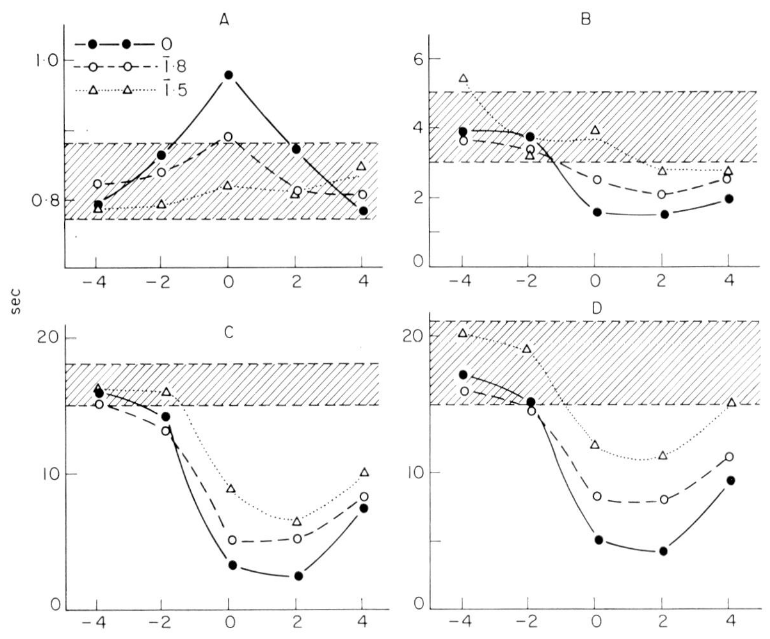

FIG. 15. The effect on the reaction released by shading the radial nerve in *Diadema antillarum*, of re-illuminating the nerve at various intensities and at several positions. The shaded area shows the maximum range of variation in control reactions during which no light was re-admitted after shading. Abscissae in all cases show the position at which light was re-admitted: position 0 is mid-way between ambitus and periproct, the positive and negative values indicating linear separation (in mm) from this in oral and aboral (–) directions. Ordinates: *A*, reaction time in seconds; *B*, height of first contraction in arbitrary units; *C*, total number of contractions recorded; *D*, duration of reaction in seconds. Relative intensities (in arbitrary logarithmic units) are shown by the conventions indicated at top left. Reproduced with permission from Millott, N. and Yoshida, M. (1960b).

by projecting the two light spots one after the other on to the outside surface. The shadow reaction is therefore moulded by events in separated receptive regions, both in the skin and radial nerve.

Complementary indications of central and peripheral sites of interaction were obtained by Yoshida (1962), who showed that during the latent period, the threshold of photic inhibition changed abruptly (Fig. 16). This suggests a dual mechanism and it is but reasonable to suspect that the initial mechanism may occur in the radial nerve and the later one in the superficial (dermal) nervous system. Yoshida also studied the effect of background lighting on the inhibitory threshold and showed, most significantly, that sensory adaptation was manifest even in areas screened from the adapting light (Yoshida, 1966). This re-emphasizes the importance of nervous interplay.

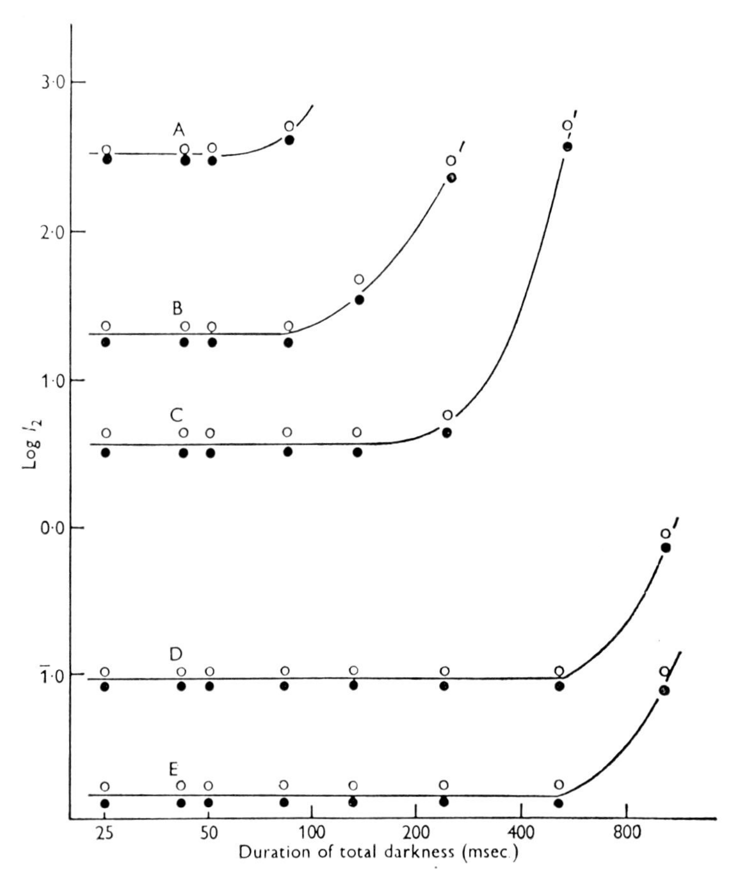

FIG. 16. The relationship between the inhibitory threshold and the time of re-illumination of the radial nerve in *Diadema setosum*. Five examples are shown in which the radial nerve was illuminated at different intensities prior to shading. Abscissae, time of re-admitting the inhibiting light in millisecs, scaled logarithmically. Ordinates, intensity of light re-admitted, in arbitrary logarithmic units. Open circles show the minimum intensities of re-admitted light required to inhibit a response. Filled circles show the maximal sub-threshold intensity (at which the shadow reaction just appeared). Reproduced with permission from Yoshida, M. (1962).

Simultaneously, Millott and Takahashi (1963) revealed other effects of interaction between photically excited nerve centres. They showed that the direction of the initial movement of a spine was determined by the interplay of excitation from superficial centres in the skin, and reflex excitation emanating from the deeply seated radial nerve. The two interact at the base of the responding spine. They also showed that the spine motor mechanism when reflexly excited by photic means is driven by thermo-sensitive centres in the radial nerves, which control the frequency and regularity of the spine oscillation in accordance with the parameters of lighting, but that the latency of the reaction although similarly influenced, is determined by a different mechanism in the radial nerve.

More recently N. Millott and H. Okumura (unpublished) have obtained electrical evidence of inhibition in the radial nerves of *Diadema*. Electrical stimulation of these nerves elicits two propagated potentials (presumably group discharges), which are distinct (Fig. 17) being fast (small) and slow (large). Moreover the latter has a relatively high threshold. It is not responsible for exciting spine movement because

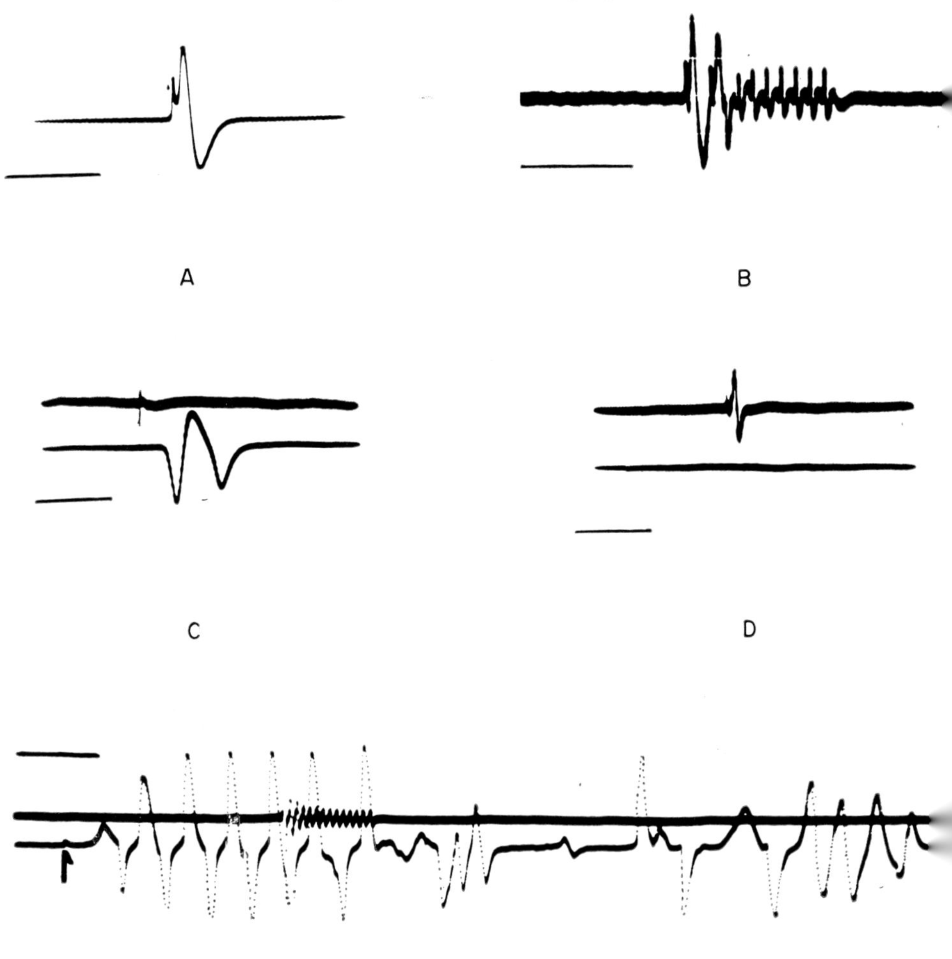

FIG. 17

simultaneous recording of the electrical responses in the radial nerve and the behaviour of the subjacent spines shows that its threshold is far higher than that of spine movement, which appears to be related to the fast potential (Fig. 17C). Repetitive electrical stimulation given immediately following the cessation of photic stimulation of the radial nerve, or a little later while the resultant waving of the spines is still in progress, exerts a sustained inhibitory action on the shadow response, which is partially or totally suppressed. In the former case, the frequency, amplitude, duration and regularity of the spine oscillations are diminished (Fig. 17E). These effects are related to the slow potential because little inhibition is evident until the threshold for the slow response is reached. The relation of such potentials to those excited by light is uncertain and is being investigated.

The resemblance of the shadow response of *Diadema* to an off-effect such as is seen in visual systems with complex receptors is obvious, for they have the same basic ingredients—inhibition coupled with nervous interplay. The resemblance receives substantial support from the striking electrophysiological demonstration of "on" and "off" discharges following photic stimulation of the isolated radial nerve, provided by Takahashi (1964) (Fig. 5).

FUNCTION OF THE DERMAL LIGHT SENSE

The dermatoptic sense appears to serve its possessors by informing them of the presence of light, its direction and of changes in intensity.

FIG. 17. Propagated potentials in the radial nerve of *Diadema antillarum*, recorded extracellularly. Records should be read from left to right.

A, Fast (small) and slow (large) potentials in an isolated radial nerve resulting from a single electrical stimulus indicated by the stimulus artifact immediately preceding the fast response. Time Scale: 250 millisec.

B, Fast (small) and slow (large) potentials in an isolated radial nerve following repetitive electrical stimulation at 10 cycles per sec. Time Scale: 1·0 sec.

C and *D*, The relation of the potentials to spine contraction. The upper record in each case shows the electrical responses of a radial nerve *in situ*, following an electrical stimulus. The lower record in each case shows the behaviour of a single inter-ambulacral spine arising alongside the stimulated nerve. Time scales: 1·0 sec. Note that in *C*, the fast potential is followed by spine movement; in *D*, the slow potential does not elicit spine contraction.

E, The inhibitory effect of the slow potential on the shadow response. In this experiment the radial nerve was illuminated when *in situ* for three minutes. At the arrow, the illumination was abruptly cut off. The vigorous response that follows in an underlying spine is recorded in the left-hand portion of the lower trace. During this response the radial nerve was stimulated repetitively so as to elicit the slow potentials which are visible in the upper trace. Note the inhibition of the photic response which follows. Time Scale: 1·0 sec.

It may perhaps also function in the less obvious and sophisticated fashion discussed below. As already indicated, it may be the sole photic sense or its functions may be discharged in conjunction with complex eyes. An exhaustive discussion of function would necessarily involve consideration of responses to photic stimuli which are far outside the scope of this account. We must therefore limit ourselves to generalities, some particulars and a few suggestions.

In detecting the mere presence of light the sense may inform its possessor of day and night or seasonal rhythms with which organismal behaviour may be synchronized. The reproductive activities of various echinoids are a case in point (see Boolootian, 1966; Millott, 1966). Again the effect of light may inform the possessor of exposure. Alternatively by mediating phototactic or photokinetic responses, the sense may serve to bring animals into the conditions of illumination best suiting their mode of life or to maintain them there. Or, as in the covering reactions of sea urchins, the light-induced coordinated activity of effectors may result automatically in protection or concealment (Millott, 1956).

However it is as a detector of sudden illumination and particularly of shading, that the function of the sense achieves its most characteristic expression. Exposed bottom-dwellers (and among such forms the sense is widely developed), are almost inevitably subjected to shading when a predator moves into the vicinity. The event signals danger, but it may also signal food or a potential host (Kennedy, 1963b). Such events may be signalled at a distance, for sensitivity may be unexpectedly high. Thus Kennedy (1960) has estimated that the internal siphonal surface of *Spisula* is sensitive to 6×10^{-2} erg/sec $\times$ cm^2.

There is more to the matter than this for the significance of inhibition must be reckoned with. Responsiveness to photic transients is no monopoly of the dermal light sense and the situation in the skin and radial nerve of *Diadema* recalls that already revealed in complex eyes as well as in other sensory systems, the significance of which has been cogently discussed several times (Hartline *et al.*, 1961; Ratliff, 1961). It is therefore possible that the nervous interplay mentioned above forms a means of emphasizing contrast existing at the boundaries of light and shade, especially if they are constantly changing as would be the case if the shadow were produced by a moving predator.

The enhanced awareness of presence and movement would no doubt pay dividends toward survival by appropriate and timely responsiveness. In *Diadema*, for example, we can judge the appropriateness, for the response induced by a reasonably large shadow first causes a phalanx of the tallest spines to be arched over a highly vulnerable ambulacrum and then swept repeatedly to and fro over the threatened area. Moreover, such action in creating flicker effects on the photosensitive skin, might thereby sustain sensitivity to the boundaries of shadows in a way analogous to the restoration of stabilized retinal images (Ditchburn, 1963).

Similarly we can judge its effectiveness from the elaborate means of predation evolved by *Cassis*, the helmet conch, which resorts to spraying the photosensitive skin of its victim with a salivary neurotoxin (Cornman, 1963). The instance of *Diadema* is but one among many, a fact which endows photic off-responses with an altogether wider significance in relation to survival (Kennedy, 1963b).

Interpreting fully the significance of the dermal light sense will doubtless call for more subtle explanations. Thus Kennedy (1963a) has suggested that the photoreceptive neurons in the ventral nerve cord of *Cambarus* (which are also secondary neurons for the tactile sense), may function by raising the level of background activity and in this way improve the discrimination of inhibitory signals fed into the system by the stimulation of mechanoreceptors. It is interesting to note however, that these photosensitive properties are retained in blind cavernicolous crayfish which rely on tactile and proprioceptive information for orientation and locomotion (Larimer, 1966). Two blind species, *Cambarus setosus* and *Orconectes pellucidus australis* have retained light sensitive caudal photoreceptors in spite of the regression of their eyes (Larimer *et al.*, 1966). But another blind species *Cambarus ayersii*, in common with *Cambarus setosus*, does not respond by locomotory reflexes to photic stimulation of the caudal photoreceptors, although it responds to illumination of the head region (Wells, 1959). It therefore appears as if here, the photosensitive neurons have lost functional connexion with the motor neurons of the pereiopods.

The dermal light sense, because of its diffuseness and the lack of structurally elaborate photoreceptors, has been widely regarded as primitive and an evolutionary survival. Such a view now appears too facile, for not only may the sensitivity displayed exceed that of some specialized photoreceptors, but the organization of the receptive surface may be complex in other respects (p. 12). Moreover, the nervous organization behind it shows some sophistication (p. 22) and is not primitive in the sense defined by Gregory (1967), who suggests that visual systems, in the course of their evolution have taken over part of the pattern-touch neural system. In *Diadema* the nerve pathways of the dermal light sense and those of the dermal touch sense are in part tangled, but they are nevertheless distinct (Millott, 1954; Cornman, 1963; Bullock, 1965).

References

Arvanitaki, A. and Chalazonitis, N. (1949). Réactions biolélectriques neuroniques à la photoactivation spécifique d'une hème-protéine et d'une carotène-protéine. *Archs Sci. physiol.* **3**: 27–44.

Arvanitaki, A. and Chalazonitis, N. (1961). Excitatory and inhibitory processes initiated by light and infra-red radiations in single identifiable nerve cells (giant ganglion cells of *Aplysia*). In *Nervous inhibition*: 194–231. Florey, E. (ed.), Oxford: Pergamon Press.

Ball, E. G. and Cooper, O. (1949). Echinochrome, its absorbtion spectra; pK′ value and concentration in eggs, amoebocytes and test of *Arbacia punctulata*. *Biol. Bull. mar. biol. Lab. Woods Hole* **97**: 231–232.

Barlow, H. B. (1961). The coding of sensory messages. In *Current problems in animal behaviour*: 331–360. Thorpe, W. H. and Zangwill, O. L. (eds.). Cambridge: University Press.

Benoit, J. and Ott, L. (1944). External and internal factors in sexual activity. *Yale J. Biol. Med.* **17**: 27–46.

Boltt, R. E. and Ewer, D. W. (1963). Studies on the myoneural physiology of Echinodermata. IV. The lantern retractor muscle of *Parechinus*: responses to stimulation by light. *J. exp. Biol.* **40**: 713–726.

Boolootian, R. A. (1966). Reproductive physiology. In *Physiology of Echinodermata*: 561–613. Boolootian, R. A. (ed.). New York: Interscience.

Bruno, M. S. and Kennedy, D. (1962). Spectral sensitivity of photo-receptor neurons in the sixth abdominal ganglion of the crayfish. *Comp. Biochem. Physiol.* **6**: 41–46.

Buddenbrock, W. von (1930). Untersuchungen über den Schattenreflex. *Z. vergl. Physiol.* **13**: 164–213.

Bullock, T. H. (1965). Comparative aspects of superficial conduction systems in echinoids and asteroids. *Am. Zool.* **5**: 545–562.

Chalazonitis, N. (1964). Light energy conversion in neuronal membranes. *Photochem. Photobiol.* **3**: 539–559.

Chalazonitis, N., Chagneux-Costa, H. and Chagneux, R. (1966). Ultra-structure des "grains" pigmentés du cytoplasme des neurones d'*Aplysia depilans*. *C.r. Séanc. Soc. Biol.* **160**: 1014–1017.

Cobb, J. L. S. and Laverack, M. S. (1966). The lantern of *Echinus esculentus* (L.). III. The fine structure of the lantern retractor muscle and its innervation. *Proc. R. Soc.* (B) **164**: 651–658.

Cornman, I. (1963). Toxic properties of the saliva of *Cassis*. *Nature, Lond.* **200**: 88–89.

Crozier, W. J. and Arey, L. B. (1919). Sensory reactions of *Chromodoris zebra*. *J. exp. Zool.* **29**: 261–310.

Dhainaut-Courtois, N. (1965). Sur la présence d'un organe photorécepteur dans le cerveau de "*Nereis pelagica*" L. (annelide polychaète). *C.r. hebd. Séanc. Acad. Sci., Paris* **281**: 1085–1088.

Ditchburn, R. W. (1963). Information and control in the visual system. *Nature, Lond.* **198**: 630–632.

Dodt, E. (1963). Photosensitivity of a localized region of the frog diencephalon. *J. Neurophysiol.* **26**: 752–758.

Eakin, R. M. (1963). Lines of evolution of photoreceptors. In *General physiology of cell specialization*: 393–425. Mazia, D. and Tyler, A. (eds.). New York: McGraw-Hill.

Eakin, R. M. (1965). Evolution of photoreceptors. *Cold Spring Harb. Symp. quant. Biol.* **30**: 363–370.

Fox, D. L. and Hopkins, T. S. (1966). The comparative biochemistry of pigments. In *Physiology of echinodermata*: 277–300. Boolootian, R. A. (ed.). New York: Interscience.

Frisch, K. von (1911). Beitrage zur Physiologie der Pigmentzellen in der Fischhaut. *Pflügers Arch. ges. Physiol.* **138**: 319–387.

Granit, R. (1947). *Sensory mechanisms of the retina.* Oxford: University Press.

Granit, R. (1955). *Receptors and sensory perception.* New Haven: Yale University Press.

Gregory, R. L. (1967). Origin of eyes and brains. *Nature, Lond.* **213**: 369–372.

Hama, K. (1961). A photoreceptor-like structure in the ventral nerve cord of crayfish, *Cambarus virilus. Anat. Rec.* **140**: 329–336.

Hartline, H. K., Ratliff, F. and Miller, W. H. (1961). Inhibitory interaction in the retina and its significance in vision. In *Nervous inhibition*: 241–284. Florey, E. (ed.). Oxford: Pergamon Press.

Hartline, H. K., Wagner, H. G. and MacNichol, E. F. Jnr. (1952). The peripheral origin of nervous activity in the visual system. *Cold Spring Harb. Symp. quant. Biol.* **17**: 125–141.

Hecht, S. (1918). The physiology of *Ascidia atra* Lesueur. II. Sensory physiology. *J. exp. Zool.* **25**: 229–299.

Hecht, S. (1934). Vision II. The nature of the photoreceptor process. In *A handbook of general experimental psychology*: 704–828. Murchison, C. (ed.). Worcester, Mass.: Clark University Press.

Hermann, H. T. and Stark, L. (1962). Prerequisites for a photoreceptor structure in the crayfish tail ganglion. *Q. Prog. Rep. Res. Lab. M.I.T.* **65**: 238–247.

Hess, C. (1914). Untersuchungen über den Lichtsinn bei Echinodermen. *Pflügers Arch. ges. Physiol.* **160**: 1–26.

Kawaguti, S. and Kamishima, Y. (1964). Electron microscopic structure of iridophores of an echinoid, *Diadema setosum. Biol. J. Okayama Univ.* **10**: 13–22.

Kawaguti, S., Kamishima, Y. and Kobashi, K. (1965). Electron microscopy on the radial nerve of the sea-urchin. *Biol. J. Okayama Univ.* **11**: 87–95.

Kennedy, D. (1960). Neural photoreception in a lamellibranch mollusc. *J. gen. Physiol.* **44**: 277–299.

Kennedy, D. (1963a). Physiology of photoreceptor neurons in the abdominal nerve cord of the crayfish. *J. gen. Physiol.* **46**: 551–572.

Kennedy, D. (1963b). Inhibition in visual systems. *Scient. Am.* **209**: 1, 122–130.

Larimer, J. L. (1966). A functional caudal photoreceptor in blind cavernicolous crayfish. *Nature, Lond.* **210**: 204–205.

Larimer, J. L., Trevino, D. L. and Ashby, E. A. (1966). A comparison of spectral sensitivities of caudal photoreceptors of epigeal and cavernicolous crayfish. *Comp. Biochem. Physiol.* **19**: 409–415.

Light, V. E. (1930). Photoreceptors in *Mya arenaria* with special reference to their distribution, structure and function. *J. Morph.* **49**: 1–42.

McCullough, C. G. (1962). Modification of *Hydra* pacemaker activity by light. *Am. Zool.* **2**: 201.

Médioni, J. (1961). Le problème de l'existence d'un sens dermatoptique chez les insectes. *Revue Psychol. Fr.* **6**: 11–20.

Messenger, J. B. (1967). Parolfactory vesicles as photoreceptors in a deep-sea squid. *Nature, Lond.* **213**: 836–838.

Millott, N. (1953). Light emission and light perception in species of *Diadema*. *Nature, Lond.* **171**: 973–974.

Millott, N. (1954). Sensitivity to light and the reactions to changes in light intensity of the echinoid *Diadema antillarum* Philippi. *Phil. Trans. R. Soc.* (B) **238**: 187–220.

Millott, N. (1956). The covering reaction of sea urchins. I. A preliminary account of covering in the tropical echinoid *Lytechinus variegatus* (Lamarck), and its relation to light. *J. exp. Biol.* **33**: 508–523.

Millott, N. (1957a). Animal photosensitivity, with special reference to eyeless forms. *Endeavour* **16**: 19–28.

Millott, N. (1957b). Naphthaquinone pigment in the tropical sea urchin *Diadema antillarum* Philippi. *Proc. zool. Soc. Lond.* **129**: 263–272.

Millott, N. (1964). Pigmentary system of *Diadema antillarum* Philippi. *Nature, Lond.* **203**: 206.

Millott, N. (1966). The enigmatic echinoids. In *Light as an ecological factor*. Bainbridge, R., Evans, G. C. and Rackham, O. (eds.). *Symp. Br. ecol. Soc.* No. 6: 265–291.

Millott, N. and Manly, B. M. (1961). The iridophores of the echinoid *Diadema antillarum*. *Q. Jl microsc. Sci.* **102**: 181–194.

Millott, N. and Takahashi, K. (1963). The shadow reaction of *Diadema antillarum* Philippi. IV. Spine movements and their implications. *Phil. Trans. R. Soc.* (B) **246**: 437–469.

Millott, N. and Vevers, H. G. (1955). Cartotenoid pigments in the optic cushion of *Marthasterias glacialis* (L). *J. mar. biol. Ass. U.K.* **34**: 279–287.

Millott, N. and Yoshida, A. (1956). Reactions to shading in the sea urchin, *Psammechinus miliaris* (Gmelin). *Nature, Lond.* **178**: 1300.

Millott, N. and Yoshida, M. (1957). The spectral sensitivity of the echinoid *Diadema antillarum* Philippi. *J. exp. Biol.* **34**: 394–401.

Millott, N. and Yoshida, M. (1959). The photosensitivity of the sea urchin *Diadema antillarum* Philippi: responses to increases in light intensity. *Proc. zool. Soc. Lond.* **133**: 67–71.

Millott, N. and Yoshida, M. (1960a). The shadow reaction of *Diadema antillarum* Philippi. I. The spine response and its relation to the stimulus. *J. exp. Biol.* **37**: 363–375.

Millott, N. and Yoshida, M. (1960b). The shadow reaction of *Diadema antillarum* Philippi. II. Inhibition by light. *J. exp. Biol.* **37**: 376–389.

Motte, I. de la (1964). Untersuchungen zur vergleichenden physiologie der lightempfindlichkeit geblendeter fische. *Z. vergl. Physiol.* **49**: 58–90.

Nagel, W. A. (1896). *Der Lichtsinn augenloser Tiere*. Jena: Fischer.

Nicol, J. A. C. (1950). Responses of *Branchiomma vesiculosum* (Montagu) to photic stimulation. *J. mar. biol. Ass. U.K.* **29**: 303–320.

Nishioka, R. S., Hagadorn, I. R. and Bern, H. A. (1962). Ultrastructure of the epistellar body of the octopus. *Z. Zellforsch. mikrosk. Anat.* **57**: 406–421.

Nishioka, R. S., Yasumasu, I. and Bern, H. A. (1966). Photoreceptive features of vesicles associated with the nervous system of cephalopods. *Nature, Lond.* **211**: 1181.

Nishioka, R. S., Yasumasu, I., Packard, A., Bern, H. A. and Young, J. Z. (1966). Nature of vesicles associated with the nervous system of cephalopods. *Z. Zellforsch. mikrosk. Anat.* **75**: 301–316.

North, W. J. and Pantin, G. F. A. (1958). Sensitivity to light in the sea anemone *Metridium senile* (L): adaptation and action spectra. *Proc. R. Soc. Lond.* **148**: 385–396.

Passano, L. M. and McCullough, C. B. (1962). The light response and the rhythmic potentials of *Hydra*. *Proc. natn. Acad. Sci. U.S.A.* **48**: 1376–1382.

Prosser, G. L. (1934). Action potentials in the nervous system of crayfish. 2. Responses to illumination of the eye and caudal ganglion. *J. cell. comp. Physiol.* **4**: 363–377.

Ratliff, F. (1961). Inhibitory interaction and the detection and enhancement of contours. In *Sensory communication*: 183–203. Rosenblith, W. A. (ed.). Boston, Mass.: M.I.T. Press and New York: Wiley.

Rockstein, M., Cohen, J. and Hausman, S. A. (1958). A photosensitive pigment from the dorsal skin and eyespots of the starfish, *Asterias forbesi*. *Biol. Bull. mar. biol. Lab., Woods Hole* **115**: 361.

Rockstein, M. and Finkel, A. (1960). Stellarin, a photosensitive pigment from the dorsal skin of the starfish, *Asterias forbesi*. *Anat. Rec.* **138**: 379.

Rockstein, M. and Rubenstein, M. (1957). The biochemical basis for positive photokinesis of the starfish, *Asterias forbesi*. *Biol. Bull. mar. biol. Lab., Woods Hole* **113**: 353–354.

Rushton, W. A. H. (1963). Increment threshold and dark adaptation. *J. opt. Soc. Am.* **53**: 104–109.

Sarasin, G. F. and Sarasin, P. B. (1887). Augen und Integument der Diadematiden. *Ergebn. naturw. Forsch. Ceylon* **1**: 1.

Singer, R. H., Rushforth, N. B. and Burnett, A. L. (1963). The photodynamic action of light on hydra. *J. exp. Zool.* **154**: 169–173.

Steven, D. M. (1951). Sensory cells and pigment distribution in the tail of the ammocoete. *Q. Jl microsc. Sci.* **92**: 233–247.

Steven, D. M. (1963). The dermal light sense. *Biol. Rev.* **38**: 204–240.

Takahashi, K. (1964). Electrical responses to light stimuli in the isolated radial nerve of the sea urchin, *Diadema setosum* (Leske). *Nature, Lond.* **201**: 1343–1344.

von Uexküll, J. (1900). Die Wirkung von licht und Schatten auf die Seeigel. *Z. Biol.* **40**: 447–476.

Vevers, H. G. (1966). Pigmentation. In *Physiology of Echinodermata*: 267–275. Boolootian, R. A. (ed.). New York: Interscience.

Vevers, H. G. and Millott, N. (1957). Cartotenoid pigments in the integument of the starfish *Marthasterias glacialis* (L). *Proc. zool. Soc. Lond.* **129**: 75–80.

Viaud, G. (1948). Le phototropisme et les deux modes de la photoréception. *Experientia* **4**: 81–88.

Wallenfels, K. (1941). Complex compounds of echinochrome and related hydroxynaphthaquinones. *Ber. dtsch. chem. Ges.* **74**B: 1598–1604 (cited from *Chem. Abstr.* **37**: 371–372).

Wells, P. H. (1959). Responses to light by cave crayfishes. *Occ. Pap. natn. speleol. Soc.* **4**: 3–15 (cited from Larimer *et al.*, 1966).

Welsh, J. H. (1934). Caudal photoreceptor and responses of the crayfish to light. *J. cell. comp. Physiol.* **4**: 379–388.

Willem, V. (1891). Sur les perceptions dermatoptiques. *Bull. Sci. Fr. Belg.* **23**: 329–346.

Wolken, J. J. (1962). Photoreceptor structures: their molecular organization for energy transfer. *J. theoret. Biol.* **3**: 192–208.

Yoshida, M. (1956). On the light response of the chromatophore of the sea urchin, *Diadema setosum* (Leske). *J. exp. Biol.* **33**: 119–123.

Yoshida, M. (1957). Spectral sensitivity of chromatophores in *Diadema setosum* (Leske). *J. exp. Biol.* **34**: 222–225.

Yoshida, M. (1960). Further studies on the chromatophore response in *Diadema setosum* (Leske). *Biol. J. Okayama Univ.* **6**: 169–173.

Yoshida, M. (1962). The effect of light on the shadow reaction of the sea urchin *Diadema setosum* (Leske). *J. exp. Biol.* **39**: 589–602.

Yoshida, M. (1966). Photosensitivity. In *Physiology of Echinodermata*: 435–464. Boolootian, R. A. (ed.). New York: Interscience.

Yoshida, M. and Millott, N. (1959). Light sensitive nerve in an echinoid. *Experientia* **15**: 13–14.

Yoshida, M. and Millott, N. (1960). The shadow reaction of *Diadema antillarum* Philippi. III. Re-examination of the spectral sensitivity. *J. exp. Biol.* **37**: 390–397.

Yoshida, M. and Ohtsuki, H. (1966). Compound ocellus of a starfish: its function. *Science, N.Y.* **153**: 197–198.

Young, J. Z. (1935). The photoreceptors of lampreys. Light sensitive fibres in the lateral line nerves. *J. exp. Biol.* **12**: 229–238.

Symp. zool. Soc. Lond. (1968) No. 23, 37–62.

THE STRUCTURE OF MOLLUSC STATOCYSTS, WITH PARTICULAR REFFRENCE TO CEPHALOPODS

V. C. BARBER*

Department of Anatomy, University College, University of London, England

SYNOPSIS

Information on the structure of mollusc statocysts is reviewed. Most molluscs have a pair of statocysts comprising a simple sac composed of hair cells and supporting cells with a statolith, but cephalopod statocysts are much more complex. Features of the electron microscopic structure of the statocyst of the lamellibranch, *Pecten maximus*, and of the cephalopods, *Octopus vulgaris* and *Sepia officinalis*, are described. Stereoscan studies on *Eledone moschata* are included. Particular reference is made to the arrangement and orientation of the ciliary groups in the macula and crista of the statocyst of *Octopus*. Each hair cell in the *Octopus* statocyst bears numerous kinocilia with a 9 + 2 internal filament content. The cilia have basal bodies, basal feet and roots. Each cilium is orientated with regard to the next, and it is postulated that passive movement of the cilia in the direction of their basal feet is excitatory (causing depolarization), whereas the reverse is inhibitory. Evidence for the theory is considered together with information about transduction mechanisms in general.

INTRODUCTION

In recent years the structure of the acoustico-vestibular systems of a variety of vertebrates have been studied with the electron microscope (see Flock, 1965; Wersäll, Flock and Lundquist, 1965, for reviews). However, only a few fine-structural studies of equivalent invertebrate structures have been made, such as those on the ctenophore statocyst (Horridge, 1965a), the statocyst of *Octopus* (Barber, 1965, 1966a, b), and the locust ear (Gray, 1960). Such studies have been so few that it has not been possible to present a complete picture of statocyst structure in any invertebrate Phylum (see Bullock and Horridge, 1965, and the review by Kolmer, 1926).

Molluscs are a particularly suitable group for such a comprehensive study as statocysts in the phylum range from the very simple to the very complex, such as in the decapod cephalopods. Because of the work on cephalopods published by me, and various studies in progress, it is now possible to present a reasonable account of statocyst structure in the molluscs.

* Present address: Department of Zoology, University of Bristol, Bristol 8.

STRUCTURE OF THE STATOCYSTS IN MOLLUSCS OTHER THAN CEPHALOPODS

The majority of these statocysts are comparatively simple in structure (see Bullock and Horridge, 1965, and Charles, 1966). Many light microscope studies have been made and in this study the structure of some representative ones will be reviewed. (The classification of Morton, 1958, which is based on Thiele, will be used in this account.)

Most molluscs have a pair of statocysts. These are usually situated near the pedal ganglion, although they send their nerves to the cerebral ganglion, passing alongside the cerebro-pedal connective. The only class of molluscs that lacks statocysts, as one might expect, are the fairly sedentary Amphineurans (Bullock and Horridge, 1965; although Morton, 1958, says that they are present in a rudimentary form). However the related *Monoplacophora* (for example *Neopilina galathea*) have a pair of statocysts (Lemche and Wingstrand, 1959). These have a duct separate from the static nerve.

The statocysts of Lamellibranchs, such as that of *Pecten inflexus* (Buddenbrock, 1915), consist of a simple sac containing hair cells, supporting cells and a central statolith (Fig. 1). Because of the flattened

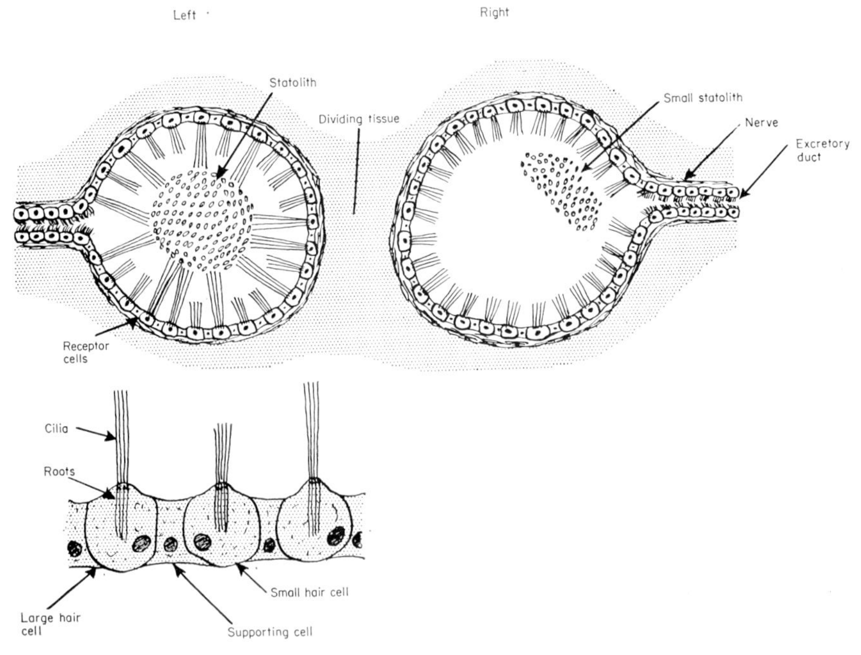

FIG. 1. The statocysts of the lamellibranch *Pecten inflexus* (based on Buddenbrock, 1915). Note that the left statocyst is better developed.

body form of the animal the statocysts are not symmetrical, the left being more complex than the right. The left has two sorts of hair cells, one with longer and the other with shorter cilia. Another difference is that the right statocyst has a smaller statolith than the left. A ciliated excretory duct leads off from the static sac of both the left and right statocysts. This duct passes through the centre of the static nerve. In some lamellibranchs (for example the Mytilidae) the excretory duct opens narrowly to the exterior. In others (the Ostreidae) statoliths are said to be absent (Morton, 1958). The Scaphopoda have statocysts similar to the lamellibranchs.

Studies of the fine structure of the statocyst of *Pecten maximus* are in progress. (For details of electron microscopical preparative methods used see Barber, Evans and Land, 1966). The static sac wall is composed of hair cells and supporting cells. The hair cells bear cilia (with a 9 + 2 filament content) at their distal ends and a few microvilli. (Fig. 2). These cilia actually pass into the statolith mass, unlike the cilia in cephalopods. The hair cells have a similar organelle content to those of

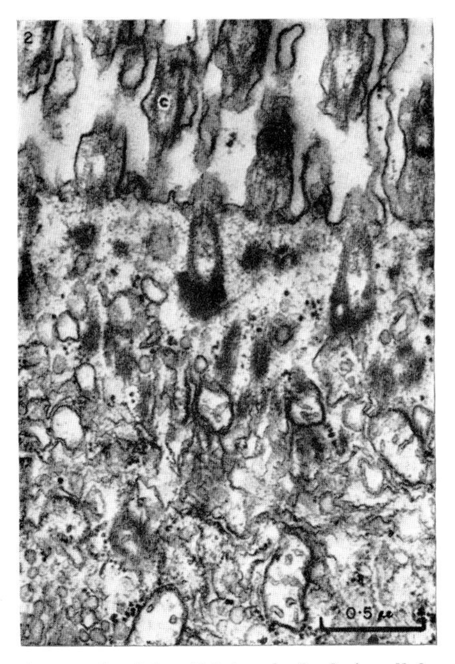

Fig. 2. Electron micrograph of the distal end of a hair cell from the right statocyst of *Pecten maximus* to show the cilia (*c.*). (All specimens shown in electron micrographs in this paper have been fixed with phosphate buffered osmium tetroxide and were stained with lead citrate. Embedded in Araldite.)

cephalopods, a particular feature being numerous large granules and multivesicular bodies. Profiles having the structural features of synapses are present below the cells in the small plexus. So, if these strucsures are functional synapses, some synaptic interaction probably occurs, either between adjacent hair cells or between hair cells and centrifugal fibres from the cerebral ganglion. It can be seen even in this early stage of the study, that in essence the structural elements in this simple statocyst, and in the more evolved statocysts of cephalopods, are the same.

The statocysts of most of the orders of Gastropoda are similar to those in *Pecten*, namely, being composed of a simple sac with hair cells and supporting cells. Descriptions of, for example the statocysts of *Helix pomatia* (Pfeil, 1922; Baecker, 1932), suggest that there are very large hair cells. However, electron microscopical studies show that these cells probably produce the statolith and that the hair cells are much smaller (M. S. Laverack, unpublished). Most Gastropod statocysts have an excretory duct, although *Helix* is reported not to have one (Pfeil, 1922). In some species, such as the pulmonate *Planorbis corneus*, there is a peculiar arrangement where the statolith crystals actually appear to pass down this duct (see Schmidt, 1912) (Fig. 3).

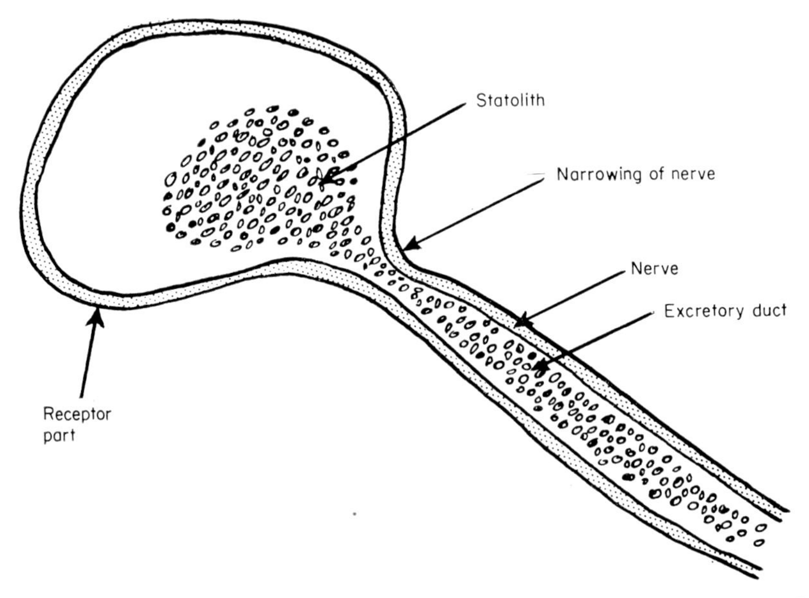

FIG. 3. A statocyst of the gastropod *Planorbis corneus* to show how the statolith mass can pass into the excretory duct (based on Schmidt, 1912).

In some Gastropods, those in the order Heteropoda, the statocysts are much more complex. These species, such as *Pterotrachea coronata*, are pelagic, so it is hardly surprising that the statocysts are more complex (Tschachotin, 1908). The statocyst in *Pterotrachea* has a macula and antimacula. The macula contains two sorts of hair cells, a large central cell, with its large pericentral supporting cell, and smaller hair cells, with other smaller supporting cells. The antimacula, which forms the rest of the sac contains other ciliated cells. In the middle of the sac is a large statolith (Fig. 4). It seems reasonable to speculate that

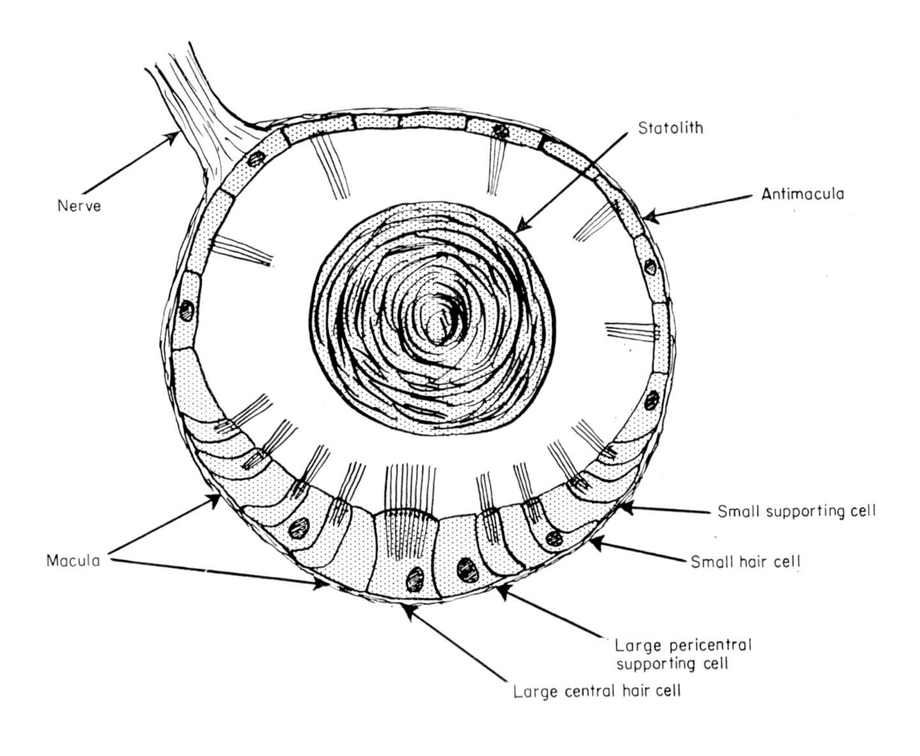

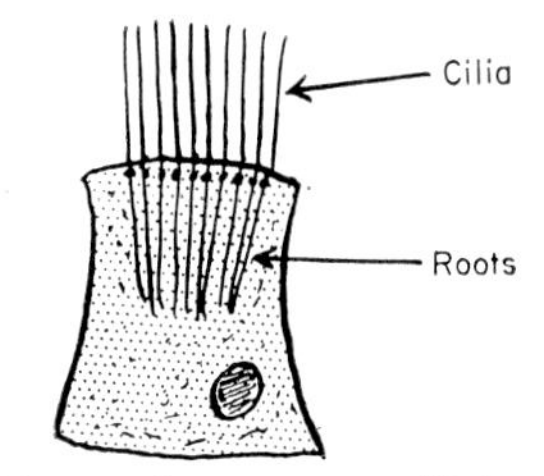

FIG. 4. A statocyst of the pelagic heteropod gastropod *Pterotrachea coronata*. Note how it is divided into a macula and anti-macula (based on Tschachotin, 1908).

cephalopod statocysts must have passed through such a stage in their evolution from simple to complex structures.

STRUCTURE OF CEPHALOPOD STATOCYSTS: AMMONOIDEA AND NAUTILOIDEA

The class Cephalopoda can be divided into three sub-classes, the Coleoidea, containing all living forms except the genus *Nautilus*; the Ammonoidea, all extinct; and the Nautiloidea, represented today only by the five living species of *Nautilus* (classification based on Morton, 1958).

Little is known about the soft parts of the extinct Ammonoids or Nautiloids. Except for *Nautilus macromphalus*, *N. pompilius*, *N. scrobiculatus*, *N. repertus*, and *N. stenomphalus*, all other species of Ammonoids and Nautiloids are extinct, so any discussion of the likely structure of their statocysts must be based on the structure as found in living nautiluses (see Stenzel, 1964, for a modern account of nautiloid structure). As far as is known there are no major differences in structure in the soft parts of the five living nautilus species and this is presumably true for the statocysts. The statocyst consists of a simple sac, probably composed of a large number of hair cells and supporting cells. There is an excretory duct, termed Kolliker's canal, and this opens directly to the exterior (Young, 1965) (Fig. 5). In general it can be seen that this statocyst is very similar to those found in other molluscs, and this probably represents the ancestral condition that was present in the extinct forms.

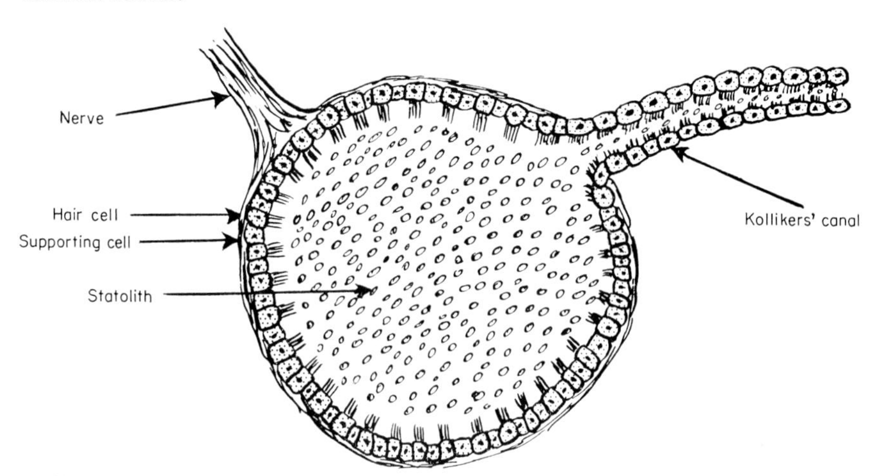

FIG. 5. A statocyst of a nautiloid cephalopod *Nautilus macromphalus*. Note how simple is its structure compared with that of other cephalopods (based on Young, 1965).

STRUCTURE OF CEPHALOPOD STATOCYSTS: COLEOIDS

The Coleoids have quite complex statocysts, with a macula and statolith, and a crista arranged approximately in three planes. There are distinct morphological differences between the three orders of coleoids, the Decapoda, the Octopoda, and the Vampyromorpha, and they are quite distinct from the nautiloids. The structure of cephalopod statocysts have been adequately described either by light microscopy (Hamlyn-Harris, 1903; Ishikawa, 1924; Klein, 1931; Young, 1960) and by electron microscopy (Barber, 1965, 1966a, b).

Octopoda

As is normal for molluscs there is a pair of statocysts in Octopods, such as for example *Octopus vulgaris* or *Eledone moschata*, and in fact the statocysts in all Octopods are rather similar.

Each statocyst consists of a cartilaginous capsule filled with perilymph. Suspended from the dorsal end of this capsule is the static sac. The sac is filled with endolymph and contains the receptor part of the statocyst, the macula and crista. The crista, a ridge of hair cells, is

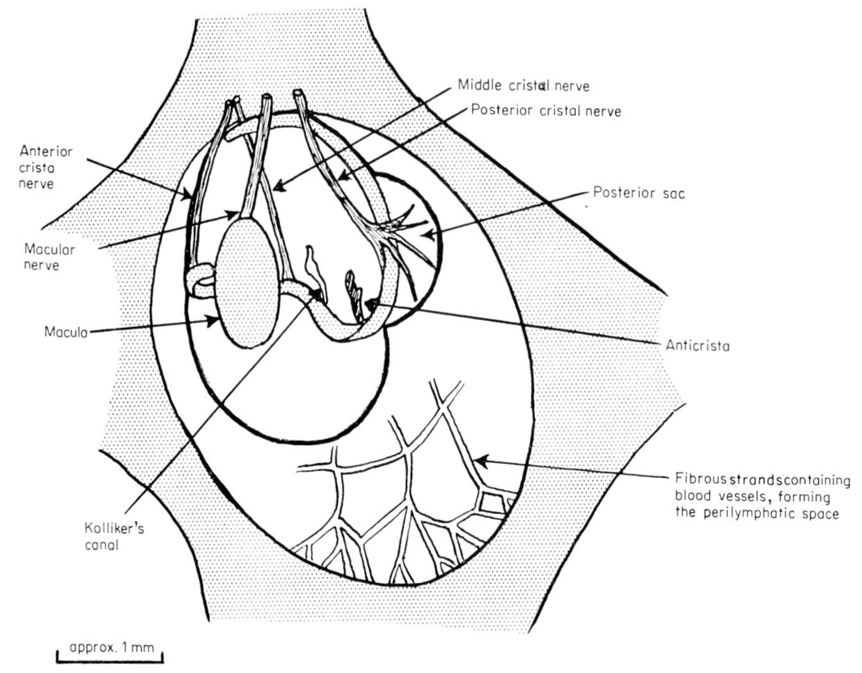

FIG. 6. A medial view of the statocyst of the octopod cephalopod *Octopus vulgaris*. A single macula and crista are present. Note the perilymphatic space, which is absent in decapods (based on Young, 1960).

orientated approximately in three main planes, each plane being composed of three sections, giving a total of nine sections. The macula, a flat plate of hair cells, is arranged vertically at the midline end of the sac. A statolith is suspended above the macula (Fig. 6) (see Young, 1960). There is one anticrista in the statocyst, lying in front of the vertical crista. Lying in the cartilage in the lateral wall of the statocyst above the longitudinal crista, is Kolliker's canal, an elongated flask-shaped sac. It opens into the static sac by a small pore and ends blindly by the crista. The macula receives its own nerve, the macular nerve. The three portions of the crista receive separate nerves. The anterior crista nerve innervates the transverse horizontal portion, the posterior crista nerve innervates the vertical crista and also the posterior sac portion of the static sac, and the middle crista nerve innervates the longitudinal horizontal portion. It must be noted that there are ciliated cells, possibly receptor cells, in the general wall of the static sac, particularly in the posterior sac portion. These are often seen in scanning electron microsopical ("Stereoscan").

The fine structure of the crista and macula are very similar, although the crista is a ridge of cells and the macula is a plate of cells. Both contain hair cells and supporting cells, and below each is a nerve plexus and this probably contains another type of cell. The whole is supported by a layer of cartilage, which contains numerous chondroblasts and blood vessels (Barber, 1965, 1966a) (Fig. 7). The blood vessels consist of a discontinuous endothelium (there are also amoebocytes in the lumen), a continuous basal membrane (basal lamina), and outside the lamina is a continuous layer of pericytes (see Barber and Graziadei, 1965, 1966a, 1967a, b, for details of the structure of cephalopod blood vessels).

The hair cells are primary receptor cells that send axons to the central nervous system. At their distal ends the hair cells bear numerous kinocilia (as many as 200), and also many short microvilli. The cilia have a 9+2 internal filament content, have basal bodies with basal feet, and have long striated roots. The roots have a periodicity of 700 Å, which is similar to the periodicity of roots found in the sensory cells in the arms of *Octopus* (Graziadei, 1964), and in other molluscs, such as in ciliated cells in the mussel (*Anodonta*) (750 Å, Gibbons, 1961).

Each cilium from a group is oriented with regard to other cilia in the group so that the two central fibres of each cilium are in line with the long axis of the group. The nine double peripheral fibres (with tails) are orientated so that filaments 1, 2, 3, 8, and 9 are on one side of the line of

FIG. 7. A greatly simplified portion of the macula of *Octopus vulgaris* as seen with the electron microscope. The crista is similar (based on Barber, 1965).

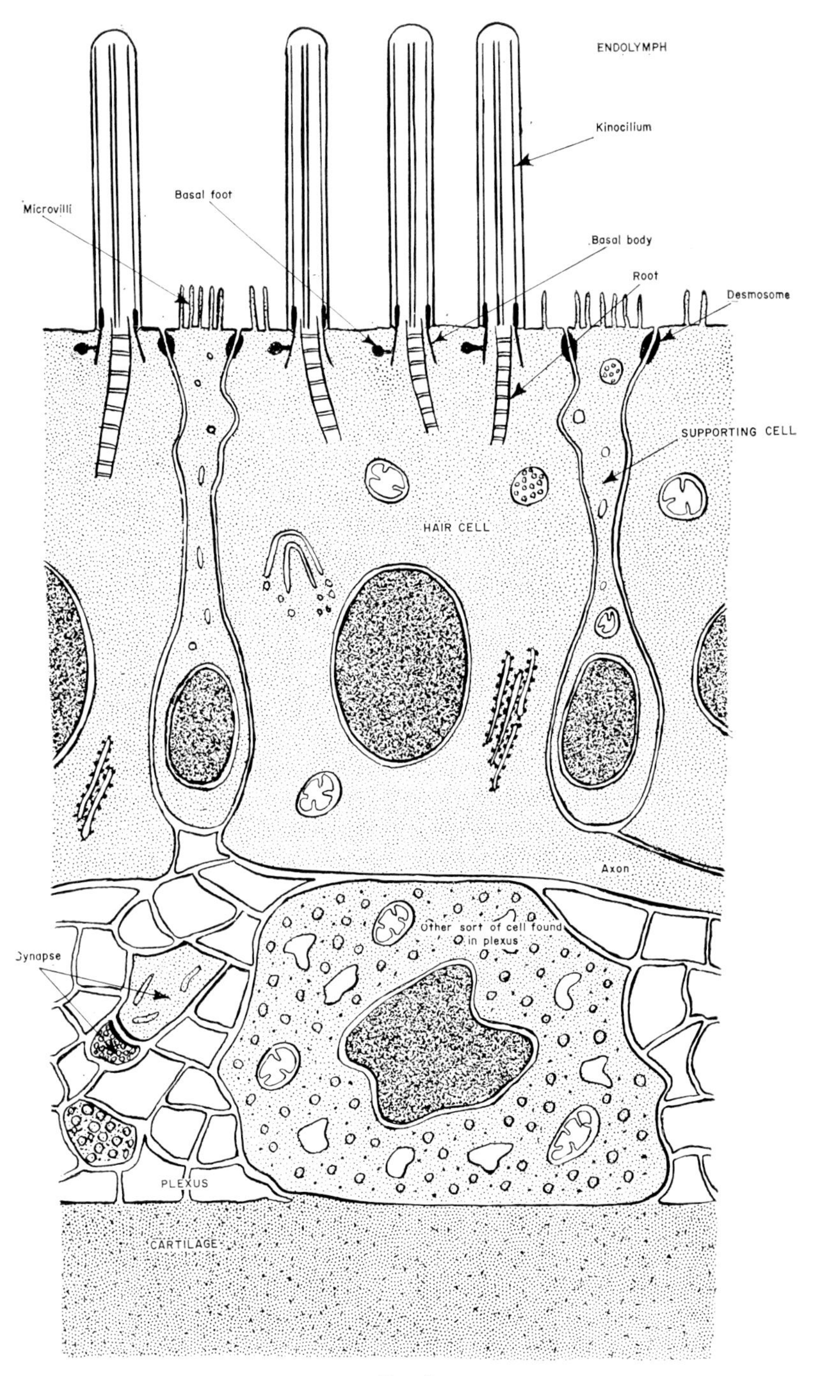
ENDOLYMPH
Kinocilium
Basal foot
Microvilli
Basal body
Root
Desmosome
SUPPORTING CELL
HAIR CELL
Axon
Other sort of cell found in plexus
Synapse
PLEXUS
CARTILAGE

Fig. 7

the two central fibres, and fibres 4, 5, 6, and 7 are on the other (numbered according to Afzelius, 1959). Another feature of the orientation is that the basal feet of the cilia are all orientated at right angles to the long axis of the group (and hence the line of the central fibres) and are on the side where four outer ciliary fibres are present (fibres 4, 5, 6, and 7). So each group of cilia is organized as summarized in Fig. 8.

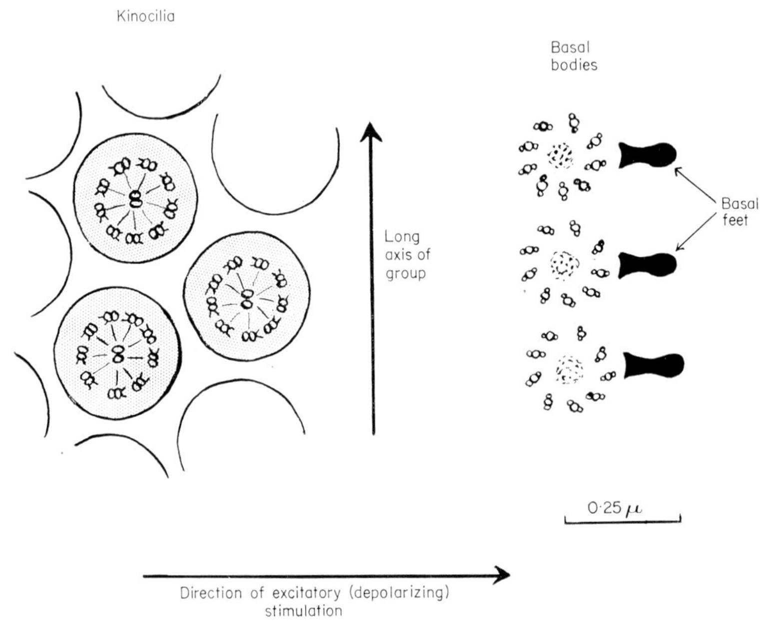

FIG. 8. Summary of the ultrastructural features of the cilia of the hair cells in *Octopus vulgaris* (as viewed from outside the cell). Excitatory stimulation is thought to occur if the cilia are bent towards the basal feet (and hence peripheral fibres 5 and 6).

Each group is also organized with regard to each other group. In the crista the long axis of each group coincides with the long axis of the crista. Another complication, at least in some of the segments of the crista, is that the basal feet of some groups point in one direction whilst others point in the opposite direction (Fig. 9). So there can be at least three rows pointing in one direction and two in the other. Young (1960) suggested that the cilia in the crista were connected in some fashion to form actual flaps (cupulae). However, Stereoscan studies on the crista of *Eledone moschata* give no evidence of this, as all the ciliary groups can be seen to be separate. Another complication is the presence of so-called single and double rows of large hair cells in different cristal segments (Young, 1960). Stereoscan studies confirm that these do

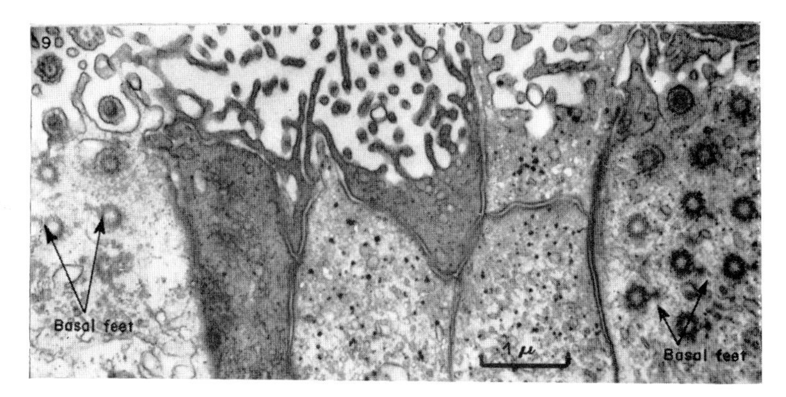

FIG. 9. Electron micrograph of a transverse section across the distal ends of two cristal hair cells. The basal feet of the cilia of the two cells point in opposite directions, hence these two cells are thought to be excited by stimuli coming from opposite directions.

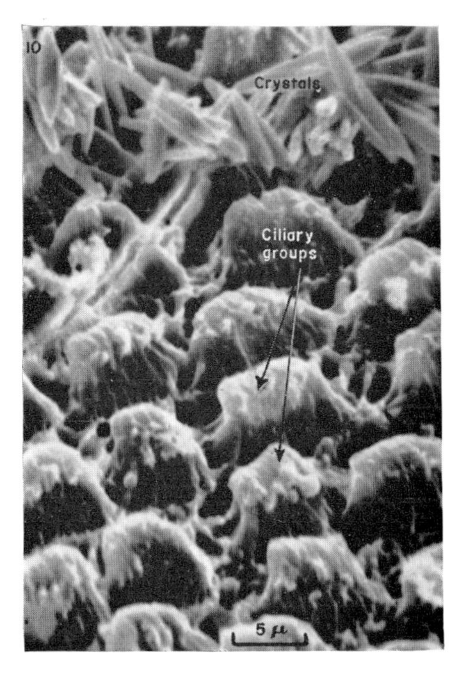

FIG. 10. Stereoscan picture of a segment of the macula of *Eledone moschata*. Note the ciliary groups arranged in rows. Some statolith crystals can also be seen (fixed phosphate buffered osmium tetroxide, dehydrated in acetone, air dried, and vacuum carbon-coated).

occur. There are several other rows in addition to these large cells, so that the significance of this is not clear.

The ciliary groups are also orientated in the macula. The long axis of each group is orientated so that each group lies along a circumferential line in the macula forming rings of hair cells (Fig. 10). All the basal feet from the groups point (maximum of nineteen cells in a row checked) *away* from the centre of the macula. It is possible that there are local variations within the macula but this is the general arrangement.

The statolith is a compact structure and is composed of quite large crystals as is shown in Fig. 10. The cilia are not embedded in the statolith, but in life may be loosely attached to it, perhaps by mucus.

A plan of the ciliary arrangement in the macula and crista is shown in Fig. 11. An attempt to map out the ciliary orientation of the crista and macula in its entirety is in progress.

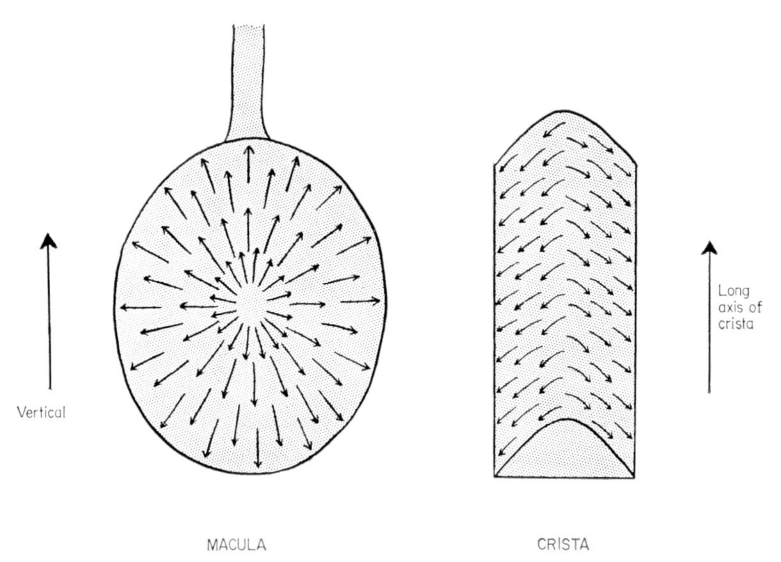

FIG. 11. Summary of the ciliary orientation in the macula and crista of *Octopus vulgaris*. The arrow represents the direction in which the basal feet from a group of cilia point.

The hair cell bodies contain the usual organelles, mitochondria, endoplasmic reticulum and agranular reticulum (which is often organized into great whorls of membranes), and multivesicular bodies. Large cytolysome granules (up to $2{\cdot}5\mu$ in diameter) are also present. The supporting cells are slender and are tightly interposed between the hair cells. At their distal ends the supporting cells bear numerous short microvilli and are attached to the hair cells by typical desmosomes. The supporting cell perikaryon is very densely packed with organelles and can easily be distinguished from the hair cell perikaryon (Barber, 1966a).

The interpretation of the plexal layer has proved very difficult. There are profiles in this layer that have all the criteria generally expected for cephalopod synapses (see Gray and Young, 1964; Barber and Graziadei, 1966b), so there is synaptic interaction in the plexus. There is some degeneration evidence that there are centrifugal fibres from the brain (Young, 1960), and it seems probable that some endings are from these fibres, but others may represent interaction between adjacent hair cells. There are several differences between the plexus in the crista and macula. In the crista, endings on the bases of hair cells or on large axons are found (see Barber, 1966a; Barber and Graziadei, 1966b), but these are not found in the macula. Post-synaptic processes projecting into the hair cells are found in the macula but these have not yet been found in the cristal plexus. Another problem is in the probable presence of another sort of cell in the plexus. Whether these are peripheral neurons (as suggested by Klein, 1931; and Young, 1960), or are a sort of glial cell (as suggested by Barber, 1966a), or are some modified supporting cell is uncertain at this stage. These cells are not easy to find with the electron microscope. They have a particular organelle content (hence the name "dark" cells used in Barber, 1966a). The plexus obviously warrants further study including some degeneration work.

There are two other particular features of note. The first is the presence of ciliated cells in the general static sac wall. Whether or not these cells are sensory is not known but seems possible. The second feature is Kolliker's canal. With electron microscopy this can be seen as a ciliated tube. The cilia have a $9 + 2$ internal filament structure, have roots, and in life can be seen to beat. The presence of numerous vesicles in these cells suggest that the canal is probably secretory, and in any case it is difficult to think of any other use for it. It could possibly secrete endolymph.

Decapoda

There are three main differences between the statocysts of octopods and decapods. The perilymphatic space is absent in decapods, the static sac being directly applied to the cartilage; there are a variable number of anticristae present in decapods (generally 11 in *Sepia* although 12 have been reported, Hamlyn-Harris, 1903), whereas octopods have only one (Fig. 12); and the last difference is that in decapods the macula is divided into sections (often three).

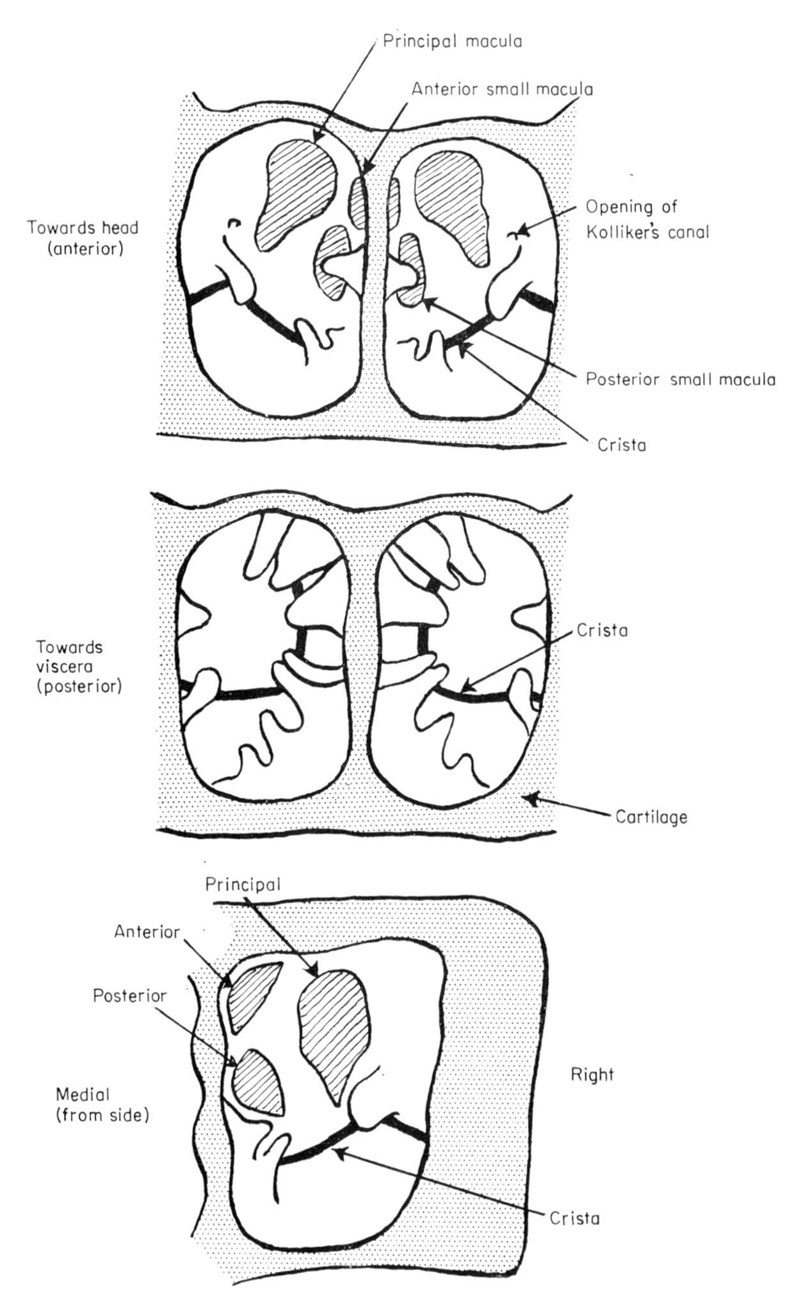

FIG. 12. Various views of the statocysts of the decapod cephalopod *Sepia officinalis*. The macula consists of three sections, and there is a crista. There is no perilymphatic space and many anticristae (based on Hamlyn-Harris, 1903).

Although the fine structure of the statocyst of *Sepia officinalis* has not yet been examined in as much detail as that of *Octopus*, there is only one obvious structural difference between the two. This is in the presence of a granular substance at the distal end of the supporting cell (Fig. 13). This granular substance is associated with the desmosomes that attach the hair cells to the supporting cells. On the hair cell

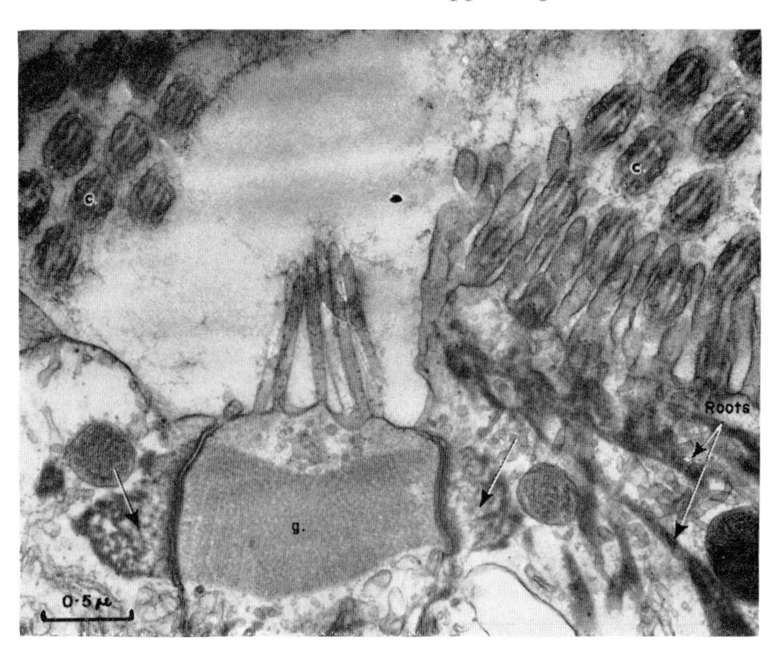

FIG. 13. Electron micrograph of a sagittal section through the distal end of hair cells and supporting cells of the principal macula of *Sepia officinalis*. Note the ground substance in the supporting cell (*g.*) and the desmosome root attachment in the hair cells (arrows). *c.* is cilium.

side of the desmosome are special root attachments. So it seems likely that the supporting cell granular substance has some sort of strengthening function. Apart from this difference there seems to be little obvious difference between the fine structure in *Sepia* and *Octopus*, although the ciliary orientation has not yet been examined in detail.

Vampyromorpha

Because of the deep sea habitat of the species in this order, such as *Vampyroteuthis infernalis* Chun, published histological studies of the animal are incomplete. Considerable damage to the animal generally occurs as it is brought up from deep water. However, some histological details of statocyst structure are now available.

It has been known since the classic studies of Pickford that the statocyst is large. For example, to quote this author, "... ear vesicle is large, in young larvae actually larger than the eyes" (Pickford, 1940). Careful dissections by Dr R. Young, of Miami, show that there is a perilymphatic space, a fact that makes the statocysts akin to those of octopods. There are differences in that there are nine anticristae (compared to one in octopods). There is a macula, and the crista is orientated in approximately three planes as in other cephalopods. Cajal-stained sections of this statocyst (from material provided by Dr Young) show some of these features but the tissue is not well enough preserved to determine much detail. No electron microscopical studies have been done.

Enough has been learnt to see that the statocyst is intermediate in some ways between the statocysts of octopods and decapods.

EVOLUTION OF MOLLUSCAN STATOCYSTS

In the majority of molluscs the statocyst pair is simple in structure, being a single sac composed of hair cells and supporting cells with a single statolith or collection of statoconia. In some species there is an excretory duct that passes through the static nerve (e.g. *Pecten*, Buddenbrock, 1915), whilst in others this duct is absent (e.g. *Helix*, Pfeil, 1922). In some the duct does not pass through the static nerve but is separate from it (e.g. *Neopilina*, Lemche and Wingstrand, 1959). In some gastropods (in the order Heteropoda (e.g. *Pterotrachea*, Tschachotin, 1908) the statocyst is more complex with a differentiation into macula and antimacula. This is the greatest complexity reached in molluscs other than cephalopods. Except for the nautiloids, such as *Nautilus macromphalus* where the statocyst is simple, all cephalopod statocysts are quite complex. The greatest complexity is reached in the decapods such as *Sepia*, where there are as many as twelve anticristal lobes. The number of anticristae has in fact been used as a means of determining a phylogenetic tree for various cephalopod genera (Ishikawa, 1924). The evolution of the molluscan statocyst can be imagined as having followed the line of starting with a simple sac-statocyst, which led on to one more like that in *Pterotrachea*, with the beginnings of a division into macula and crista, and this lead in turn to the complex statocysts of advanced cephalopods culminating in the decapods. This is summarized in Fig. 14.

PHYSIOLOGY AND FUNCTION OF MOLLUSC STATOCYSTS

Apart from a limited number of experiments on the effects of unilateral or bilateral extirpation of statocysts or section of the static

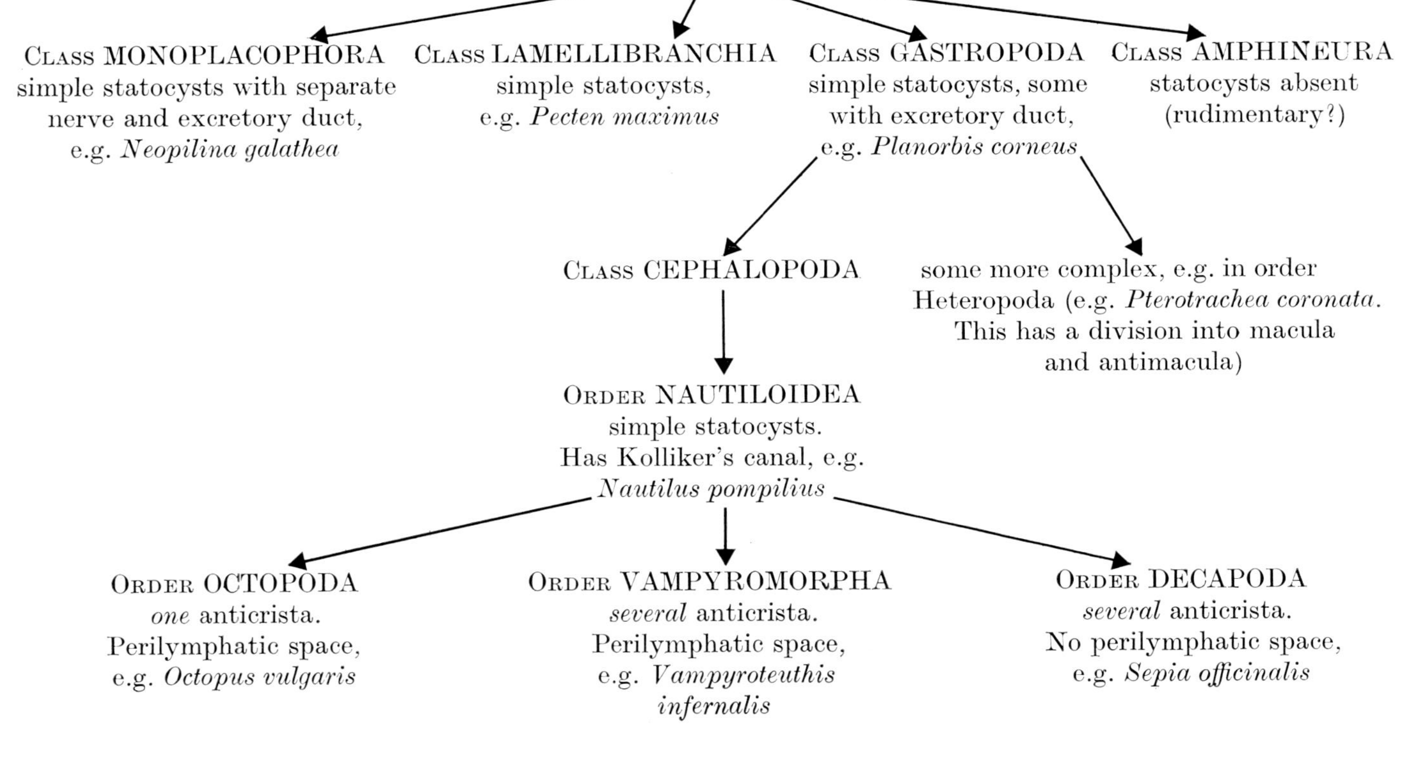

Fig. 14. Summary of the statocyst structure in the molluscs.

nerves, there are no other studies of the physiology and function of statocysts in molluscs other than cephalopods. This is not surprising considering the technical difficulties arising from the very small size of the statocysts. However, such experiments show that the statocysts do have an orientation function in Gastropods (*Lymnaea stagnalis*, Lever and Geuze, 1965; *Pterotrachea coronata*, Tschachotin, 1908), Lamellibranchs (*Pecten inflexus*, Buddenbrock, 1915), and in Cephalopods (Boycott, 1960; Dijkgraaf, 1961).

Many more experiments have been done on cephalopods than on other molluscs. Behavioural and lesion studies (Boycott, 1960; Dijkgraaf, 1961; Wells, 1960) have shown that the statocysts are important in maintaining a proper orientation. Extirpation of *both* statocysts abolishes the nystagmus reaction in *Octopus vulgaris* (Dijkgraaf, 1961) and in *Sepia officinalis* (Dijkgraaf, 1963a) and the animals are unable to maintain their eye-slits in the correct, nearly horizontal orientation (*Octopus vulgaris*, Wells, 1960). Discrete static nerve sections (Dijkgraaf, 1961) and electrophysiological recordings from the crista nerve (Maturana and Sperling, 1963) have shown that the crista responds to angular acceleration. This latter study also showed that these cristal hair cells are able to respond to low frequency vibrations, but behavioural studies have given no evidence of the ability to respond to sounds in *Octopus* (Hubbard, 1960) or in *Sepia* (Dijkgraaf, 1963b).

A THEORY OF THE FUNCTIONAL POLARIZATION OF THE HAIR CELL CILIA IN *Octopus* STATOCYSTS

Polarization of the cilia in vertebrate acoustico-vestibular organs

The correlation of directional sensitivity and some ultra-structural features of the hair cells is obviously important with regard to how the mechanical energy is transduced into electrical changes in the hair cells to signal position, movement, or vibration. In vertebrate systems a theory connecting the orientation of the kinocilia, stereocilia, basal bodies, and basal feet has emerged and it correlates well with the electrophysiological data that is available (see Flock, 1965; Wersäll *et al.*, 1965, for a review) (Fig. 15).

In the vestibular organs and the lateral-line organs the sensory hair bundle is composed of a number of stereocilia and one kinocilium (sometimes two kinocilia, see Flock, 1964). In the organ of Corti the kinocilium is lacking (Duvall, Flock and Wersäll, 1966; Flock, Kimura, Lundquist and Wersäll, 1962). The hair cells of the organ of Corti do,

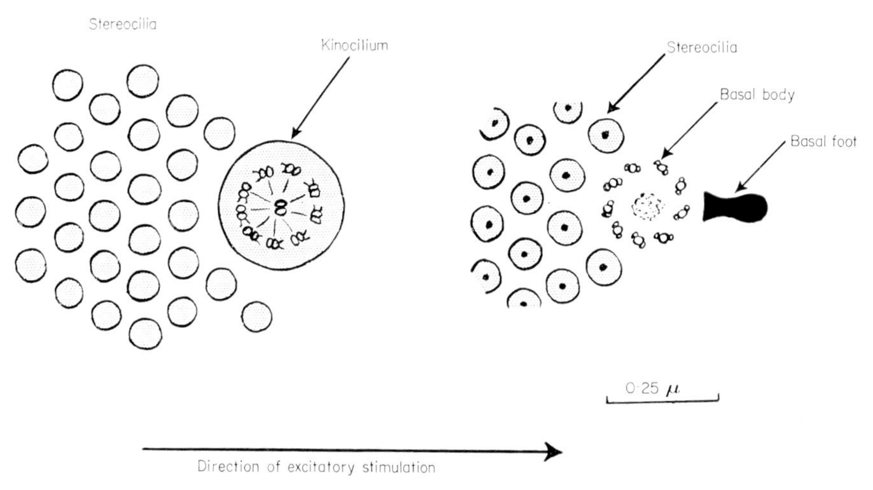

FIG. 15. Ciliary orientation in vertebrate acoustico-vestibular systems (this is not the exact arrangement for the Organ of Corti. See text.)

however, possess a centriole located in a position that corresponds to the position of the basal body of the hair cells in the vestibular receptors or in the lateral-line organ. The kinocilium is present in the organ of Corti in embryos but is shed before birth (Duvall *et al.*, 1966). Another complication is that the vestibular receptors or lateral-line receptors possess a basal foot on the kinociliary basal body. This basal foot points away from the stereocilia. A basal foot is *not* present in the organ of Corti (Duvall *et al.*, 1966). The internal filaments of the kinocilium are also orientated, so that the two central filaments of the kinocilium are at right angles to the axis of the basal foot and so that four pairs of peripheral fibres lie behind the axis of the central fibres and five pairs lie in front (Fig. 15). Lowenstein and Wersäll (1959) first suggested that (in fish) this polarization had a functional correlate. This theory was extended so that it included the basal foot (Flock and Duvall, 1965; Lowenstein, Osborne and Wersäll, 1964). The theory suggests that the sensory hair cells are depolarized when the sensory hairs are displaced towards the kinocilium (or centriole) of the cell, and also, of course, towards the basal foot. The cells are thought to be hyperpolarized (inhibited) when displacement occurs in the opposite direction. Nerve fibres innervating these cells increase their rate of firing during the depolarization phase and decrease their firing rate on hyperpolarization (see Flock, 1965, for a review).

As can be seen, one basic difference between vertebrate hair cells and those of cephalopods is that the cephalopod hair cell bears numerous kinocilia, whereas the hair cells of vertebrates bear one kinocilium and a number of stereocilia. However, the basal foot/kinocilia orientation is the same in both. So the evidence that supports the vertebrate theory is equally applicable (and in fact more so) to cephalopod hair cells as we shall see.

Polarization of motile cilia

The kinocilia in groups of motile cilia are orientated exactly as are the kinocilia in the hair cells of *Octopus*. Gibbons (1961) showed that in the lamellibranch *Anodonta cataracta*, the beating stroke was towards peripheral fibres 5 and 6 and hence towards the basal foot. However, the studies of Afzelius (1961) showed that the cilia beat towards fibre 1 in ctenophore comb plates (basal feet not described). In any case the cilia seem to beat in this plane (Fawcett and Porter, 1954). Certain ciliated protozoa are able to change their direction of beat (Okajima, 1953; Sleigh, 1960) and in these animals it has been shown that the cilia are randomly organized (i.e. each cilium is not organized with regard to the next as shown by the ciliary fibres) (in *Opalina*, Noirot-Timothée, 1959; Wessenberg, 1966) and have no basal foot (*Opalina*, Noirot-Timothée, 1959, not mentioned; Wessenberg, 1966, basal foot absent; in *Trichonympha*, *Pseudotrichonympha*, and *Holomastigotoides*, Gibbons and Grimstone, 1960). So it seems that there is a structural correlate with the direction of beat.

Polarization and physiological response in Octopus

Considering the vertebrate evidence on ciliary orientation and physiology, and the knowledge of the orientation of motile cilia, the following theory is put forward to correlate the ciliary polarization and probable physiological response in *Octopus*. The theory is that passive movement of the cilia in the crista and macula of the statocyst of *Octopus* in the direction of peripheral fibres 5 and 6, and hence in the direction of the basal foot, causes excitation and depolarization of the hair cell, whereas movement in the opposite direction causes hyperpolarization and inhibition. This is summarized in Fig. 8.

The theory would mean that because of its ciliary orientation, the vertical crista would have hair cells that respond to rotation around the vertical axis, the longitudinal crista around the longitudinal axis, and the transverse crista around the transverse axis. At least some of the cristal segments, because groups of cells are orientated in opposite directions, would be able to respond in opposite directions. Because of the orientation of hair cells in the macula, movement of the statolith up or down would stimulate some cells and inhibit others on opposite sides of the macular plate.

Physiological evidence

There are only two studies that give physiological evidence relevant to the theory.

Experimental section of the static nerves of *Octopus* so that only the posterior crista nerve was left was done to five animals in a study by Dijkgraaf (1961). This nerve innervates the vertical crista and posterior sac portion of the static sac. Four of the animals showed a weak but unmistakeable post-rotatory reflex to the side of the preceding rotation (around the vertical axis). Two of these four operated animals reacted equally well to rotation to either side. "It seems", to quote Dijkgraaf, "that the sensory endings . . . are sensitive to rotation in both directions." As can be seen these findings correlate well with the anatomy.

A restricted electrophysiological study has been made (Maturana and Sperling, 1963). Recordings were made from the middle cristal nerve, which supplies the longitudinal crista. Experiments were then made by rotation around the *vertical* axis, but surprisingly no experiments were reported of rotation around the longitudinal axis. The units that were recorded from (a) had no spontaneous activity (unlike vertebrate hair cells, although they are of course not primary sensory cells), (b) were sensitive to low frequency vibrations, (c) responded with a large discharge (100 spikes/sec) to a counter-clockwise rotation in the horizontal plane (as seen from above), and with a smaller discharge (45 spikes/sec) at the *end* of a clockwise rotation. Thus the effective stimulus was angular acceleration (or alternatively, unidirectional linear acceleration) in a counter-clockwise direction (Maturana and Sperling, 1963). As was realized by these workers and was commented on by Wells (1966), this was not the expected result. However, the hair cells were found to be unidirectional in response, which fits in with the theory, but it would be expected that the hair cells would respond to rotation around the longitudinal axis and *not* around the vertical axis. Many explanations can be formulated for this result but as no rotational experiments in other planes were reported it seems pointless to speculate. A further, more extensive study is obviously needed.

So the physiological evidence is equivocal, but at least there is some support for the theory.

Some notes on the transduction process

Exactly how cilia beat is uncertain. Recent studies however, suggest that active sliding of peripheral fibres probably produces the bending (Satir, 1965). Other studies have shown that the ciliary fibres are composed of sub-fibres, which are themselves composed of sub-units (Andre and Thiéry, 1963; Grimstone and Klug, 1966; Pease,

1963; Ringo, 1967) rather as in microtubules (Ledbetter and Porter, 1964). Some information is now accumulating about the biochemistry of these sub-units (Gibbons, 1963, 1965; Gibbons and Rowe, 1965), for example it is known that most of the adenosine triphosphatase activity in cilia (the protein "dynein") is located in the arms of the peripheral fibres (see Gibbons, 1965). Another recent study has histochemically demonstrated that the ciliary rootlet of the human retinal rod cells contains adenosine triphosphatase activity (Matsusaka, 1967). Exactly how these findings will eventually fit together to give a comprehensive explanation of ciliary movement is still uncertain.

When cilia beat electrical changes occur in the cells. Horridge (1965b) has even gone as far as to say that, "it seems reasonable to conclude that the depolarization of the cell sets off the beat of the cilia of that cell." It seems reasonable to suppose that a reverse of this can occur, namely the passive bending of the cilia producing an action potential in the cell. How this transduction occurs is not known but it has been shown (in the Pacinian corpuscle) that the initiation of electrical activity primarily involves a change in sodium permeability (see Gray, 1959). How such a membrane change could occur we do not know, but it could be that mechanical deformation on passive movement of the cilia causes changes in the molecular organization of the excited area (as suggested for the Pacinian corpuscle, Gray, 1959), or movement may activate a chemical reaction, which triggers off the process (Duncan, 1963).

Although it seems reasonable to suppose that the kinocilia and stereocilia are intimately involved in transduction in acoustico-vestibular systems, recent work on olfactory receptors must not be overlooked. (Dr M. S. Laverack pointed that out to me.) The cilia of the olfactory receptor cells were removed, with apparently little effect on the olfactory ability of the cells (Tucker, 1967). In Tucker's experiments the basal bodies were left unaffected and this suggests that they may be of critical importance for transduction. Perhaps the only function of the cilia is to transmit passively the stimulus to the basal body and basal foot where the critical transduction stage occurs.

Acknowledgements

I should like to thank Professor J. Z. Young and Professor E. G. Gray for their continued advice and encouragement; Dr P. Dohrn and the staff of the Stazione Zoologica, Naples, for all their help. The visits to Naples were partly financed by a grant to Professor J. Z. Young from the U.S. Office of Aerospace Research. Dr A. Boyde and Mr R.

Willis kindly assisted me with the Stereoscan study, using a Stereoscan electron microscope provided for him by the Science Research Council; Dr M. Land provided specimens of *Pecten* and gave much useful discussion; Mr S. Waterman provided invaluable photographic assistance and Miss B. T. Charlton much useful technical help.

References

Afzelius, B. (1959). Electron microscopy of the sperm tail. Results obtained with a new fixative. *J. biophys. biochem. Cytol.* **5**: 269–278.

Afzelius, B. (1961). The fine structure of the cilia from ctenophore swimming plates. *J. biophys. biochem. Cytol.* **9**: 383–394.

Andre, J. and Thiéry, J. P. (1963). Mise en évidence d'une sous-structure fibrillaire dans les filaments axonématiques des flagelles. *J. microsc.* **2**: 71–80.

Baecker, R. (1932). Die Mikromorphologie von *Helix pomatia* und einigen anderen Styllommatophoren. *Ergebn. Anat. EntwGesch.* **29**: 449–585.

Barber, V. C. (1965). Preliminary observations on the fine structure of the *Octopus* statocyst. *J. microsc.* **4**: 547–550.

Barber, V. C. (1966a). The fine structure of the statocyst of *Octopus vulgaris*. *Z. Zellforsch. mikrosk. Anat.* **70**: 91–107.

Barber, V. C. (1966b). The morphological polarization of kinocilia in the *Octopus* statocyst. *J. Anat., Lond.* **100**: 685–686.

Barber, V. C., Evans, E. and Land, M. F. (1966). The fine structure of the eye of the mollusc, *Pecten maximus*. *Z. Zellforsch. mikrosk. Anat.* **76**: 295–312.

Barber, V. C. and Graziadei, P. (1965). The fine structure of cephalopod blood vessels. I. Some smaller peripheral vessels. *Z. Zellforsch. mikrosk. Anat.* **66**: 765–781.

Barber, V. C. and Graziadei, P. (1966a). Blood vessels of cephalopods: their fine structure and innervation. *Symp. Electr. Activ. Blood Vessels.* (Cambridge) **1966**. (*Bibl. anat.* **8**: 66–71.) Basel/New York: Karger.

Barber, V. C. and Graziadei, P. (1966b). Cephalopod synaptic structure. *6th Int. Congr. Electron Microscopy* (Kyoto) **2**: 433–434. Tokyo: Maruzen Co. Ltd.

Barber, V. C. and Graziadei, P. (1967a). The fine structure of cephalopod blood vessels. II. The vessels of the nervous system. *Z. Zellforsch. mikrosk. Anat.* **77**: 147–161.

Barber, V. C. and Graziadei, P. (1967b). The fine structure of cephalopod blood vessels. III. Vessel innervation. *Z. Zellforsch. mikrosk. Anat.* **77**: 162–174.

Boycott, B. B. (1960). The functioning of the statocysts of *Octopus vulgaris*. *Proc. R. Soc.* (B) **152**: 78–87.

Buddenbrock, W. von. (1915). Die Statocyst von *Pecten*, ihre Histologie und Physiologie. *Zool. Jb.* (Zool.) **35**: 301–356.

Bullock, T. H. and Horridge, G. A. (1965). *Structure and function in the nervous system of invertebrates.* 2 Vols. San Francisco and London: W. H. Freeman.

Charles, G. H. (1966). Sense organs (less cephalopods). In *Physiology of Mollusca* **2**: 455–521. Wilbur, K. M. and Yonge, C. M. (eds.). New York and London: Academic Press.

Dijkgraaf, S. (1961). The statocyst of *Octopus vulgaris* as a rotation receptor. *Pubbl. Staz. zool. Napoli* **32**: 64–87.

Dijkgraaf, S. (1963a). Nystagmus and related phenomena in *Sepia officinalis*. *Experientia* **19**: 29.

Dijkgraaf, S. (1963b). Versuche über Schallwahrnehmung bei Tintenfischen. *Naturwissenschaften* **50**: 50.

Duncan, C. J. (1963). Excitatory mechanisms in chemo- and mechanoreceptors. *J. theor. Biol.* **5**: 114–126.

Duvall, A. J., Flock, Å. and Wersäll, J. (1966). The ultrastructure of the sensory hairs and associated organelles of the cochlear inner hair cells, with reference to directional sensitivity. *J. Cell Biol.* **29**: 497–505.

Fawcett, D. W. and Porter, K. R. (1954). A study of the fine structure of ciliated epithelia. *J. Morph.* **94**: 221–282.

Flock, Å. (1964). Structure of the macula utriculi with special reference to directional interplay of sensory responses as revealed by morphological polarization. *J. Cell Biol.* **22**: 413–431.

Flock, Å. (1965). Transducing mechanisms in the lateral line canal organ receptor. *Cold Spring Harb. Symp. quant. Biol.* **30**: 133–145.

Flock, Å. and Duvall, A. J. (1965). The ultrastructure of the kinocilium of the sensory cells in the inner ear and lateral line organs. *J. Cell Biol.* **25**: 1–8.

Flock, Å., Kimura, R., Lundquist, P-G. and Wersäll, J. (1962). Morphological basis of directional sensitivity of the outer hair cells in the Organ of Corti. *J. acoust. Soc. Am.* **34**: 1351–1355.

Gibbons, I. R. (1961). The relationship between the fine structure and direction of beat in gill cilia of a lamellibranch mollusc. *J. biophys. biochem. Cytol.* **11**: 179–205.

Gibbons, I. R. (1963). Studies on the protein components of cilia from *Tetrahymena pyriformis*. *Proc. natn. Acad. Sci. U.S.A.* **50**: 1002–1010.

Gibbons, I. R. (1965). Chemical dissection of cilia. *Archs Biol., Paris* **76**: 317–352.

Gibbons, I. R. and Grimstone, A. V. (1960). On flagellar structure in certain flagellates. *J. biophys. biochem. Cytol.* **7**: 697–716.

Gibbons, I. R. and Rowe, A. J. (1965). Dynein, a protein with adenosine triphosphatase activity from cilia. *Science, N.Y.* **149**: 424–426.

Gray, E. G. (1960). The fine structure of the insect ear. *Phil. Trans. R. Soc.* (B) **243**: 75–94.

Gray, E. G. and Young, J. Z. (1964). Electron microscopy of synaptic structure of *Octopus* brain. *J. Cell Biol.* **21**: 87–103.

Gray, J. A. B. (1959). Initiation of impulses at receptors. In *Handbook of physiology* I: 123–145. Section I. *Neurophysiology*. Field, J. (ed.). Baltimore: Waverley Press.

Graziadei, P. (1964). Electron microscopy of some primary receptors in the sucker of *Octopus vulgaris*. *Z. Zellforsch. mikrosk. Anat.* **64**: 510–522.

Grimstone, A. V. and Klug, A. (1966). Observations on the substructure of flagellar fibres. *J. Cell Sci.* **1**: 351–362.

Hamlyn-Harris, R. (1903). Die Statocysten der Cephalopoden. *Zool. Jb.* (Anat.) **18**: 327–358.

Horridge, G. A. (1965a). Relations between nerves and cilia in ctenophores. *Am. Zool.* **5**: 357–375.

Horridge, G. A. (1965b). Intracellular action potentials associated with the beating of the cilia in ctenophore comb plate cells. *Nature, Lond.* **205**: 602.

Hubbard, S. J. (1960). Hearing and the octopus statocyst. *J. exp. Biol.* **37**: 845–853.

Ishikawa, M. (1924). On the phylogenetic position of the cephalopod genera of Japan based on the structure of statocysts. *J. Coll. Agric. Imp. Univ. Tokyo* **7**: 165–210.

Klein, K. (1931). Die Nervenendigungen in der Statocyste von Sepia. *Z. Zellforsch. mikrosk. Anat.* **14**: 481–516.

Kolmer, W. (1926). Statoreceptoren. Bau der statischen Organe. In *Handbuch der normalen und pathologischen Physiologie.* **11**: 767–790. Receptionsorgane I. Bethe, A., von Bergmann, G., Embden, G. and Ellinger, A. (eds.). Berlin: Julius Springer.

Ledbetter, M. C. and Porter, K. R. (1964). Morphology of microtubules of plant cells. *Science, N.Y.* **144**: 872–874.

Lemche, H. and Wingstrand, K. G. (1959). The anatomy of *Neopilina galathea* Lemche, 1957 (Mollusca, Tryblidiacea). *Galathea Rep.* **3**: 9–73.

Lever, J. and Geuze, J. J. (1965). Some effects of statocyst extirpations in *Lymnaea stagnalis. Malacologia* **2**: 275–280.

Lowenstein, O., Osborne, M. P. and Wersäll, J. (1964). Structure and innervation of the sensory epithelia of the labyrinth in the thornback ray (*Raja clavata*). *Proc. R. Soc.* (B) **160**: 1–12.

Lowenstein, O. and Wersäll, J. (1959). Functional interpretation of the electron microscopic structure of the sensory hairs in the cristae of the elasmobranch *Raja clavata* in terms of directional sensitivity. *Nature, Lond.* **184**: 1807–1808.

Matsusaka, T. (1967). ATP'ase activity in the ciliary rootlet of human retinal rods. *J. Cell Biol.* **33**: 203–208.

Maturana, H. R. and Sperling, S. (1963). Unidirectional response to angular acceleration recorded from the middle cristal nerve in the statocyst of *Octopus vulgaris. Nature, Lond.* **197**: 815–816.

Morton, J. E. (1958). *Molluscs.* London: Hutchinson University Library.

Noirot-Timothée, C. (1959). Recherches sur l'ultrastructure d'*Opalina ranarum. Ann. Sci. nat.* (Zool.) (12) **1**: 265–281.

Okajima, A. (1953). Studies on the metachronal wave in *Opalina*. I. Electrical stimulation with the micro-electrode. *Jap. J. Zool.* **11**: 87–100.

Pease, D. C. (1963). The ultrastructure of flagellar fibrils. *J. Cell Biol.* **18**: 313–326.

Pfeil, E. (1922). Die Statocyste von *Helix pomatia* L. *Z. wiss. Zool.* **119**: 79–113.

Pickford, G. E. (1940). The Vampyromorpha, living-fossil cephalopoda. *Trans. N.Y. Acad. Sci.* (2) **2**: 169–181.

Ringo, D. L. (1967). The arrangement of sub-units in flagellar fibres. *J. Ultrastruct. Res.* **17**: 266–277.

Satir, P. (1965). Studies on cilia. II. Examination of the distal region of the ciliary shaft and the role of filaments in motility. *J. Cell Biol.* **26**: 805–834.

Schmidt, W. (1912). Untersuchungen über die Statocysten unserer einheimischen Schnecken. *Jena. Z. naturwiss.* **48**: 515–562.

Sleigh, M. A. (1960). The form of beat in cilia of *Stentor* and *Opalina. J. exp. Biol.* **37**: 1–10.

Stenzel, H. B. (1964). Living nautilus. In *Treatise on invertebrate paleontology.* Part K. Mollusca. 3. K59–K93. The Geological Society of America and the University of Kansas Press.

Tschachotin, S. (1908). Die Statocyste der Heteropoden. *Z. wiss. Zool.* **90**: 343–422.

Tucker, D. (1967). Olfactory cilia are not required for receptor function. *Fedn Proc. Fedn Am. Socs exp. Biol.* **26**: 544. Abstract 1609.

Wells, M. J. (1960). Proprioception and visual discrimination of orientation in *Octopus*. *J. exp. Biol.* **37**: 489–499.

Wells, M. J. (1966). Cephalopod sense organs. In *Physiology of Mollusca* **2**: 523–545. Wilbur, K. M. and Yonge, C. M. (eds.). New York and London: Academic Press.

Wersäll, J., Flock, Å. and Lundquist, P-G. (1965). Structural basis for directional sensitivity in cochlear and vestibular sensory receptors. *Cold Spring Harb. Symp. quant. Biol.* **30**: 115–132.

Wessenberg, H. (1966). Observations on cortical ultrastructure in *Opalina*. *J. microsc.* **5**: 471–492.

Young, J. Z. (1960). The statocysts of *Octopus vulgaris*. *Proc. R. Soc.* (B) **152**: 3–29.

Young, J. Z. (1965). The central nervous system of *Nautilus*. *Phil. Trans. R. Soc.* (B) **249**: 1–25.

NOTE ADDED IN PROOF

The receptor cells of *Pterotrachea* statocysts bear cilia and microvilli (V. C. Barber and Dilly, in preparation), and those of *Hermissenda* bear cilia (R. M. Eakin, unpublished data). The ciliated cells in *Helix* statocysts are said to be the receptor cells (Quattrini, 1967, *Boll. Soc. ital. Biol. sper.* **43**: 785), which does not agree with the description by Laverack (1968, *Symp. zool. Soc. Lond.* No. 23). The scanning electron microscopical structure of the statocysts of *Eledone* and *Loligo* have been described (Barber and Boyde, 1968, *Z. Zellforsch. mikrosk. Anat.* **84**: 269; and in preparation). The structure of the sense organs of *Nautilus* have been described (Barber, 1967, *J. microsc.* **6**; 1067). The receptor cells in the statocysts of an annelid, *Protodrilus*, are said to bear cilia and microvilli (Merker and Harnack, 1967, *Z. Zellforsch. mikrosk. Anat.* **81**; 221). The evidence for the sliding filament theory of ciliary bending (Satir, 1967, *J. gen. Physiol.* **50**: 241) and for the modes of control of ciliary motion (Kinosita and Murakami, 1967, *Physiol. Rev.* **47**: 53) have been reviewed.

Symp. zool. Soc. Lond. (1968) No. 23, 63–74.

SOME ASPECTS OF THE EFFECT OF LIGHT ON VISUAL PIGMENTS

T. I. SHAW

Queen Mary College, University of London, England

SYNOPSIS

Reviewing some of the literature concerned with changes undergone by the visual pigments after they have been illuminated, particular attention is paid to the evaluation of evidence that the protein component of the pigments undergoes a configurational change. The nature of that change is discussed.

INTRODUCTION

Those visual pigments which have been most carefully investigated have proved to be proteins with a lipid group associated with the chromophore. They are extractable from retinae by detergents. The lipid of the chromophore is retinal, the aldehyde of vitamin A. It occurs in one of two forms retinal$_1$ or retinal$_2$ which are based on corresponding forms of vitamin A_1 and vitamin A_2. The difference between these forms is an extra conjugated double bond which is present in the second form. The retinal$_1$ based pigments form visual purple or rhodopsin as well as at least one cone pigment known as iodopsin, the retinal$_2$ based pigments give visual violet or porphyropsin. From variations in the absorption spectra between rhodopsins extracted from different species it seems that the protein component, known as opsin, must differ, at least slightly, from animal to animal. The nature of the linkage between opsin and retinal has been discussed by several authors but the most widely held current view is based on the work of Morton and Pitt (1955) which indicated that the aldehydic group of the retinal forms a Schiff's base with a free amino group of the opsin by the exclusion of a molecule of water.

CHANGES IN THE ISOMERISM OF RETINAL

The structure of retinal permits the existence of a number of isomeric forms. In particular, it has a ring structure attached to a chain of nine carbon atoms. Within the chain there are alternating double bonds. Such double bonds can, in principle, permit of the existence of isomers by occurring in *cis* or *trans* forms, although owing to steric hindrance by other groupings, not all double bonds may be able to occur in each configuration. Hubbard and Wald (1952) showed that two of the isomers of $retinal_1$ would combine with opsin *in vitro* to give light sensitive pigments. One of these isomers, now known as 11 *cis* $retinal_1$ formed rhodopsin. It is apparently in the 11 *cis* form that retinal occurs in the visual pigments. Another isomer, the 9 *cis*, formed a light sensitive pigment, different in its absorption spectrum from rhodopsin and called iso-rhodopsin. Under the influence of long wavelength light the retinal is converted to a form which is no longer capable of combining with opsin to generate rhodopsin. Hubbard and Wald showed that this inactive form of retinal contained all the double bonds of the carbon chain in the *trans* form. The retinal of the visual pigment is converted from the 11 *cis* to the all *trans* form during the bleaching of the visual pigment. Apparently the change occurs very early in the processes of bleaching and is probably the primary photochemical event, for as Kropf and Hubbard (1958) showed light can re-isomerize the all *trans* retinal while it is still attached to opsin generating rhodopsin and iso-rhodopsin.

INTERMEDIATE COMPOUNDS FORMED IN THE BLEACHING OF VISUAL PIGMENTS

The process of bleaching in pigment extracts is by no means instantaneous, the initial photochemical changes are followed by a series of thermally hastened reactions in which rhodopsin is converted first to prelumirhodopsin (Yoshizawa and Wald, 1963) then to lumirhodopsin which is subsequently transformed to metarhodopsin I and then to metarhodopsin II (Matthews, Hubbard, Brown and Wald, 1963). Subsequently in a rather slow reaction the retinal splits off from the opsin moiety. The various intermediates have been recognized by their absorption spectra, using stabilization both by low temperatures, which slow down the thermal reactions and by the use of dried films in which the decay of the lumirhodopsin and particularly metarhodopsin I is delayed (Wald, Durrell and St George, 1950). It is clear that as a result of illumination more reactions than the simple isomerization of retinal proceed in the visual pigment molecules. Various authors (Wald and Brown, 1952; Radding and Wald, 1956; Hubbard, Brown and Kropf, 1959; Erhardt, Ostroy and Abrahamson, 1966), recognizing this have suggested that there are configurational changes that take place in the opsin. It is the purpose of the present article to gather together the evidence that such changes do occur.

THIOL GROUP EXPOSURE ON BLEACHING

A number of reagents react stoichiometrically with the free thiol groups in protein and Wald and Brown (1952) undertook the estimation of such groups in rhodopsin by titration with silver nitrate in the presence of ammonia. The end point of the titration was determined amperometrically, the electric current flowing through a platinum electrode being recorded. Free thiol groups were detected in the unbleached rhodopsin solution and more became accessible to the silver nitrate after exposure to light. When further data on the concentration of rhodopsin used in the experiments became available, they calculated that between about two and three sulphydryl groups were freed during the bleaching of each rhodopsin molecule (Wald and Brown, 1953). This was found to be true for a number of different species. The point has recently been explored more fully by Erhardt *et al.* (1966) using cattle rhodopsin. They confirmed that sulphydryl groups became accessible to titration on bleaching and in alkaline solutions they found three such groups appeared in each rhodopsin molecule. By cooling the solutions to $-29°$, when metarhodopsin I is stable, they showed that it was only subsequent to the formation of this intermediate that the thiol groups were liberated.

The overall picture is by no means simple, however, for if slightly different conditions are used for the titration of the thiol groups the stoichiometry changes so that at neutral pH a total of four groups are exposed in each rhodopsin molecule. In a later paper Ostroy, Rudney and Abrahamson (1966) found that the results were dependent upon the exact conditions used and that in particular the intermediates depended upon the titrant present in the solutions.

It is certainly attractive to suppose that the exposure of sulphydryl groups arises from some configurational change undergone by the opsin in which groups prior to bleaching are held deep within the protein molecule but that they are brought to the surface after the rhodopsin has been exposed to light. Perhaps, however, the exposure might arise from a partial detachment of the retinal from the opsin which allows the titrants to reach groups previously protected from reacting. As Dartnall (1957) has pointed out there are likely to be many points of attachment of the chromophore group and it is possible that these are successively broken during the thermal reactions which succeed the initial photochemical event.

HYDROGEN ION BINDING PROPERTIES

Solutions of visual pigments tend to become alkaline on bleaching by light. Normally the buffers added to the solutions markedly restrict the pH change, but the effect is pronounced when they are in low concentration. Radding and Wald (1956) who observed the change found that in the first 30 seconds after illumination one hydrogen ion binding site became available in each rhodopsin molecule. There was also a much slower change in which acid was released. It seems clear that even the rapid change does not take place in the earliest stage of rhodopsin bleaching for Erhardt *et al.* (1966) noted that it failed to occur at $-29°C$ in 50% aqueous glycerol solutions. They used cattle pigment and metarhodopsin I was formed at this temperature. Apparently it is subsequent to the formation of metarhodopsin I that the hydrogen ion binding group becomes exposed. Indeed both Matthews *et al.* (1963) and Abrahamson and Erhardt (1964) find that it is during the conversion of metarhodopsin I to metarhodopsin II that the group is revealed. The nature of the acid binding group of metarhodopsin II was studied by investigating the effect of pH on the equilibrium between the two metarhodopsin forms. The results that have been obtained indicate a pK of 6·4 and with this value in view Matthews *et al.* (1963) have suggested that the group involved might be an imidazole group of histidine.

It is noteworthy that the tendency to take up hydrogen ions on bleaching, originally demonstrated in solutions of visual pigment, occurs also in isolated rod suspensions (Falk and Fatt, 1966). Here again the same stoichiometry is observed and the reaction occurs very soon after illumination.

As in the case of thiol group exposure it is difficult to be sure that the change in acid binding is not simply due to a partial detachment of retinal from the opsin exposing the effective group rather than some alteration in the configuration of the protein being responsible for the change.

ABSORPTION IN THE ULTRAVIOLET

Two of the absorption bands of rhodopsin and porphyropsin seem to be attributable to the retinal component of the molecules. These bands are the obvious (A) absorption band in the visible and a rather small band known as the *cis* peak, or B band, which occurs in the near ultraviolet. Other absorption bands are present, however, at shorter wavelengths. Notable among these bands is one occurring at about 278 nm. It is known as the C band. This is apparently due to the opsin and is characteristically placed for the absorption by aromatic amino acids such as tyrosine and tryptophan. The absorption in this region has been investigated by Takagi (1963) who noted that a series of changes took place in this region of the absorption spectrum when cattle rhodopsin solutions were bleached. The band occurring at 278 nm was shifted to a shorter wavelength and so too were other bands occurring at 231, 286·5 and 292 nm. These bands again are probably connected with aromatic amino acids. The results were uninfluenced by pH at least between 4·7 and 9·7 or by ionic strength and neither ageing nor treatment with glycerol had any effect. Surprisingly, however, the changes were reduced by concentrated sucrose solutions and increased by concentrated urea. It was not only in pigment extracts that the effects were noted but they were also shown to occur in the isolated outer segments of rods when these were bleached by light. Using cooled solutions and dried films Takagi found that the changes seemed to occur after the formation of metarhodopsin I and were therefore rather late in the sequence of reactions involved in rhodopsin bleaching.

The changes in ultraviolet absorption by rhodopsin on bleaching cannot, as a whole, be explained in any obvious and simple picture of structural change but they do seem to imply that the opsin itself, in particular the aromatic amino acids, are involved in the alterations undergone after light is absorbed by the pigment. Takagi suggested that slight configurational changes like denaturation occurring in the protein were responsible for the observed results.

COTTON EFFECTS

In view of the evidence that some changes occur in the protein component of the visual pigments during bleaching it seemed important that other techniques should be used, in particular techniques which might reveal the nature of the changes. Late in the last century Cotton had shown that optically active molecules could usefully be studied by measuring the rotation of the plane of polarized light at various wavelengths through an absorption band, a study known as optical rotatory dispersion. In such a study information about various parts of the molecule becomes available. If the chromophore group responsible for the band is dissymmetric; that is it lacks a centre, plane or axis of symmetry, or if it is located in an asymmetric environment, then the rotation varies markedly and characteristically with wavelength. Such a band is said to show a Cotton effect. A second, but related phenomenon, was also discovered by Cotton who found that optically active absorption bands preferentially absorbed either right or left circularly polarized light and are said to show circular dichroism. The two effects of optical rotatory dispersion and circular dichroism are closely related and the results from either can be used to predict the other.

Studies of these effects are particularly relevant to proteins which can be optically active both by virtue of their amino acids which contain an asymmetric carbon atom and by their configuration, for they may exist as helical structures. Helices are themselves necessarily dissymmetric and must contribute to the optical activity of the whole molecule. We can notice in passing that retinal has no asymmetric carbon atom and, indeed, has a plane of symmetry through the molecule. We might thus expect that the retinal would contribute nothing to measurements on optical rotatory dispersion or circular dichroism studies on rhodopsin.

There are two groups of Cotton effects which have been studied on polypeptides and proteins and which are relevant to the present discussion. The first was noted by Stryer and Blout (1961) who found that when an optically inactive dye was attached to a polypeptide molecule the visible absorption band of the dye showed a marked Cotton effect. This occurred if the polypeptide was in the α helical configuration. If, however, the L-polyglutamic acid, which was the polypeptide used, was in the random coil configuration the absorption band ceased to be optically active. Evidently the asymmetric environment provided by the α helix imposes itself on the otherwise symmetric dye chromophore. The second group of relevant Cotton effects relate to the absorption bands associated with the backbone chain of polypeptides and proteins. These absorption bands are located in the far ultraviolet of the spectrum. Studies on synthetic polypeptides made from L-amino acids show that when these are in the form of α helices there are three separate bands which show circular dichroism (Holzwarth and Doty, 1965). Two of the bands, located at 222 and 206 nm, are negatively dichroic, that is they preferentially absorb right circularly polarized light; the third is at 193 nm and is positive. Proteins which are presumed to contain an α helix show a similar circular dichroism (e.g. Mommaerts, 1966). However, the dichroism in the far ultraviolet depends on configuration and polypeptides having a random coil configuration show only a single, sharp, negatively dichroic band located at 200 nm (Holzwarth and Doty, 1965). A polypeptide, apparently in the β configuration showed yet a third pattern of circular dichroism (Sarkar and Doty, 1966), and a similar pattern was present in proteins which contained appreciable amounts of the β structure in their molecule. It is clear that studies on the optical rotatory dispersion or circular dichroism of rhodopsin may be expected to yield some insight into whether the opsin does change in configuration as the molecule is bleached.

The optical rotatory dispersion of rhodopsin

Two studies have been carried out upon the optical rotatory dispersion of rhodopsin. Williams (1966) found that there were changes in the rotation of the plane of polarized light in solutions of frog, *Rana pipiens*, visual purple when this was exposed to light. He used a non-optically active detergent to extract the pigment from the retinae. Studying the visible region of the spectrum he found the change on bleaching to be represented by an S-shaped curve which is the characteristic pattern observed for chromophore groupings showing a Cotton effect. The chromophore grouping principally responsible must arise in the retinal portion of the visual pigment molecule. However, as was mentioned in the preceding section retinal itself has a plane of symmetry and alone should not give an optically active band. It is clear that the asymmetry must be induced by the opsin molecule and in this connection one is reminded of the asymmetry induced in dyestuffs absorbed to polypeptides in the α helical configuration (Stryer and Blout, 1961). Williams also studied the effect of regenerating the rhodopsin. This he achieved by giving an initial bleaching flash of light followed, after a short time, by a second flash. The second burst of light re-isomerized some of the retinene which had been converted to the all-*trans* form of the molecule by the first flash. Some 11-*cis* retine was formed and regenerated rhodopsin. Undoubtedly one other isomer—the 9-*cis*—was also formed and gave rise to iso-rhodopsin but probably this formed a relatively small fraction of the total pigment. Studying the optical rotatory dispersion of the regenerated pigment Williams found that it was as asymmetric as the original rhodopsin. It is clear therefore that the changes in opsin configuration are reversible.

Another study on the changes in the optical rotatory dispersion of rhodopsin on bleaching has been made by Kito and Takezaki (1966), who used cattle visual pigment. These authors again noted changes in optical activity in the visible region of the spectrum that followed illumination. Indeed down to wavelengths as short as 350 nm the unbleached pigment was more dextrotatory than the bleached form. At still shorter wavelengths the opposite was the case and the curves for the rotation of bleached and unbleached rhodopsin crossed at between 330–340 nm. Of particular interest was their observation that changes occurred not only in solutions of rhodopsin but also in rod particles which had been broken down by sonic treatment. The authors pointed out that the results suggested a change in the helical structure of the opsin arising from the bleaching of the pigment.

The circular dichroism of visual pigment

The optical activity of the chromophore operating in the visible region of the absorption spectrum of visual pigments was first indicated in preliminary studies by Crescitelli and Shaw (1964) and the experiments have subsequently been amplified and extended to the ultraviolet region of the spectrum by Crescitelli, Mommaerts and Shaw (1966). A variety of pigments have been studied, the rhodopsins of the frog, *Rana pipiens*, the bullfrog, *Rana catesbiana* and of cattle, together with the golden visual pigment (chrysopsin) of the conger eel *Conger conger* and the porphyropsin of the carp *Cyprinus carpio*.

It was at first thought that the conger eel pigment was negatively dichroic but it was later shown that this was an error and that in conformity with the other pigments studied it absorbed more left circularly polarized light and was therefore positively dichroic. The difference in absorption between right and left circularly polarized light is quite small. Indeed a solution with a normal optical density of 1·0 at the maximum of the visible absorption band shows a difference in optical density between right and left circular light of only about 5×10^{-4}. Specialized instruments are now available for such measurements.

The pigments were generally extracted by digitonin which is itself optically active and it seemed possible that the digitonin might be imposing an asymmetry on the chromophore. However, this was ruled out by the use of synthetic detergents as extractants. The dichroism persisted in solutions of visual pigments made with cetyltrimethyl ammonium bromide or Triton X-100. Apart from the visible absorption band, visual pigments generally display an absorption band, the B band, located at about 350 nm. As mentioned earlier this is commonly known as the *cis* peak and is associated with the occurrence of retinal in the *cis* configuration. Like the absorption band in the visible, the *cis* peak was also found to show a positive circular dichroism which was as strong as that occurring in the visible band although the ordinary absorption of the *cis* peak is weak. This was true for at least cattle and frog rhodopsin, but the *cis* peak failed to show circular dichroism in the porphyropsin from carp retinae. When any of the pigments were bleached the dichroism in the visible and near ultraviolet (to 320 mn) was lost. The loss occurred whether hydroxylamine was added to the solution or not. In the former case the retinene was detached from the opsin while in the latter case it would have remained in association with it.

The occurrence of a Cotton effect in the visible absorption band of visual pigments, has already been discussed in connection with the results of Williams (1966) in the preceding section. The opsin of the unbleached pigments might be partially in the α helical configuration, thereby imposing an asymmetry on the retinal. If that is so then on bleaching the α helix, or at least that part of it close to the retinene, is presumably lost. To examine this possibility further, the dichroism of digitonin extracts of frog rhodopsin and carp porphyropsin have been measured in the far ultraviolet. These extracts show a negatively dichroic band at 220 nm and, as far as one can penetrate owing to light absorption, a second is indicated at about 210 nm. These are bands characteristic of polypeptides and proteins containing an α helical configuration but it must be remembered that some of the dichroism may be due to protein impurities present in the extracts and need not be entirely attributed to the opsin present. The noteworthy point is that, on bleaching, the dichroism in the far ultraviolet is reduced, although it is not abolished. The reduction amounts to about $\frac{1}{6}$ of the total circular dichroism of frog rhodopsin at 220 nm. Although the same absolute reduction occurs in carp porphyropsin when this is of such a concentration that it shows the same normal absorption in the visible region of the spectrum, the fraction of dichroism lost on bleaching is only about a half of that shown in the frog. It may be that more protein impurities were present in the preparation of carp porphyropsin that was used. The reduction in circular dichroism that occurs in the far ultraviolet when visual pigments are bleached, again suggests that there is a loss of right handed α helix (or a gain of left handed α helix) from opsin during the bleaching. As in all interpretations of studies on the Cotton effect in visual pigments it must be remembered that knowledge on the effects in proteins and polypeptides is still in its infancy and the conclusions must to some extent be regarded as tentative.

It is of some interest to try and estimate how extensive a loss of α helix would be needed to explain the foregoing results. Such calculations have been used previously for proteins (e.g. Mommaerts, 1966). Those α helical polypeptides that have been studied show, at 220 nm, residual elipticities of $3{\cdot}5$–4×10^4 deg cm^2/decimole (Holzwarth and Doty, 1965). This corresponds to circular dichroic absorptions of $10{\cdot}6$–$12{\cdot}1$ optical density units for molar solutions having 1 cm light path length. The change in u.v. circular dichroic absorption shown by frog rhodopsin on bleaching amounted to 5×10^{-3} optical density units when a 1 cm path length of light was used and the rhodopsin was present at such a concentration that the ordinary absorption of the unbleached pigment amounted to $1{\cdot}0$ optical density unit, measured at the maximum of the visible absorption band. The rhodopsin solution, according to the data of Wald and Brown (1953) must have had a concentration of 0·025 mM; the amino acids lost from the α helical state apparently amounted to 0·47–0.41 mM. It would seem therefore that between 17 and 19 amino acids are involved in the change. The calculation however must be regarded with great caution particularly as Sarkar and Doty (1966) have suggested that hydration of groups may influence the circular dichroism.

DISCUSSION

There has been gathered together a number of observations which indicate that apart from a change in the isomerism of retinal brought about by light there are other alterations in the visual pigment molecule which occur on bleaching. Changes in accessible thiol groupings, in hydrogen ion binding and in ultraviolet spectra, all suggest that the opsin is undergoing a configurational change. The suggestion is supported by observations on the Cotton effect which must be related to the protein rather than to the retinal present in visual pigments, though the changes indicated may be unrelated to those indicated by the other techniques. The Cotton effect changes are those one would expect for a reversible partial denaturation of the opsin when a portion of the helical structure is lost. Very tentative calculations suggest that 17–19 amino acid residue in each rhodopsin molecule are involved in the change, at least in frog visual pigment. Hubbard (1954) estimated the molecular weight of visual purple to be about 40,000 which implies that it must contain some hundreds of amino acid residues. The change in configuration indicated by the circular dichroism studies therefore means that rather a small fraction of the molecule is involved in the loss of α helix. However a chain of 18 amino acids arranged as an α helix would extend over 27 Å in length whereas Courtauld models of retinal indicate that even in the all-*trans* form it would only have a length of 17·5 Å. Thus the change in protein configuration would have to be somewhat larger than what one would expect for a change triggered by a prosthetic group as small as retinal.

At the moment it is not clear what role, if any, the configurational changes of the opsin during bleaching play in stimulating the eye. They could be involved in generating an enzymatically active form of opsin or perhaps in influencing the permeability of a membrane. However, many of the results have been obtained in experiments which have only been carried out on extracts of the visual pigments, and only in a few cases have the effects been shown to occur when bleaching takes place in the rods themselves.

Indeed the question arises as to whether the changes occur sufficiently rapidly after the initial photochemical event for them to play any role in stimulating the eye. Such evidence as there is here, from hydrogen ion binding, thiol group exposure and ultraviolet absorption changes indicate that major changes in protein configuration occur after the formation of metarhodopsin I and are therefore too slow to be involved in stimulating the retinal (Wald, Brown and Gibbons, 1963). However, kinetic studies (Erhardt *et al.*, 1966) suggest an internal configuration change involving an ordering process in the region of the prosthetic group is involved in the conversion of lumirhodopsin to metarhodopsin I. As yet there is no data from Cotton effect studies to show whether perhaps these indicate configurational changes occurring in the earliest stages of bleaching. Finally it should be noted that even if the changes discussed above are not involved in stimulating the eye in vision they may be involved in determining the behaviour of the retina which is profoundly affected in its response by the amount of uncombined opsin present (Rushton, 1965).

References

Abrahamson, E. W. and Erhardt, F. (1964). Protein configuration changes in the photobleaching of rhodopsin. *Fedn Proc. Fedn Am. Socs exp. Biol.* **23**: 384.

Crescitelli, F., Mommaerts, W. F. H. M. and Shaw, T. I. (1966). Circular dichroism of visual pigments in the visible and ultraviolet spectral regions. *Proc. natn Acad. Sci. U.S.A.* **56**: 1729–1734.

Crescitelli, F. and Shaw, T. I. (1964). The circular dichroism of some visual pigments. *J. Physiol., Lond.* **175**: 43P–45P.

Dartnall, H. J. A. (1957). *The visual pigments.* London: Methuen.

Erhardt, F., Ostroy, S. E. and Abrahamson, E. W. (1966). Protein configuration changes in the photolysis of rhodopsin. I. The thermal decay of cattle lumirhodopsin *in vitro*. *Biochim. biophys. Acta* **112**: 256–264.

Falk, G. and Fatt, P. (1966). Rapid hydrogen ion uptake of rod outer segments and rhodopsin solutions on illumination. *J. Physiol., Lond.* **183**: 211–224.

Holzwarth, G. and Doty, P. (1965). The ultraviolet circular dichroism of polypeptides. *J. Am. chem. Soc.* **87**: 218–228.

Hubbard, R. (1954). The molecular weight of rhodopsin and the nature of the rhodopsin-digitonin complex. *J. gen. Physiol.* **37**: 381–399.

Hubbard, R., Brown, P. K. and Kropf, A. (1959). Action of light on visual pigments. *Nature, Lond.* **183**: 442–448.

Hubbard, R. and Wald, G. (1952). *Cis-Trans* isomers of vitamin A and retinene in the rhodopsin system. *J. gen. Physiol.* **36**: 269–315.

Kito, Y. and Takezaki, M. (1966). Optical rotation of irradiated rhodopsin solutions. *Nature, Lond.* **211**: 197–198.

Kropf, A. and Hubbard, R. (1958). The mechanism of bleaching rhodopsin. *Ann. N.Y. Acad. Sci.* **74**: 266–280.

Matthews, R. G., Hubbard, R., Brown, P. K. and Wald, G. (1963). Tautomeric forms of metarhodopsin. *J. gen. Physiol.* **47**: 215–240.

Mommaerts, W. F. H. M. (1966). Ultraviolet circular dichroism of myosin. *J. molec. Biol.* **15**: 377–380.

Morton, R. A. and Pitt, G. A. J. (1955). Studies on rhodopsin. 9 : pH and the hydrolysis of indicator yellow. *Biochem. J.* **59**: 128–134.

Ostroy, S. E., Rudney, H. and Abrahamson, E. W. (1966). The sulphydryl groups of rhodopsin. *Biochim. biophys. Acta* **126**: 409–412.

Radding, C. M. and Wald, G. (1956). Acid base properties of rhodopsin and opsin. *J. gen. Physiol.* **39**: 909–922.

Rushton, W. A. H. (1965). Bleached rhodopsin and visual adaptation. *J. Physiol., Lond.* **181**: 645–655.

Sarkar, P. K. and Doty, P. (1966). The optical rotatory properties of the β configuration in polypeptides and proteins. *Proc. natn. Acad. Sci. U.S.A.* **55**: 981–989.

Stryer, L. S. and Blout, E. R. (1961). Optical rotatory dispersion of dyes bound to macromolecules. Cationic dyes: Polyglutamic acid complexes. *J. Am. chem. Soc.* **83**: 1411–1418.

Takagi, M. (1963). Studies on the ultraviolet spectral displacements of cattle rhodopsin. *Biochim. biophys. Acta* **66**: 328–340.

Wald, G. and Brown, P. K. (1952). The role of sulphydryl groups in the bleaching and synthesis of rhodopsin. *J. gen. Physiol.* **35**: 797–821.

Wald, G. and Brown, P. K. (1953). The molar extinction of rhodopsin. *J. gen. Physiol.* **37**: 189–200.

Wald, G., Brown, P. K. and Gibbons, I. (1963). The problem of visual excitation. *J. opt. Soc. Am.* **53**: 20–35.

Wald, G., Durrell, J. and St. George, R. C. C. (1950). The light reaction in the bleaching of rhodopsin. *Science, N.Y.* **111**: 179–181.

Williams, T. P. (1966). Induced asymmetry in the prosthetic group of rhodopsin. *Vision Res.* **6**: 293–300.

Yoshizawa, T. and Wald, G. (1963). Prelumirhodopsin and the bleaching of visual pigments. *Nature, Lond.* **197**: 1279–1286.

Symp. zool. Soc. Lond. (1968) No. 23, 75–96.

FUNCTIONAL ASPECTS OF THE OPTICAL AND RETINAL ORGANIZATION OF THE MOLLUSC EYE

M. F. LAND*

Department of Physiology, University College, University of London, England

INTRODUCTION

It is well known that the usual effect of external energy (e.g. light) on an appropriate sense cell is to cause excitation. This excitation takes the form of a depolarizing receptor potential and leads to the initiation of impulses in the sense cell itself or, by synaptic activity, in a second order neurone. However, it has been suspected for a long time now that in molluscs, and particularly in the bivalves, there exist primary visual receptors whose response pattern is the opposite of this, with light causing inhibition of any existing responses, and evoking activity in the cells and their axons only at its termination (Hartline, 1938; Kennedy, 1960). The situations in which these receptors have been found are, not surprisingly, ones in which the behaviourally meaningful stimulus is the removal of light, as for example in the shadow responses of bivalves. The more respectable kinds of receptor, those which are excited by light, are also found in molluscs, but in another context. They occur in places like the cephalic eyes of gastropods, where what the animal requires is information about the actual light intensity, and not just about decreases in it.

The purpose of the present article is to assess the evidence, behavioural, anatomical and physiological, for the existence of these two kinds of receptor in gastropod and bivalve molluscs.

Orientation behaviour and shadow responses

Gastropods and bivalves exhibit two easily distinguished kinds of visual behaviour. Firstly, they can use light to orientate, moving either towards or more commonly away from regions of higher light intensity. These responses usually take the form of a tropotaxis (in the terminology of Fraenkel and Gunn, 1961) although every variant from klinokinesis in the bivalve *Lasaea rubra* (Morton, 1960) to telotaxis in the gastropod *Littorina littorea* (Newell, 1958) has been reported. Secondly, many species can respond to sudden decrease of light intensity, in bivalves by

* Present address: Department of Physiology, University of California, Berkeley, California, U.S.A.

closing the shell and withdrawing the siphons, and in gastropods by withdrawing the body into the shell and adhering tightly to the substratum. The first kind of behaviour is clearly concerned with habitat selection, and the second with defence against those predators which cast shadows.

Orientation behaviour is best known in the gastropods (see review by Charles, 1966) but it is not confined to that group. In spite of the absence of paired cephalic eyes, the usual pre-requisite for phototaxis, many bivalves are capable of precisely directed visual orientation. Different species of *Pecten*, for example, perform either positive or negative phototaxes in which the pallial eyes take part (Buddenbrock and Moller-Racke, 1953) and even in eyeless forms like *Lima hians* (Crozier, 1921) and *Lasaea rubra* (Morton, 1960) orientation by tropotaxis, in the sense that direction finding is immediate "without trial movements" seems to occur. The organ implicated as both receptor and effector appears, in these cases, to be the foot.

Conversely, shadow responses have been most studied in the bivalves (Braun, 1954a, b; Braun and Job, 1965). A few species like *Mya arenaria* respond by withdrawing their siphons to lightening as well as darkening (Hecht, 1918) but this is unusual. Like orientation behaviour, however, shadow responses are not confined to the one group, and gastropods like *Helix pomatia* (Föh, 1932), *Chromodoris zebra* (Crozier and Arey, 1919) and *Onchidium floridanum* (Arey and Crozier, 1921) respond to slight decreases in light intensity in various behaviourally meaningful ways. The situation in *Helix* is interesting. The tentacular eyes are undoubtedly concerned in orientation (Buddenbrock, 1919), but Föh (1932) found that the eyes were completely insensitive to shadow and that the part of the animal most sensitive was the region of the mantle just in front of the shell. This can only mean that different groups of receptors are involved in the two kinds of response. Crozier and Arey (1919) reached similar conclusions for *Chromodoris*.

Types of receptor

Since, in phototaxis, the central nervous system requires to be kept continuously informed about the relative intensities of light coming from different directions, it would be reasonable to suppose that the eyes contain receptors whose discharge is tonic, and whose firing frequency is a function of light intensity; they might, in other words, have characteristics similar to those of *Limulus* ommatidial cells (Hartline and Graham, 1932).

In shadow responses, on the other hand, one might expect to find cells which respond phasically, and only to decreases in intensity. It is not necessary, however, to postulate a special kind of receptor to detect dimming in this way, as "off" responses of a similar sort could be produced by second- or later-order neurones as a result of synaptic inhibition from the axons of "on" responding primary receptors. In fact in barnacles (Gwilliam, 1963), insect dorsal ocelli (Ruck, 1961) and also in *Limulus* (Wilska and Hartline, 1941) this is how "off" responses are produced. Nevertheless, in spite of these findings from arthropods, there are good reasons for believing that in molluscs fundamentally different types of receptor *are* involved in the mediation of the two kinds of behavioural response. The evidence for this statement comes from a number of sources, but it is least equivocal in the case of the receptors in the eyes of the scallop, *Pecten*. This evidence will be considered separately on p. 77, and compared with results from other species on p. 85.

THE SITUATION IN *Pecten*

The remarkable pallial eyes of *Pecten* and related genera (*Chlamys*, *Spondylus*, *Amussium*, see Dakin, 1928) contain two retinae lying one above the other. The earlier authors all agreed that the cells in the two

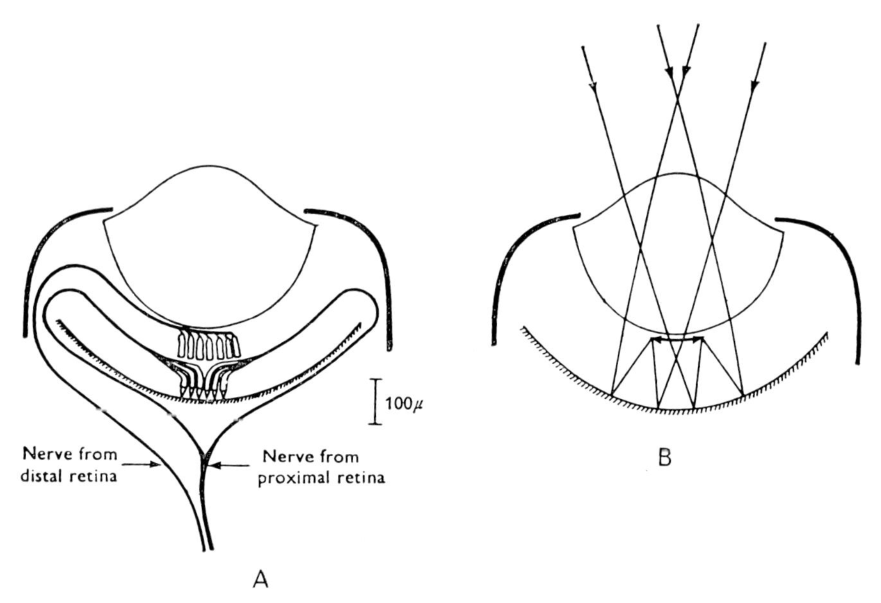

FIG. 1. Diagrams of A, innervation and B, optical system of the eye of *Pecten*. In A, only the central region of the retina is shown: the retina actually extends right across the back of the eye. From Land (1966a).

layers were histologically different, but they often disagreed in their interpretation of the differences (the older literature is reviewed in Miller, 1958). Each retina gives rise to a separate branch of the optic nerve (Fig. 1A), and these two branches remain separate for a short distance before joining and running in through the mantle to the visceral ganglion. There are approximately 5000 receptors in each retina in *P. maximus*, and each gives rise to a fibre in the optic nerve. The optic nerve fibres are the axons of the retinal cells by two criteria: (i), the nerves from both retinae grow out *from* the eye during development (Butcher, 1930) and hence have their cell bodies in the retina itself; (ii), in electron micrographs the axons are seen to be in cytoplasmic continuity with the cells of the two retinae (Barber. Evans and Land, 1967; and W. H. Miller, personal communication).

Optic nerve responses

Hartline (1938) recorded from single fibres in the optic nerve of *Pecten irradians*. He found that individual fibres responded to the onset of illumination, or to the cessation, but not to both. By cutting the nerve from either the proximal retina (nearest the back of the eye) or the distal retina, he showed that whether a fibre gave "on" or "off" responses depended on which retina it came from. Cells of the distal retina gave "off" responses and cells of the proximal retina "on" responses.

The proximal cell responses begin with a burst of impulses at "on", settling to a tonic discharge whose frequency is a function of light intensity. In *P. irradians* the response continues for the duration of illumination (Fig. 2). These responses are very similar to those recorded from *Limulus* optic nerve fibres, and in Hartline's words "are typical of sense organ activity." In the distal fibres there is no response to the onset of, or during illumination, but cessation of illumination is followed, after a latency which may be as long as two seconds at threshold, by a burst of impulses. Hartline (1938) stressed the importance of the preceding period of illumination in the generation of the "off" response: "The higher the intensity of the preceding period of illumination, and the longer its duration (up to one or two minutes), the shorter is the latent period of the 'off' response, the higher its frequency, and the longer does it last."

This statement is equally true of the "off" response in *P. maximus* (Land, 1966a). Thus any proposed mechanism for the "off" response must account for the ability of the distal cells to integrate light intensity over a long period of time, without producing a response in the nerve during this period.

The question that Hartline's investigation (1938) raised, and which he clearly stated, was whether the distal cells were a new kind of receptor sensitive only to the removal of light, or were ganglion

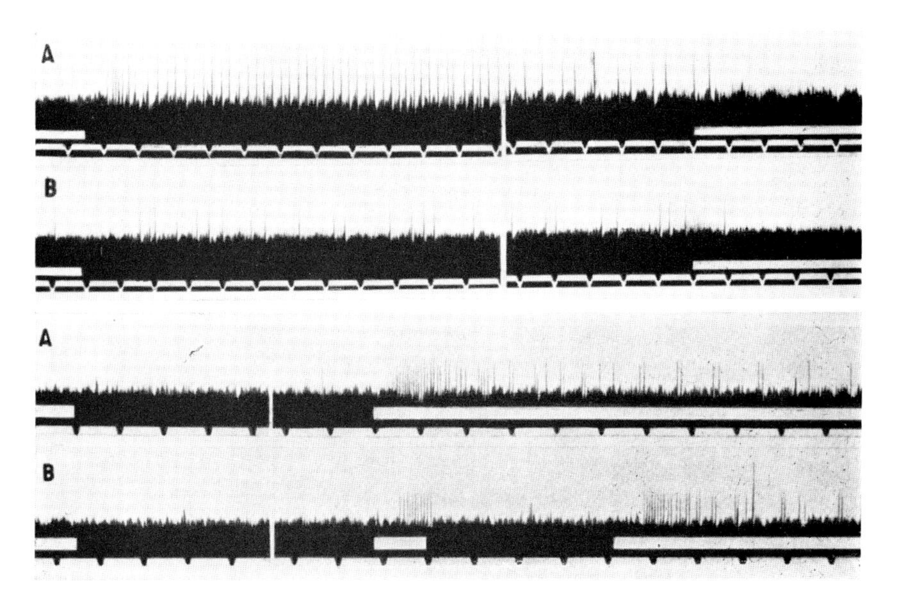

FIG. 2. Responses of single fibres from the retinae of *Pecten irradians*. (From Figs 3 and 4, H. K. Hartline, *J. cell. comp. Physiol.* 11: 471 and 473.) The upper pair of records shows "on" responses in a fibre from the proximal retina. The intensity in A is 100 times that in B, and the amplification in B slightly less than in A. Illumination ends 30 sec after onset.

The lower pair shows "off" responses in a fibre from the distal retina. B shows the effect of re-illumination at the height of an "off" response. Illumination ends 20 sec after onset. In all records the signal marking illumination is blackening of the white line above the time marker. Time in $\frac{1}{5}$ sec intervals.

cells producing "off" responses as a rebound response to the cessation of inhibitory synaptic activity from the proximal cells. The latter explanation would make the distal cells analogous to the second-order cells in barnacles and insect dorsal ocelli (see p. 77).

Electron microscopy

At the time of Hartline's paper, contemporary anatomical opinion favoured the idea that the distal cells were primary receptors since they contained cilium-like appendages which Hesse (1908) suggested had a photoreceptive function. Recent electron microscope studies (Miller, 1958; Barber *et al.*, 1967) have confirmed the existence of these

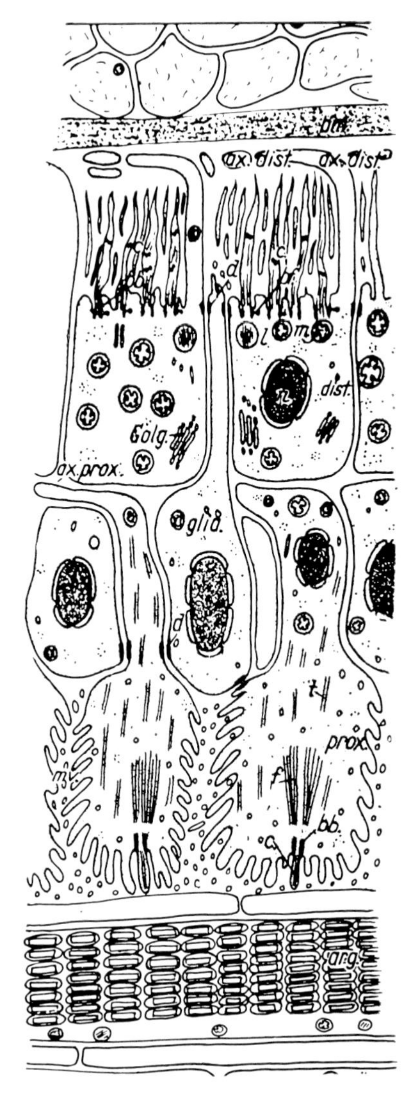

FIG. 3. Diagram of a section through the retina of *Pecten maximus*, based on electron microscopy. From Barber *et al.* (1967).

Below the lens is a septum or basement membrane (*bm*) beneath which lie the distal cells (*dist.*), bearing cilia (*c*) at their septal ends. The cells contain cytolysomes (*l*), mitochondria (*m*), Golgi apparatus (*Golg.*) and are attached to the interstitial cells (*glia*) and other distal cells by desmosomes (*d*). The distal cell axons (*ax. dist*) pass up to, and eventually through the septum. The cilia of the distal cells have basal bodies (*bb.*, with basal feet (*bf*). In the proximal cell layer (*prox*) the cells contain one or two cilia with basal bodies, but the receptive surface is composed of microvilli (*mv*). The cells contain microtubules (*t*) and filaments. They bear axons distally (*ax. prox*). The reflecting argentea (*arg.*) lies below the retina.

appendages. Miller found that in *P. irradians* the septal edge of the distal cells (nearest the front of the eye) contained a number of oval structures, each composed of the coiled, tightly apposed membranes of modified cilia. In *P. maximus* the picture is similar except that the cilia are often not coiled, but extend straight as flattened sacs from the cell body to the septum which runs beneath the lens (Fig. 3). The cilia which form these sacs have a $9+0$ filament content, and the individual fibres, which diverge slightly after leaving their basal body, can often be traced into the sac for a considerable distance. There are of the order of 100 of these modified cilia in each cell, and they form a characteristic, rather regular, lamellar array. The axon arises from the same end of the cell as the cilia, and adjacent to them.

The proximal cells have, in contrast, a structure much more typical of other invertebrate photoreceptors (see Eakin, 1963, 1965a). The presumed receptive surface is situated at the end of the cell furthest from both the axon and the light, and is composed of a rather irregular array of microvilli (Miller, 1960; Barber *et al.*, 1967). The microvilli surround a roughly conical extension of the cell body which contains one or sometimes two short cilia. The vast majority of invertebrate photoreceptors which have so far been examined contain a microvillous surface. This may be a tightly packed "rhabdom" structure in arthropods, or a rather looser array as in most other phyla.

Optics

Physiological evidence that the distal cells are primary receptors came from an investigation of the eye's optics (Land, 1965, 1966b). Although the eye possesses a lens, its focal point lies so far behind the eye that the image produced by it lies nowhere near either set of retinal cells—in fact the image would be about 1200 μ from the lens centre, whereas the proximal retina is only about 300 μ, and the distal retina 200 μ from the lens centre. A real image *is* formed in the eye, not by the lens, but by reflection at the argentea—a multilayer structure composed of guanine crystals which functions as a highly efficient specular reflector for blue-green light (Land, 1966b). The argentea lines the whole of the back of the eye, and as the back of the eye is spherical, or very nearly so, it produces a real, inverted image which is easily visible to an observer looking into the pupil. The image falls on the region of the retina occupied by the ciliary lamellae of the distal cells (Fig. 1B). The optical function of the lens seems to be to provide a certain amount of correction for the spherical aberration inherent in reflection at the spherical surface, in a way rather similar to the action of the corrector plate in a Schmidt astronomical telescope.

Movement responses

If the distal cells are receptors, rather than ganglion cells, the existence of this image should mean that individual cells of the distal retina will respond to decreases in light intensity in small parts of the visual field; in particular they should respond to moving objects. If these cells are "off" receptors, they should respond to the trailing edges of light objects, and the leading edges of dark ones. This was found to

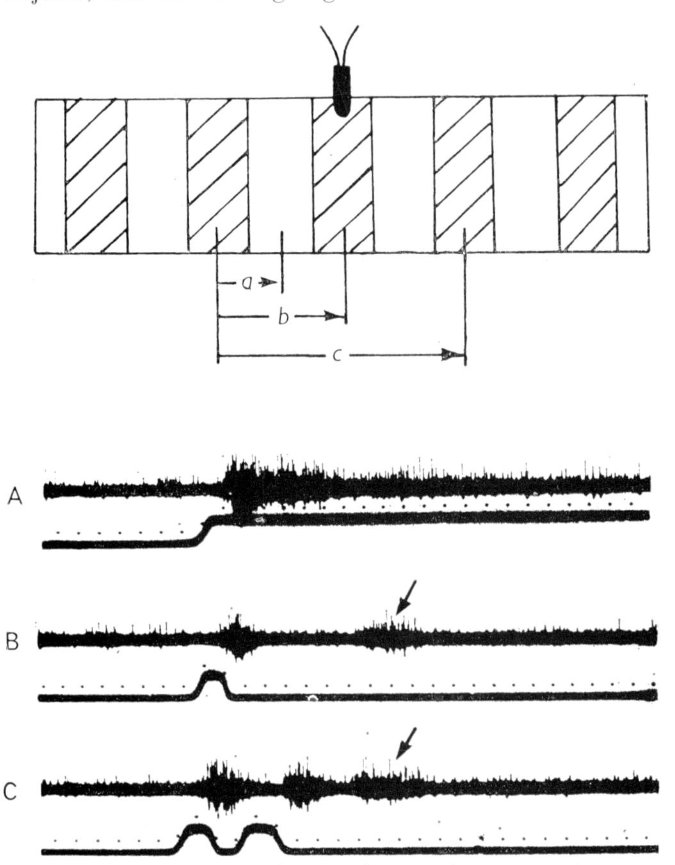

FIG. 4. Records from *Pecten maximus* optic nerve in response to movement of a pattern of equal dark and light stripes displayed on a screen. From Land (1966a).

In each record the pattern was held stationary for 1 min, and then moved through $\frac{1}{2}$, 1, and 2 "wavelengths" of the pattern respectively. Note that responses only occur when previously illuminated regions of the pattern, and hence of the distal retina, are darkened. In B and C, the responses are rapidly suppressed as the responding cells are re-illuminated, in A, this does not occur. The arrows show weak (long latency) bursts probably due to the transient lightening and then darkening of the previously darkened areas.

Time marks: 0·1 sec. Upward movement of the lower trace indicates lightening of photodiode in centre of the screen. Each dark stripe subtends 16° at the eye.

be the case (Land, 1966a). Dark stripes moved in steps through the eye's field of view elicited responses in the optic nerve not only when they entered the field, but also after each movement. Conversely, light stripes only elicited responses provided they had remained stationary in the visual field for a period of time (10 sec or more depending on the brightness of the stripe). In both cases it is removal of light from a particular part of the retina which acts as the stimulus. "Off" responses from the whole distal retina are suppressed by re-illumination: similarly re-illumination of a responding *part* of the retina, achieved by suitably manipulating a pattern of stripes, causes suppression of the response. Figure 4 shows an experiment in which this was done using a grating, movement of which caused no net increase or decrease of light intensity at the eye.

These results show that movement responses have the characteristics of "off" responses, and because there is no image in the region of the proximal cells (which are in fact in contact with the reflector), these "off" responses must be the result of photochemical activity in the distal cells themselves.

Two features of the "off" response are light-dependent.

(i) The build-up of excitability during illumination, which is represented in the "whole field" response by the dependence of the response parameters on the duration of previous illumination, and in movement responses by the time that a light stripe must remain stationary before it will elicit a response on being moved.

(ii) The actual inhibition of action potentials while the distal cells are illuminated.

Both these features are "image dependent", and so both must originate in the distal cells. However, whether both have their origin in a single photochemical process, or represent two photochemically separate antagonistic processes—as Kennedy (1960) has shown to be the case in the rather similar "off" responses from *Spisula* (see below)—cannot at present be answered.

Behaviour

The optical and electrophysiological picture of the *Pecten* eye is now sufficiently detailed to account for the "receptor end" of most of the species' known visual behaviour. Buddenbrock and Moller-Racke (1953) divide the stimuli to which *Pecten* respond into: (*a*) the distribution of brightness in the surrounding environment; (*b*) reduction of light intensity by shadowing; (*c*) movements in the optical environment.

These stimuli result in different patterns of behaviour. The overall distribution of light and dark controls the direction in which the animals swim, and prior to this the direction in which the tentacles are extended. Different species prefer different environments: *P. varius* swims to the darker part of the environment, *P. jacobaeus* to the lighter. According to Buddenbrock and Moller-Racke the animals always swim to the overall lighter or darker region, irrespective of whether that part is a uniform tone or a pattern of stripes. It is reasonable to conclude that the receptors mediating these responses are the proximal cells, firstly, because information relating to the *static* pattern of light and dark is (in *P. irradians*) only available from these receptors. Whether a particular species swims to the light or the dark need not depend on the properties of the receptors, so much as on the central interpretation of the discharge pattern; what is required in either case is continuous information about light intensity. Secondly, when the animal is actually swimming, a patterned environment would act as a very effective stimulus to the distal cells, while a uniform field would not. As scallops ignore this difference when orientating, the distal cells are not likely to be involved.

Although the proximal cells do not receive an image as such they are directionally sensitive to the extent that the pupil only permits each to receive an illuminating cone of about 50°, so there should be some directional element in the discharge pattern from any one eye, as well as in the complete pattern from all 60 or so eyes. There does not therefore seem to be a conflict between the absence of a resolved image on the proximal retina, and Buddenbrock and Moller-Racke's statement that the animals will orientate to a light or dark region by "telo-taxis".

All *Pecten* species respond in much the same way to shading and movement, by withdrawing the tentacles and closing the valves. The response may be rapid and involve both fast and slow muscles, or slower without the fast muscle participating (M. F. Land, unpublished). The two valves are usually not shut tight, although repeated stimuli can cause total, long-lasting closure. Shading and movement do not cause swimming in *P. maximus*, although other stimuli like dilute acid on the tentacles or air in the mantle cavity often do.

Buddenbrock and Moller-Racke used two kinds of test to determine the sensitivity to dimming. In one they used two lights in different positions with respect to the animal, and produced dimming by turning one light off. In the other test the same arrangement was used except that a diffusing screen was placed between the lamps and the animal. Thus the latter method caused the same relative amount of dimming from all directions, while the former produces complete extinction in one part of the visual field and no change in another. They found that the most sensitive species, *P. varius*, responded to a non-directional decrease in intensity of 5·4% but, with the same animal in the two-light situation it responded to a decrease of 0·52%. They rightly interpreted this 10-fold difference in sensitivity as indicating the existence of an image. In *P. maximus* optic nerve responses have been recorded to non-directional decreases of 10% (Land, 1966a) and behavioural responses to 5% decreases.

Buddenbrock and Moller-Racke found that dark objects would cause closure of *P. varius* if they moved through approximately 1° relative to the animal, and in *P. jacobaeus* through 2°. In comparison, optic nerve responses were recorded from *P. maximus* to 2° movements. Behavioural and electrophysiological estimates of sensitivity are thus in excellent agreement.

Actually, the distance on the image which corresponds to a 2° movement in a large (1 mm) eye of *P. maximus* is about 9 μ, i.e. the diameter of a single distal receptor. In view of the fact that in this eye light has already passed, unfocused, through the distal retina before forming the reflected image, thereby reducing the image contrast by a factor of two, the preservation of acuity as good as the grain of the retina is a notable achievement.

Conclusions

Two aspects of the visual environment are important to scallops: the overall pattern of light and dark, and small changes in total or local illumination caused by shadows or movement. Information relating to these two aspects is extracted by separate receptor systems in the eyes, one responding to on-going and sustained illumination, and the other to decrease only. The two kinds of receptors are anatomically distinct, the "on" responding cells containing microvilli, and the "off" responding modified cilia. The "off" system is highly directionally sensitive by virtue of the reflected image, and therefore able to detect movement; the "on" system is less directional, but nevertheless capable of providing the basis for orientation behaviour.

OTHER MOLLUSCS

This section will be devoted to a review of recent anatomical and electrophysiological work on light receptors in other molluscs, and will attempt to compare the structure and functional properties of these receptors with those of the two kinds of cell found in *Pecten*.

Receptors concerned in orientation

The receptors in the paired cephalic eyes of a number of gastropods have been studied by electron microscopy. In *Helix pomatia* (Röhlich and Török, 1963; Schwalbach, Lickfeld and Hahn, 1963), *Helix aspersa* (Eakin, 1965b), *Viviparus maleatus* (Clark, 1963), *Littorina littorea* (Owen and Charles, 1966) and the nudibranch *Hermissenda crassicornis* (Eakin, Westfall and Dennis, 1967) the structure of the receptors is basically similar to that of the proximal cells of *Pecten*. The presumed photoreceptive surface is a tuft of microvilli, regularly arranged in some species, less so in others, at the opposite end of the cell from the origin of the axon. *Littorina*, *Viviparus* and during development *Helix aspersa* cells contain a cilium, as in *Pecten*, the others do not. There is a geometrical difference between these cells and *Pecten* proximal cells in that in all the gastropod eyes the microvillous ends of the cells are nearest the lens, while in *Pecten* they are furthest away. It is difficult to know what significance to attach to this as the layout of the *Pecten* eye is unique anyway.

The only electrophysiological study of such receptors so far published* is the elegant work on the nudibranch *Hermissenda crassicornis* begun by Barth (1964) and continued after Barth's death by Dennis (1965, 1967a, b). Barth succeeded in recording from single receptors in this eye—there are only five—with microelectrodes. He found that individual cells' responses to light could be either a depolarization accompanied by an increase in the rate of discharge of action potentials, or a hyperpolarization accompanied by a cessation of discharge. In the second case the end of hyperpolarization usually led to a rebound "off" response. More complex "on/off" responses were also seen. Dennis succeeded in penetrating pairs of cells simultaneously, and showed that lateral inhibition occurred between them. From a number of lines of evidence, he concluded that, irrespective of whether a particular cell gives "on" or "off" responses, the initial effect of light is to depolarize the cell. A spike in one cell causes an inhibitory post-synaptic potential in adjacent cells; these can summate, leading to hyperpolarization and the inhibition of spiking. Whether a cell gives "on" or "off" responses Dennis attributes mainly to variations from cell to cell in the relative influence of the excitatory (light induced) and inhibitory (post-synaptic) components. Cells can change from being mainly "on" responding to being "off" responding in the course of a single penetration. Dennis suggests that the cell to cell variations, and the changes with time, are due to differences in the state of adaptation to light of the various cells. If this interpretation is correct, *Hermissenda* presents a similar situation to that in *Limulus*, where "on", "off" and "on/off" responses can be elicited from single axons by varying the state of adaptation of adjacent ommatidia, and hence the pattern of

* See note added in proof, p. 96.

lateral inhibition affecting the ommatidium being recorded from Ratliff and Mueller, 1957). It is probably fair to conclude that the primary response to light in *Hermissenda* is in all cases a depolarization, as it is assumed to be in *Pecten* proximal cells, but as Dennis points out the possibility of some primary inhibition occurring as well has not been rigorously excluded.

M. J. Dennis (personal communication) has also recorded from the cephalic eyes of the littoral pulmonate *Onchidium damelli*. The potential recorded from a microelectrode in a receptor showed a depolarizing response to light with an initial phasic peak which declined to a tonic plateau maintained as long as the light was on. The optic nerve response reflected this pattern: an initial burst of impulses followed by a sustained response at a lower frequency.

Although the number of species so far investigated is too small to permit a definite generalization, the evidence so far suggests that gastropod and bivalve receptors of the microvillous type give excitatory, depolarizing responses to light, and a discharge pattern with phasic and tonic components which may or may not be modified by lateral inhibition.

Receptors concerned in shadow responses

The only electron microscope studies of receptors, other than the distal cells of *Pecten*, which are thought to mediate shadow responses are those of Yanase and Sakamoto (1965) on the dorsal eyes of the pulmonate *Onchidium verruculatum*, and Barber and Land (1967) on the siphonal eyes of the cockle, *Cardium edule*.

Some species of *Onchidium* possess in addition to a pair of cephalic eyes a number of eye-like structures 1–200 μ in diameter on their dorsal surface. There are between 18 and 46 of these structures in *O. verruculatum*, arranged on papillae projecting from the animal's back (Stantschinsky, 1908; Hirasaka, 1922). Unlike the cephalic eyes, which are structurally similar to those of other gastropods, the retina in the dorsal eyes is inverted, with the optic nerve penetrating the retina in the centre. The microstructure of the receptor cells is strikingly similar to that of the distal cells of *Pecten*. Each possesses an "outer segment" furthest from the light, containing a number of lamellar structures composed of whorls of flattened cilia; as in *Pecten* the lamellae are joined to the rest of the cell by the ciliary stalks. Each whorl of cilia contains a granular object at its centre.

There is at present no behavioural study of *O. verruculatum*, but Arey and Crozier (1921) investigated the behaviour of *O. floridanum*, a species without dorsal eyes. They found that the dorsal region was extremely sensitive to shadow, but that the cephalic eyes were not. It is difficult to think up a function for the dorsal eyes in *O. verruculatum* except as rather more sophisticated shadow detectors; in fact the possibility arises that they may possess a reasonable image (the animals browse in air, and the eyes have a corneal surface of approximately the right radius for image formation) and could therefore respond to movement as in *Pecten*.

Recently Fujimoto, Yanase and Iwata (1966) have recorded "E.R.G.'s" from the corneae of both the cephalic and dorsal eyes. The former give a cornea-negative potential for the duration of illumination, with no "off" effect; the latter give a much weaker cornea-negative

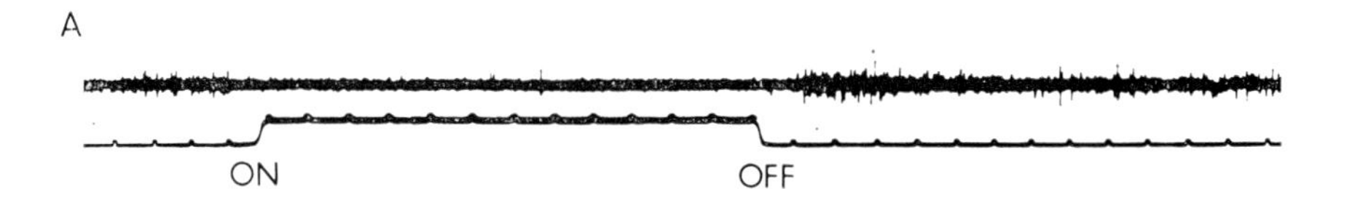

FIG. 5. "Off" responses from *Cardium edule* external pallial nerve. From Barber and Land (1967).

A shows the response to illumination of the eyes after 5 min in darkness. Note spontaneous dark activity is abolished by light, and "off" discharge.

B shows shorter latency "off" response after 5 min in light: "off" response abolished by re-illumination.

Time marks: 0·1 sec. Light intensity 10^3 lumens/m^2.

potential during illumination, but a pronounced cornea-positive response at "off". These findings are certainly compatible with the suggestion that the effect of light on the receptors of the dorsal eyes is inhibitory (the orientation of the cells is opposite to that in the cephalic eyes) with the cells depolarizing only at "off". A comparison with the distal cells in *Pecten* is certainly inviting.

In *Cardium edule* approximately 60 small eyes are borne at the apices of tentacles which arise from around the base of each siphon. Each eye contains a small number of receptors (12–20) whose axons run down the optic tentacles and join the external pallial nerve. Each receptor bears a large number of cilia which form lamellar arrays, not quite as regular as in the distal cells of *Pecten* but nevertheless very similar (Barber and Land, 1967). Recordings from bundles of fibres in the external pallial nerve, near the origin of the optic nerves, showed that they contained only "off" responding fibres (Fig. 5). They gave a weak discharge in the dark, inhibited by illumination, but recommencing at a higher frequency at "off". The latency and frequency of the "off" discharge were, as in *Pecten*, dependent on the duration and intensity of the previous period of illumination. The behavioural response mediated by these eyes is a typical lamellibranch shadow response: withdrawal of the siphons and closure of the shell. In spite of the superficial resemblance of the optical system of these eyes to the mirror system in *Pecten*—*Cardium* eyes are also lined with a multilayer reflector—the small size of these eyes appears to preclude much movement detection. Movements will only evoke behavioural responses if accompanied by a decrease in intensity of about 10%, and unlike *Pecten* movement of a grating, which causes no overall change in illumination, is not an effective stimulus.

The most comprehensive study of a molluscan "off" receptor so far performed is Kennedy's (1960) paper on the single "off" responding neurone in the pallial nerve of the lamellibranch *Spisula sollidissima*. Kennedy found that this fibre was spontaneously active in the dark at a frequency of about 5/sec, and that this activity was inhibited by illumination, but recommenced at "off" with a burst of impulses. He showed that the response had two co-existent components, one excitatory and most sensitive to longer wavelengths, the other inhibitory and more sensitive to shorter wavelengths. In a normal response evoked by white light the inhibitory component dominates during illumination, but by selectively adapting the nerve to blue light Kennedy was able to "expose" the excitatory component, so that the fibre fired to "on" as well as "off". Kennedy was of the opinion that the two pigments which mediated the excitatory and inhibitory processes were both contained in the responding axon. However, his evidence for both excitation and inhibition being primary processes was based on the histological absence of other cell bodies in that part of the nerve, and, as he admits, it is not conclusive. No histologically identifiable ending was found in the nerve, so we do not know whether or not this receptor is anatomically similar to those of *Pecten*, *Onchidium* and *Cardium*.

Kennedy (1960) also recorded from bundles of fibres in the siphonal nerves of *Mya arenaria* and *Venus mercenaria*, and obtained responses to both "on" and "off" with little sustained discharge during either light or dark. It is not, however, known whether *single* fibres respond to both "on" and "off", or to one or the other. The detailed anatomy of the receptors is not known, although Light (1930) has described their light-microscopic appearance. *Spisula*, *Venus* and *Mya* all give behavioural responses to shadow.

Finally, Arvanitaki and Chalazonitis (1961) have recorded light induced responses from central ganglion cells in *Aplysia*. Although these cells have high thresholds for light stimuli, and their situation in the body is such that they cannot be considered to be receptors in the usual sense, they nevertheless show responses which are qualitatively similar to many of those just discussed. Some cells are depolarized by light, others hyperpolarize and give "off" responses when light is removed. In some of the "off" responding cells, hyperpolarization can be seen to be due to summation of inhibitory post-synaptic potentials (cf. *Hermissenda*) and the light response of these cells is presumably secondary. However, in others hyperpolarization is smooth and immediate, and presumably the pigment mediating it is actually in the responding cell. Similar, but more pronounced hyperpolarizations (without i.p.s.p.'s) followed by "off" responses were produced in other cells by infra-red radiation (0·7–4 μ). What the behavioural consequences (if any) of these responses are, is not known.

"Off" response mechanisms

No-one has yet succeeded in inserting a microelectrode into *Pecten* distal cells, or, with the exception of *Aplysia*, into any of the other probable "off" receptors mentioned, so it is not known exactly what electrical changes in the cells are responsible for the "off" responses in the nerve. A retinal mass response has been recorded from *Pecten* (Ratliff, 1956, and my own records) using an electrode inserted into the front of the eye, but the responses obtained were only to on-going and sustained illumination. Possibly the contribution of the distal cells is completely masked by that from the proximal cells.

It would be tempting to suppose that the two aspects of the "off" response in *Pecten*—the build-up of excitability during illumination, and the inhibition of action potentials—are analogous to the excitatory and inhibitory processes which Kennedy found in *Spisula*. However, the crucial experiments of selective adaptation have not yet been performed on *Pecten*, and there are sufficient differences in detail between the stimulus/response characteristics of the *Spisula* and *Pecten* "off" responses (Kennedy, 1960; Land, 1966a) to make one suspicious of a direct comparison. Cronly-Dillon (1966) measured an action spectrum for the closing response of *Pecten maximus*, by finding the intensity at which extinction of the light would just elicit closing at each wavelength. He found that the curve had two peaks, a main one at 475 nm, and a second at 540 nm, and this certainly suggests that two pigments may be involved, as in *Spisula*. However, Cronly-Dillon's use of a white adapting light in addition to the monochromatic source, and the fact that the reflecting argentea exerts a colour-selective effect on the light which reaches the distal receptors (Land, 1966b), mean that this action spectrum is unlikely to be quite that of the primary photochemical process (or processes), so that the involvement of two pigments in the *Pecten* response is not yet certain.

One possibility that can probably be ruled out, in *Pecten* anyway, is that the "off" response is the result of lateral inhibition, as in some of the *Hermissenda* cells (Dennis, 1967a). In the first place there are no anatomically identifiable synapses between the distal cells, or their axons in the vicinity of the eye (Barber, Evans and Land, 1967). Secondly, no "on" responses are recorded in the distal nerve, and it seems hardly likely that the whole retina could be subjected to lateral inhibition without an impulse being fired first.

Primary inhibition, i.e. directly light-induced hyperpolarization or membrane clamping, is almost certain to play a rôle in the *Pecten* "off" response, but until a microelectrode study coupled with spectral sensitivity measurements is made, it will not be possible to say whether or not other processes are also involved.

No explanation can at present be offered for the association, which it has been the main purpose of this article to substantiate, between the highly ciliated structure of these cells, and the production of primary "off" responses.

THE OPTICAL DESIGN OF ORIENTATION EYES AND MOVEMENT DETECTORS

The cephalic eyes of most marine gastropods lack the optical requirements for image formation. Many have spherical or nearly spherical "lenses", but these are usually situated too close to the receptors to form an image. As Pumphrey (1961) pointed out, the focal length of a homogeneous sphere made of dry protein (refractive index 1·53—probably the highest index available for biological lens manufacture) would be about 4 lens radii in water, and this should be the distance of the retina from the lens centre. By suitably decreasing the refractive index of such a lens from the centre to the periphery, the focal length can be reduced to a minimum of 2·5 radii, and this is the condition in the eyes of fish and cephalopods. However, in those marine gastropods whose eyes do possess lens-like structures, e.g. *Murex brandaris* (Hilger, 1885), *Pterocera lambis* (Prince, 1955) there is virtually no space at all between the lens and the retina, and hence no possibility of image formation. The only exceptions to this are the carnivorous heteropods like *Pterotrachea* with retina to lens-centre distances of about 3 lens radii (Hesse, 1900); however, the heteropods presumably use their highly specialized eyes in hunting rather than merely orientation. In other aquatic gastropods the absence of an image means that information about small objects and movements would not be provided by the receptors, although the "wide pinhole" structure of most of these eyes would permit a spatially averaged pattern of light and dark—suitable for orientation purposes—to be transmitted to the CNS. Terrestrial and semi-aquatic gastropods do rather better because the cornea provides an additional refracting surface. *Littorina littorea*, in air, probably does have an image as good as the retina can use (Newell, 1965) but then each receptor subtends about 7° at the lens centre, and the acuity can hardly be considered good.

The argument that in an orientation eye a good image is unnecessary does not apply to eyes used in movement detection. Here the detectability of a moving object will depend on how good individual retinal cells are at responding to intensity changes in as small a region of the environment as possible, and a good image, or other method of obtaining the required spatial resolution would be important. On the whole, molluscs do not seem to have exploited the movement detecting potentialities of "off" receptors by providing them with images, but when they have the results have been spectacular, and optically much more adventurous than the corresponding achievements in cephalic vision.

Pecten's solution to the problem of obtaining good resolution in a reasonably compact space has been to use a mirror; the only other example known to me of an eye where a useful image is probably formed by reflection is in the median eye of the large deep-sea ostracod *Gigantocypris* (Hardy, 1956, p. 229 and plate 12). Another equally ingenious solution is in the compound eyes of the lamellibranch *Arca noae*. These structures, which may number up to 200 on a single specimen (Patten, 1886), consist of 10–80 sense cells embedded in pigment. Each cell has a minute lens-like facet, and each points in a slightly different direction. They enable the animal to respond to shadows and movement (Braun, 1954a) but their microanatomy and electrophysiology have not been examined. They are particularly interesting in that they are paralleled by a very similar structure in a filter-feeding annelid *Branchiomma vesiculosum*. This worm bears similar compound eyes at the ends of its feeding tentacles, and these are also thought to mediate a withdrawal response to shadow (Nicol, 1950). But what is most intriguing from our present point of view, these visual cells in *Branchiomma* contain highly ciliated structures very similar to those of the distal cells of *Pecten* (Krasne and Lawrence, 1966). This is the only example known at present of such a structure with a definitely photoreceptive rôle, outside the Mollusca.

CONCLUSIONS

In gastropods and bivalves behavioural responses to visual stimuli may consist of orientation behaviour of various kinds, or defensive responses to shadow and movement. It is suggested that these two kinds of behaviour are usually mediated by different types of receptor which can be distinguished by anatomical and physiological criteria. Receptors concerned in orientation have microvillous receptive surfaces and respond to ongoing and sustained illumination: receptors concerned in shadow responses have highly ciliated receptive surfaces and give primary responses only to decrease in illumination.

These are necessarily interim statements, based on the small amount of experimental evidence currently available, but it is hoped that they may provoke further research. Inevitably most past work has concentrated on structures recognizable as eyes, but it is clear from the behavioural literature that much mollusc behaviour, orientational and defensive, is based on dermal receptors, and about these virtually nothing is known. It is hoped that future work, particularly on these receptors, may be able to amplify or deflate the present conclusions.

REFERENCES

Arey, L. B. and Crozier, W. J. (1921). On the natural history of *Onchidium*. *J. exp. Zool.* **32**: 443–502.

Arvanitaki, A. and Chalazonitis, N. (1961). Excitatory and inhibitory processes initiated by light and infrared radiations in single identifiable nerve cells (giant ganglion cells of *Aplysia*). In *Nervous inhibitions*: 194–231. Florey, E. (ed.). Oxford: Pergamon.

Barber, V. C. and Land, M. F. (1967). Eye of the cockle, *Cardium edule*: anatomical and physiological observations. *Experientia* **23**: 677.

Barber, V. C., Evans, E. M. and Land, M. F. (1967). The fine structure of the eye of the mollusc *Pecten maximus*. *Z. Zellforsch. mikrosk. anat.* **76**: 295–312.

Barth, J. (1964). Intracellular recordings from photosensory neurones in the eyes of a nudibranch mollusc (*Hermissenda crassicornis*). *Comp. Biochem. Physiol.* **11**: 311–315.

Braun, R. (1954a). Zum Lichtsinn facettenaugentragender Muscheln. *Zool. Jb.* (Zool. u. Physiol.) **65**: 91–125.

Braun, R. (1954b). Zum Lichtsinn augenloser Muscheln. *Zool. Jb.* (Zool. u. Physiol.) **65**: 194–208.

Braun, R. and Job, W. (1965). Neues zum Lichtsinn augenloser Muscheln. *Naturwissenschaften* **52**: 485.
Buddenbrock, W. von (1919). Analyse der Lichtreaktionen der Helicidien. *Zool. Jb.* (Zool. u. Physiol.) **37**: 315–360.
Buddenbrock, W. von and Moller-Racke, I. (1953). Über den Lichtsinn von *Pecten. Pubbl. Staz. zool. Napoli* **24**: 217–245.
Butcher, E. O. (1930). The formation, regeneration and transplantation of eyes in *Pecten* (*Gibbus borealis*). *Biol. Bull. mar. biol. Lab. Woods Hole* **59**: 154–164.
Charles, G. H. (1966). Sense organs (less Cephalopods). In *Physiology of Mollusca*: 455–521. Wilbur, K. M. and Yonge, C. M. (eds.). London: Academic Press.
Clark, A. W. (1963). Fine structure of two invertebrate photoreceptor cells. *J. Cell Biol.* **19**: 14A.
Cronly-Dillon, J. R. (1966). Spectral sensitivity of the scallop *Pecten maximus. Science, N.Y.* **151**: 345–346.
Crozier, W. J. (1921). Notes on some problems of adaptation: 5. The phototropism of *Lima. Biol. Bull. mar. biol. Lab. Woods Hole* **41**: 102–105.
Crozier, W. J. and Arey, L. B. (1919). Sensory reactions of *Chromodoris zebra. J. exp. Zool.* **29**: 261–310.
Dakin, W. J. (1928). The eyes of *Pecten*, *Spondylus*, *Amussium* and allied lamellibranchs, with a short discussion on their evolution. *Proc. R. Soc.* (B) **103**: 355–365.
Dennis, M. J. (1965). Lateral inhibition in a simple eye. *Am. Zool.* **5**: 651.
Dennis, M. J. (1967a). Electrophysiology of the visual system of a nudibranch mollusc. *J. Neurophysiol.* **30**: 1439–1465.
Dennis, M. J. (1967b). Interactions between the five receptor cells of a simple eye. In *Invertebrate nervous systems*. Wiersma, C. A. G. (ed.). Univ. Chicago Press.
Eakin, R. M. (1963). Lines of evolution of photoreceptors. In *General physiology of cell specialisation*: 393–425. Mazia, D. and Tyler, A. (eds.). New York: McGraw-Hill.
Eakin, R. M. (1965a). Evolution of photoreceptors. *Cold Spring Harb. Symp. quant. Biol.* **30**: 363–370.
Eakin, R. M. (1965b). Development of photoreceptoral organelles in the eye of the pulmonate snail *Helix aspersa. Am. Zool.* **5**: 249.
Eakin, R. M., Westfall, J. A. and Dennis, M. J. (1967). Fine structure of the eye of a nudibranch mollusc, *Hermissenda crassicornis. J. Cell Sci.* **2**: 349–358.
Föh, H. (1932). Der Schattenreflex bei *Helix pomatia. Zool. Jb.* (Zool. u. Physiol.) **52**: 1–78.
Fraenkel, G. and Gunn, D. L. (1961). *The orientation of animals.* New York: Dover.
Fujimoto, K., Yanase, T., Okuno, Y. and Iwata, K. (1966). Electrical response in *Onchidium* eyes. *Mem. Osaka Gakugei Univ. B.* **15**: 98–108.
Gillary, H. L. and Wolbarsht, M. L. (1967). Electrical responses from the eye of a land snail. *Revue can. Biol.* **26**: 125–134.
Gwilliam, G. F. (1963). The mechanism of the shadow reflex in *Cirripedia. Biol. Bull. mar. biol. Lab. Woods Hole* **125**: 470–485.
Hardy, A. C. (1956). *The open sea.* London: Collins.
Hartline, H. K. (1938). The discharge of impulses in the optic nerve of *Pecten* in response to illumination of the eye. *J. cell. comp. Physiol.* **11**: 465–477.
Hartline, H. K. and Graham, C. H. (1932). Nerve impulses from single receptors in the eye. *J. cell. comp. Physiol.* **1**: 277–295.

Hecht, S. (1918). Sensory equilibrium and dark adaptation in *Mya arenaria*. *J. gen. Physiol.* **1**: 545–558.
Hesse, R. (1900). Untersuchungen über die Organe der Lichtempfindung bei niederen Thieren VI. Die augen einiger Mollusken. *Z. wiss. Zool.* **68**: 379–477.
Hesse, R. (1908). *Das Sehen der niederen Tiere*. Jena: Fischer.
Hilger, C. (1885). Beiträge zur Kenntnis des Gastropodenauges. *Morph. Jb.* **10**: 351–371.
Hirasaka, K. (1922). On the structure of the dorsal eyes of *Onchidium*, with notes on their function. *Annotnes. zool. jap.* **10**: 171–181.
Kennedy, D. (1960). Neural photoreception in a lamellibranch mollusc. *J. gen. Physiol.* **44**: 277–299.
Krasne, F. B. and Lawrence, P. A. (1966). Structure of the photoreceptors in the compound eyespots of *Branchiomma vesiculosum*. *J. Cell Sci.* **1**: 239–248.
Land, M. F. (1965). Image formation by a concave reflector in the eye of the scallop, *Pecten maximus*. *J. Physiol., Lond.* **179**: 138–153.
Land, M. F. (1966a). Activity in the optic nerves of *Pecten maximus* in response to changes in light intensity, and to pattern and movement in the optical environment. *J. exp. Biol.* **45**: 83–99.
Land, M. F. (1966b). A multilayer interference reflector in the eye of the scallop, *Pecten maximus*. *J. exp. Biol.* **45**: 433–447.
Light, V. E. (1930). Photoreceptors in *Mya arenaria*, with special reference to their distribution, structure and function. *J. Morph.* **49**: 1–42.
Miller, W. H. (1958). Derivatives of cilia in the distal sense cells of the retina of *Pecten*. *J. biophys. biochem. Cytol.* **4**: 227–228.
Miller, W. H. (1960). Visual photoreceptor structures. In *The cell* **4**: 325–364. Brachet, J. and Mirsky, A. E. (eds.). London: Academic Press.
Morton, J. E. (1960). The responses and orientation of the bivalve *Lasaea rubra* Montagu. *J. mar. biol. Ass. U.K.* **39**: 5–26.
Newell, G. E. (1958). An experimental analysis of the behaviour of *Littorina littorea* (L.) under natural conditions and in the laboratory. *J. mar. biol. Ass. U.K.* **37**: 241–266.
Newell, G. E. (1965). The eye of *Littorina littorea*. *Proc. zool. Soc. Lond.* **144**: 75–86.
Nicol, J. A. C. (1950). Responses of *Branchiomma vesiculosum* (Montagu) to photic stimulation. *J. mar. biol. Ass. U.K.* **29**: 303–320.
Owen, G. and Charles, G. H. (1966). The fine structure of the retina of *Littorina littorea*. Quoted in Charles, G. H. (1966).
Patten, W. (1886). Eyes of molluscs and arthropods. *Mitt. zool. Stn Neapel* **6**: 542–756.
Prince, J. H. (1955). The molluscan eyestalk: using as an example *Pterocera lumbis*. *Tex. Rep. Biol. Med.* **13**: 323–339.
Pumphrey, R. J. (1961). Concerning vision. In *The cell and the organism*: 193–208. Ramsay, J. A. and Wigglesworth, V. B. (eds.). Cambridge: Univ. Press.
Ratliff, F. (1956). Retinal action potentials in the eye of the scallop. *Biol. Bull. mar. biol. Lab. Woods Hole* **111**: 310.
Ratliff, F. and Mueller, C. G. (1957). Synthesis of On–Off and Off discharges in *Limulus*. *Science, N.Y.* **126**: 840.
Röhlich, P. and Török, L. J. (1963). Die Feinstruktur des Auges der Weinbergschnecke (*Helix pomatia*, L.). *Z. Zellforsch. mikrosk. anat.* **60**: 348–368.

Ruck, P. (1961). Electrophysiology of the insect dorsal ocellus. *J. gen. Physiol.* **44**: 605–657.

Schwalbach, G., Lickfeld, K. G. and Hahn, M. (1963). Der mikromorphologische Aufbau des Linsenauges der Weinbergschnecke (*Helix pomatia*, L.). *Protoplasma* **56**: 242–273.

Stantschinsky, W. (1908). Über den Bau der Rückenaugen und die Histologie der Rückenregion der Oncidien. *Z. wiss. Zool.* **90**: 137–180.

Wilska, A. and Hartline, H. K. (1941). The origin of "off responses" in the optic pathway. *Am. J. Physiol.* **133**: 491–492.

Yanase, T. and Sakamoto, S. (1965). Fine structure of the visual cells of the dorsal eye in molluscan, *Onchidium verruculatum*. *Zool. Mag., Tokyo* **74**: 238–242.

NOTE ADDED IN PROOF

A comprehensive electrophysiological study of a "microvillous" type of receptor has recently been made by Gillary and Wolbarsht (1967), who recorded electrical activity in the eye of the land snail *Otala lactea*. They found that illumination of this eye evoked a negative-going "E.R.G.", and spike activity in the optic nerve. This consisted of an "on" discharge with a tonic and phasic component, but without an "off" response. They found that the action spectrum of the response was the same when the eye was adapted to both red and blue light, and concluded that a single pigment ($\lambda^{max} = 490$ nm) was involved. The responses of "microvillous" receptors in gastropods thus appear entirely similar to the corresponding kind of receptor in *Pecten*.

Symp. zool. Soc. Lond. (1968) No. 23, 97–111.

THE EYE OF THE SLUG, *AGRIOLIMAX RETICULATUS* (Müll.)

P. F. NEWELL

Westfield College, University of London. England

and

G. E. NEWELL

Queen Mary College, University of London, England

SYNOPSIS

The optical properties of the eye of *Agriolimax reticulatus* are discussed and reasons are given for the belief that it is incapable of forming clear retinal images although the dioptric system would suffice to bring light to a focus on the photoreceptors. The eye is adapted to detect changes in light intensity only and for operating at night.

The function of the accessory retina is considered and a few simple experiments are mentioned which suggest that it is an infra-red receptor.

INTRODUCTION

As is well-known, the slug, *Agriolimax reticulatus*, is mainly a nocturnal animal, dwelling below the surface during the hours of daylight, and emerging to crawl and feed on the surface of the soil just after sunset and throughout most of the night, much of its overt activity taking place in darkness or dim light. Under these conditions it is unlikely that its small eyes could be image-forming and might be expected to differ in structure from those described for some prosobranch gastropods which are active under conditions of high light intensity and which are thought to have image-forming eyes (Evans, 1961; Newell, G. E., 1965). Nor does the behaviour of slugs suggest that they respond to light other than to changes in intensity (Newell, P. F., 1966, and in press) which largely govern their diurnal migrations to and from the soil surface.

It is the purpose of this paper to discuss the structure and optical properties of the eyes of *Agriolimax reticulatus* in relation to the behaviour of the animal.

METHODS

Agriolimax reticulatus is a species of slug which is common and easily obtainable at most seasons of the year. The structure of the eye was mainly studied in serial sections of Bouin's-fixed tentacles which had been fully extended by injecting the fluid into the body cavity of slugs previously narcotized with carbon dioxide. The material was embedded in celloidin and wax by Peterfi's method and sections cut at thicknesses ranging from 2 to 7 μ. These were stained in Masson's trichrome, Mallory's trichrome or Heidenhain's iron haemotoxylin with a suitable counterstain. Some series of sections were bleached in acidified potassium chlorate before staining to remove melanin and reveal cell outlines.

Direct estimations of the focal length of excized lenses were made by placing them in dilute (0·6%) salt solution on a microscope slide, shining parallel light through them and by means of the graduations on the fine adjustment of the microscope measuring the distance of the brightest image from the front surface of the lens. For a spherical lens the focal length is then expressed as the distance from the centre of the lens to the principal focus. The refractive index of the cornea was arrived at by estimating in what glycerin/water mixture the excized cornea most nearly became invisible and then using an Abbé refractometer, the refractive index of the fluid was determined.

Sections for preparing electron-micrographs were made on excized eyes fixed either in Palade's fixative buffered at pH 7·2 or in 2% potassium permanganate in tap water (pH 8·9) and subsequently embedded in Araldite.

GENERAL STRUCTURE

The general structure of photoreceptors has been reviewed by Wolken (1962) and the structure of the stylommatophoran eye and its relation to other structures in the ocular tentacle has been made known to us through the work of Hesse (1902), Smith (1906), Yung (1911), Lane (1962), Chétail (1963) and others, whilst the fine structure of the retina of *Helix* has been described by Eakin (1963), Röhlich and Török (1963) and by Schwalbach, Lickfield and Hahn (1963). This account is solely of the eye of *Agriolimax reticulatus* but there is little reason to believe that the eye of other slugs and for that matter, of snails, depart from it in broad outline.

In brief, each eye is borne near the tip of each extended ocular tentacle beneath a thin layer of collagen and a layer of transparent epidermis (conjunctiva). The eye is a closed vesicle, ovoid in shape, with mean outside dimensions of 140 μ and 180 μ (along the optic axis), its wall being composed of a single layer of cells (Fig. 1 and Fig. 8). But these cells are by no means uniform in size and structure.

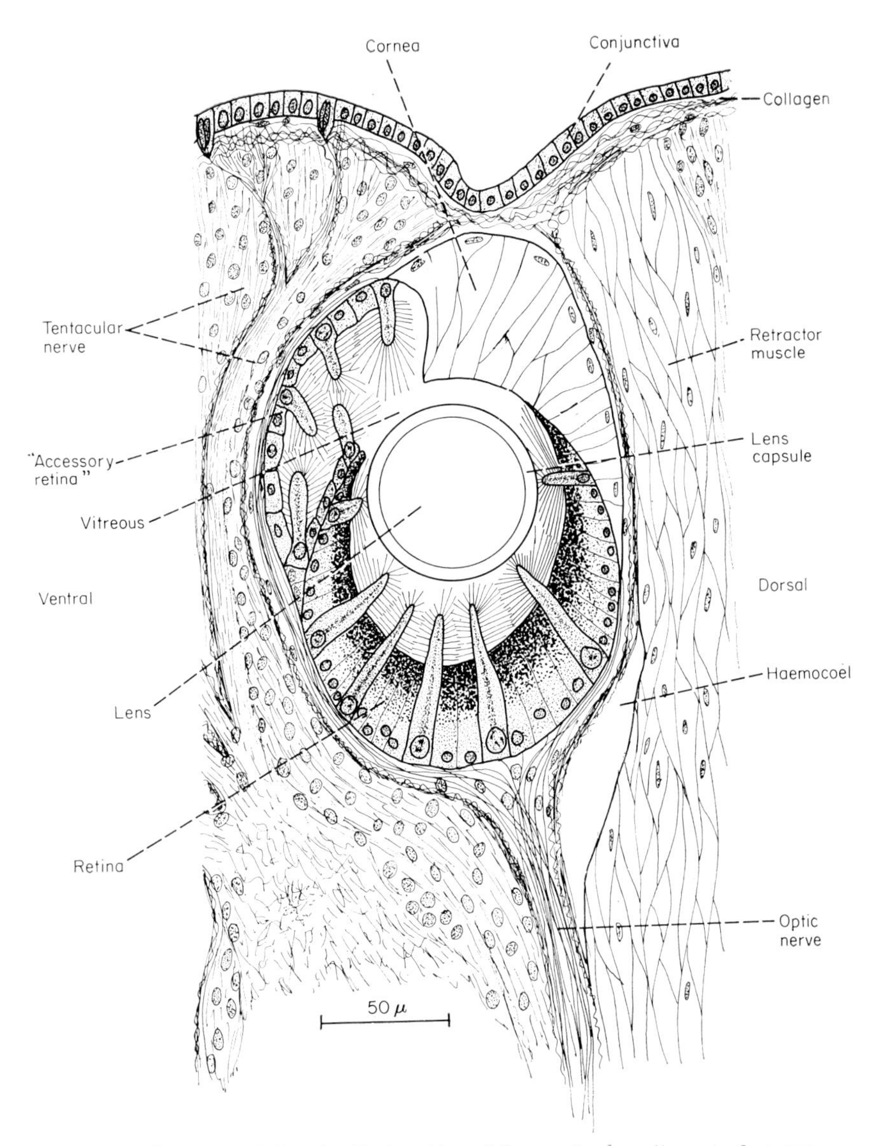

FIG. 1. Diagrammatic longitudinal section of the eye to show its main features.

At the front surface the cells are large, transparent and have a high refractive index of 1·4. They function as a first refractive surface or cornea, refracting light received through the overlying collagen and conjunctiva. In the posterior half of the eye vesicle the cells are of two sorts: photoreceptors and pigment cells, their cell bodies forming a retina about 45 μ thick near the back of the eye but thinner at the sides towards the front surface. The photoreceptor cells are large columnar cells and are few in number, only about ten appearing in any one longitudinal section of thickness 5 μ. They are easily distinguished from pigment cells, which have been bleached to remove the melanin, because each bears at its apex (i.e. at the inner end facing the lens) a stout projection, the rod. This has very numerous, radially arranged fibres or branches, clearly visible under good conditions in the light microscope (Fig. 8) and figured by Hesse (1902) and Smith (1906). These outer segments or rods project for a distance of about 15–25 μ beyond the ends of the cells, i.e. they are about half as long as the cell body itself Their structure will be referred to on p. 101. From the base or outer end of the cell arises a nerve fibre which runs in a radial fashion under the connective tissue capsule of the eye before piercing it to contribute to the optic nerve, which runs back to the cerebral ganglion, following a course near the dorsal surface of the tentacle.

The pigment cells are more numerous than the photoreceptors which they surround. Each pigment cell is packed with spherical granules of dark brown melanin throughout its inner half.

The eye capsule is a delicate layer of collagen fibres about 3 μ thick, continuous at the front of the eye with a similar layer under the conjunctiva.

Inside the eye lies an approximately spherical lens with a diameter of about 65 μ and a focal length of 100 μ (average of 12 lenses). Thus, its radius of curvature is 35 μ and the ratio of focal length/radius of curvature (Mattiessen's, 1880, ratio) is 2·8, which agrees well with that found for the lens of the eye of *Littorina littorea* (Newell, G. E., 1965). The outer layers of the lens stain darker than the rest, indicating that some sort of lens capsule is present but no sign of this was found in electronmicrographs.

Between the lens and the retina is a layer of apparently structureless vitreous humour into which the rods of the photoreceptors project.

No mention has so far been made of a remarkable structure noted by Hesse (1902) but first described by Smith (1906) in *Limax maximus* and named by him the accessory retina. This is a layer of cells lining a pocket, or diverticulum, from the main retina on its ventral side (when the optical tentacle is fully extended) and is thus below the optic axis of the eye. Its cells are of two types: supporting cells, which, unlike those of the main retina are unpigmented, and receptor cells, which resemble those of the main retina in form but which have no regular orientation and are few in number, perhaps twelve only in the whole accessory retina. The vitreous humour of the accessory retina is continuous with that of the rest of the eye.

No suggestion has yet been made as to the function of accessory retina but this question is considered in a subsequent section (p. 105).

FINE STRUCTURE

Seen under the higher powers of magnification of the light microscope the photoreceptors, as has been mentioned, appear as tall columnar cells with nuclei lying near their bases. Each photoreceptor has a distal segment or rod about 12–25 μ long and about 5–7 μ in

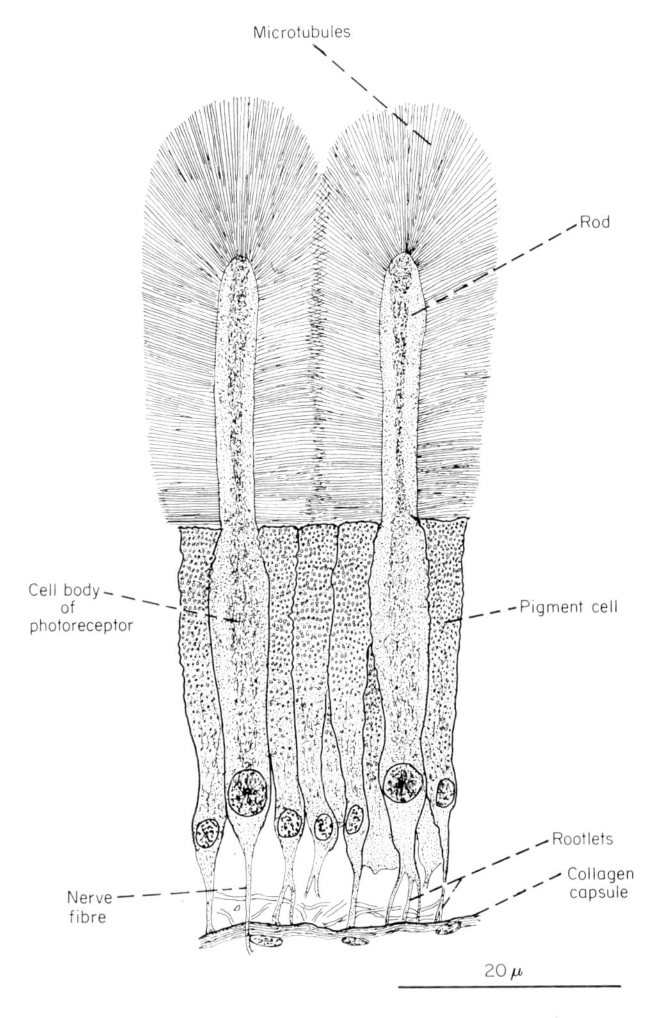

FIG. 2. Vertical section of the retina as seen in the light microscope.

diameter from which sprout very numerous radially arranged fibres giving a bottle-brush appearance, the brushes from adjacent cells overlapping with each other so that practically all the vitreous is filled with them. Under the oil immersion lens the fibres seem to have a diameter of less than 0·5 μ (Fig. 2).

Longitudinal and transverse sections of the eye, about 500 Å thick were used to prepare electronmicrographs, examples of which are reproduced in this paper. From these it is possible to amplify and modify impressions received from studies in the light microscope. In the electronmicrographs a system of radially arranged structures resembling the fibres seen in the light microscope is apparent but they prove to be very numerous branching microtubules (called microvilli in photoreceptors of other organisms by some authors) arising as evaginations of the surface membrane of the rod (Figs 6 and 7). The microtubules are between 7 μ and 12 μ long but only about 0·05 μ in diameter. Obviously they are far below the limits of resolution of a light microscope and it must be concluded that the brush-like appearance seen and described by Smith (1906) is due to groups of parallel microtubules clumping together. Most of the microtubules run a fairly straight course outward from the surface of the rod but others, much fewer in number, are coiled (Fig. 7). This radial arrangement of microtubules is in contrast with that described by Eakin (1963) for the eye of *Helix aspersa* and by Schwalbach, Lickfield and Hahn (1963) and by Röhlich and Török (1963) for *H. pomatia* and termed by them "*mikrozotten*". In these snails the brush of microtubules spring mainly from the ends of the photoreceptors and mostly lie parallel to the direction of the incident light whereas in the eye of the slug most of the tubules are at right angles to the light and in this respect resemble the eyes of cephalopods and arthropods. The microtubules are filled with a substance which has an irregular granular appearance in electron micrographs.

The core or central part of each rod is rich in large mitochondria whilst its more superficial layers are largely occupied by a much folded endoplasmic reticulum. Also within the rod are multivesicular bodies which seem to be filled with small vesicles budded off from Golgi bodies, several of which lie in the neighbourhood of the nucleus (Fig. 8). The multivesicular bodies, so-called by Macrae (1963), probably represent the material termed "*Schaumstruktur*" by Schwalbach *et al.* (1963) in the visual cells of *Helix pomatia*. Macrae described similar structures in the eye of the planarian, *Dugesia tigrina*, and suggested that the small vesicles budded off from the Golgi bodies are related to the release of nerve transmitter substances. In slug photoreceptors the multivesicular bodies are most numerous around the base of the cell and vesicles can be seen in the nerve fibre leaving the cell, evidence which supports Macrae's view of their role.

The cell body is in the form of a column with a fairly uniform diameter for most of its length but narrows towards its base below the nucleus where it is drawn out into a nerve fibre and a number of rootlets which rest on the collagen capsule of the eye. The nerve fibres, however, run radially inside the capsule and penetrate it to contribute to the optic nerve at the inner pole of the eye. In its inner third the cell body is indented at numerous places and into the hollows fit processes from the

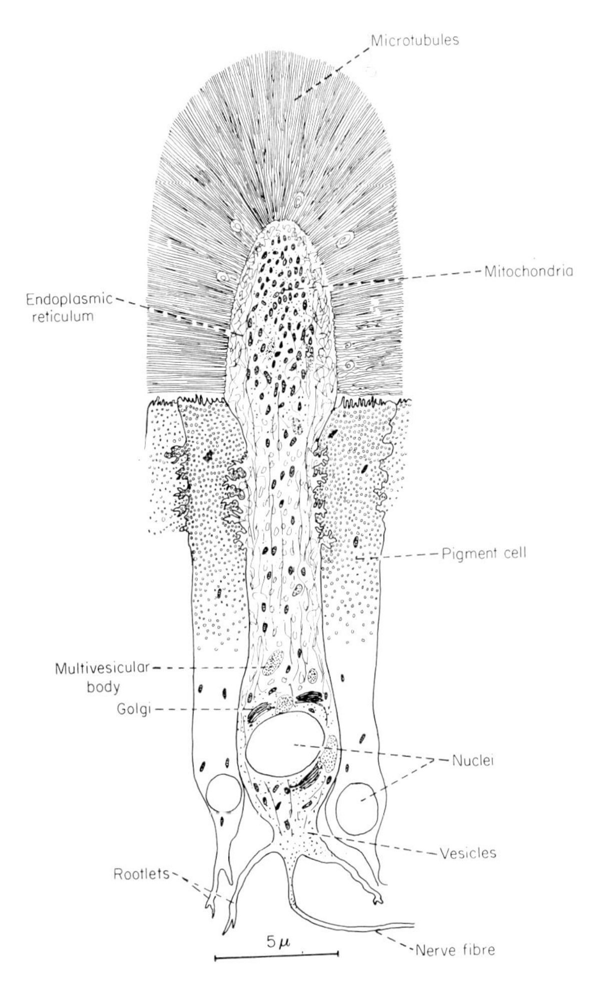

Fig. 3. Diagram to show the general structure of the retinal cells as seen in electron-micrographs. The rod of the photoreceptor is proportionately shorter than those shown in Fig. 2.

pigment cells, the regular shape of the photoreceptors being quite unlike the dendritic arrangement described by Schwalbach *et al.* (1963) for the photoreceptors of *H. pomatia* (Fig. 3). (See note added in proof, p. 111.)

OPTICAL PROPERTIES

The dioptric system consists of the transparent cornea formed by the large cells at the front of the eye vesicle. Its front surface lies just below the subepidermal layer of collagen of the epidermis of the end of the tentacle, which forms a conjunctiva. The refractive index of the cornea is 1·44 and the focal length of the lens in saline is 100 μ. From these data it is possible to draw on a scale diagram of the eye the approximate mean path of selected rays entering the eye (Fig. 4).

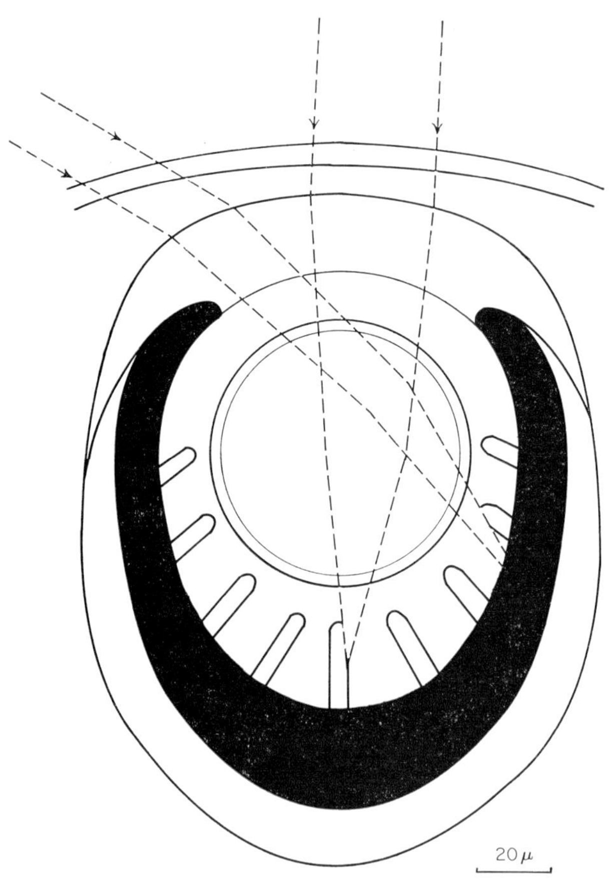

FIG. 4. Scale drawing of a longitudinal section of the eye with rays drawn to show their convergence on the retina.

The first convergence takes place at the convex front surface and this has been estimated using Snell's Law. The convergence produced by the lens was estimated by drawing rays at angles determined graphically from the convergence of rays of parallel light to the principal focus. For simplicity, refraction at the conjunctiva has been neglected though this will be significant for the most oblique rays. Moreover, it appears that the shape of the eye alters a good deal according to the state of contraction of the tentacle. For these reasons the path of the rays can only be approximate. Nevertheless, the diagram shows that in principle, light from distant objects can be brought to a focus on the retina. Yet there are good reasons for thinking that the eye is quite unsuitable for form vision. For one thing the photoreceptor cells are few in number and at the back of the eye are 20 μ apart so that the rods subtend an angle of about 15° at the centre of the lens which gives a measure of the low visual acuity. But the fringe of microtubules must be assumed to play some part in vision, probably to present an enormous surface over which visual pigment is spread and through which incoming light must pass. Since the microtubules of neighbouring photoreceptors overlap, the image must be blurred.

Altogether, the eye seems to be structurally adapted for detecting changes in light intensity only and for operating at night. The cornea and lens would then be functioning to concentrate light on the retina whilst the few photoreceptors, each with the enormous area presented by their microtubules arranged for the most part at right angles to the incident light, would also favour responses to low light intensity.

An eye of this type would be perfectly adequate for slugs to perform the simple orientating behaviour that has been described (P. F. Newell, 1966).

THE FUNCTION OF THE ACCESSORY RETINA

The fact that the accessory retina lies below the optical axis of the eye when the tentacle is fully extended means that light reaching it will not pass through the lens and hence not be concentrated on the photoreceptors. This alone suggests that it plays no part in vision for which the main retina will suffice. Yet the cells in both retinae are structurally alike and it seems plausible to assume that they also are radiation receptors, perhaps in the infra-red band as Callahan (1965a, b) has suggested for some insects. No conclusive evidence can be brought forward in support of this hypothesis but a few observations suggest that it may be correct.

1. Slugs will turn and crawl away from a black body heat emitter.

2. If the ends of the tentacles are cut off, slugs lose the ability to avoid a radiant heat source, although otherwise crawling normally.

   It follows that the receptors concerned lie in the tentacles.

3. It can be noticed that when a slug is crawling, the tip of first one and then the other tentacle is temporarily withdrawn and a study of sections of a tentacle shows that this has the effect of rotating the eye so that the accessory retina becomes displaced

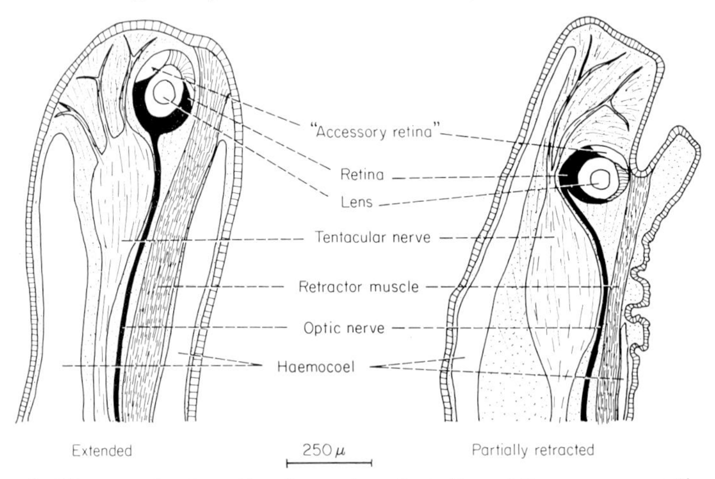

FIG. 5. Diagrams to show the change in orientation of the accessory retina as the eye is withdrawn into the tentacle.

to lie under the outer surface of the tentacle at the base of a pit, thus placing it nearer to an outside source of radiation than the main retina (Fig. 5).

SUMMARY

The eyes of the slug, *Agriolimax reticulatus* are closed, ovoid vesicles with mean outside dimensions of 140 μ and 180 μ. Each eye lies at the tip of an ocular tentacle into which it can be retracted. Inside the eye is a spherical lens with a radius of 65 μ and a focal length of 100 μ in saline so that the ratio of focal length to radius of curvature is 2·8.

Each eye has a thin capsule of collagen continuous with that underlying the transparent epidermis or conjunctiva. The front surface of the eye is composed of large, transparent cells with the high refractive index of 1·4, which function as a cornea.

The rest of the walls of the eye vesicle are the columnar cells of the retina and these are of two types, photoreceptors and pigment cells. Towards the back of the eye these have a height of about 45 μ but less

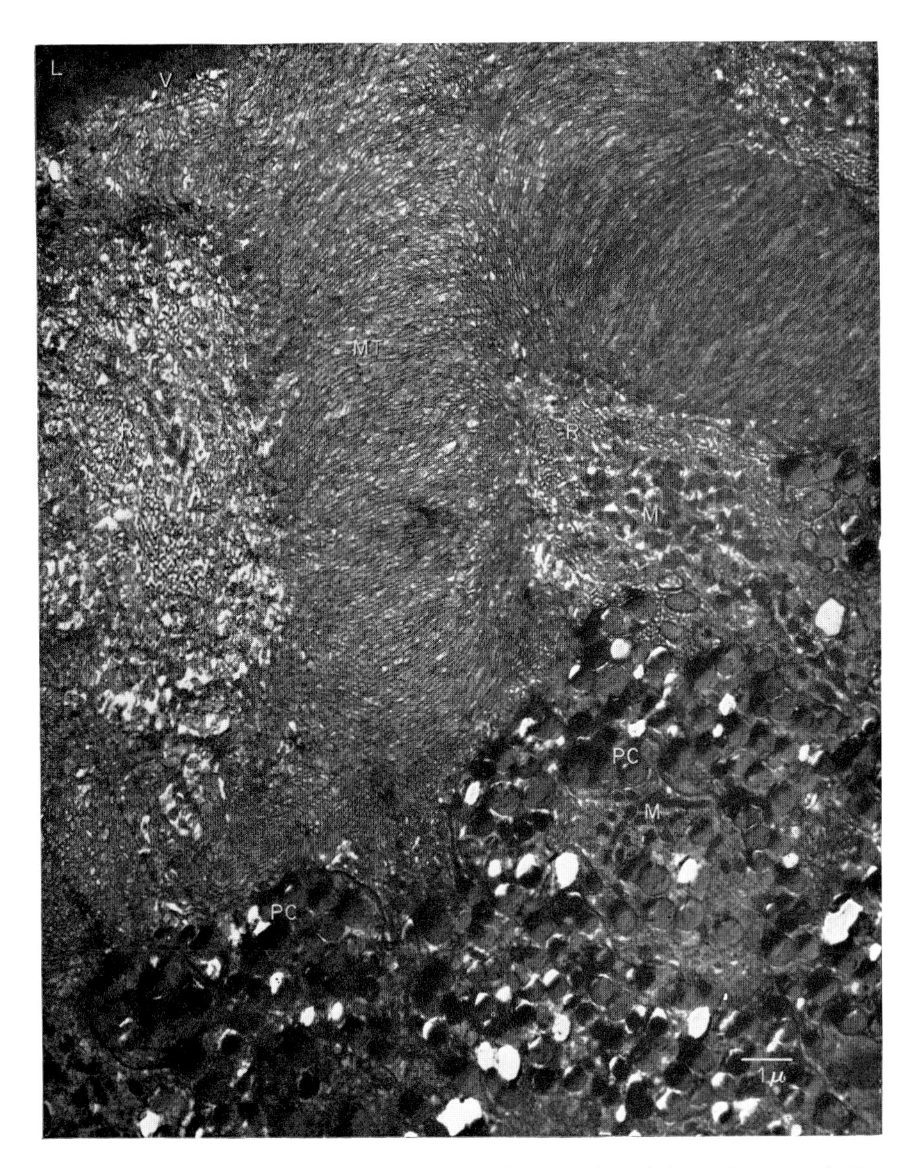

FIG. 6. Electronmicrograph of a slightly oblique section of the retina towards the upper ends of the cells. Osmium fixed. *L*—lens; *M*—mitochondria; *MT*—microtubules; *PC*—pigment cell; *R*—rod; *V*—vitreous body.

at the sides of the eye. The photoreceptors bear at the apex a rod-like projection, between 15 and 20 μ long and this has a fringe of what appear to be radially arranged fibres under the light microscope but which electron micrographs reveal to be microtubules about 7–12 μ long. Each

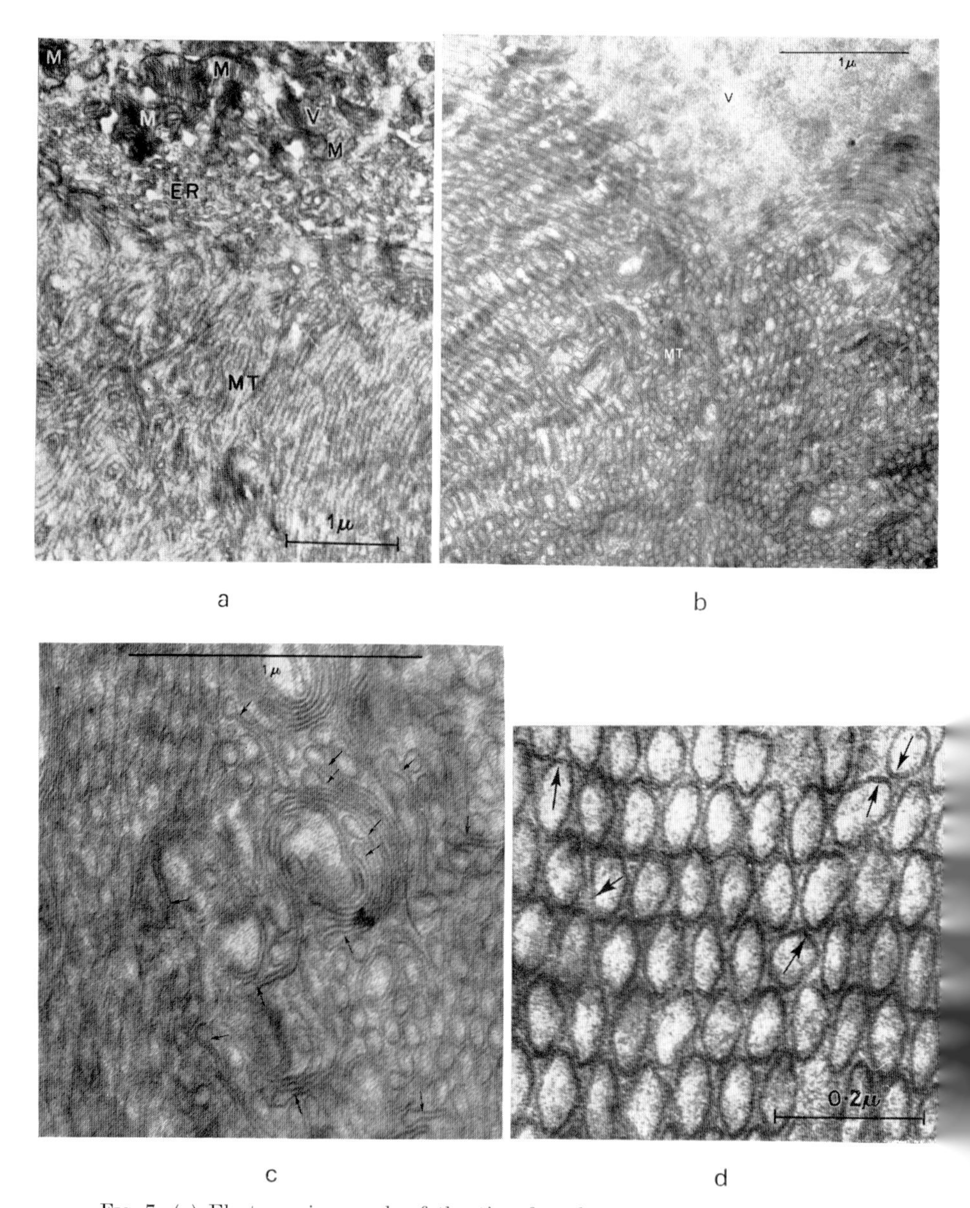

Fig. 7. (a) Electronmicrograph of the tip of a photoreceptor showing the microtubules arising from the cell body. Permanganate fixed. (b) Electronmicrograph of the microtubules. Permanganate fixed. (c) Coiled microtubles. Pointers indicate sites of possible budding of the microtubules. Permanganate fixed. (d) Electronmicrograph showing transverse sections of microtubules. Pointers indicate places where the three-layered structure of the walls are visible. Permanganate fixed. *ER*—endoplasmic reticulum; *M*—mitochondria; *MT*—microtubules; *MV*—multivesicular body; *V*—vitreous body.

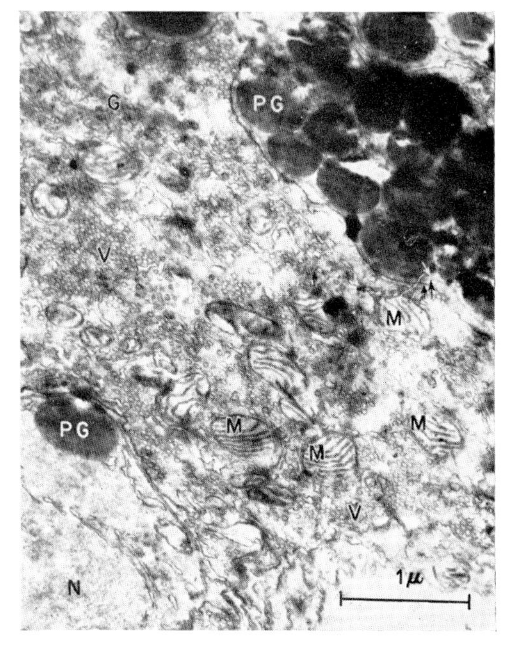

a

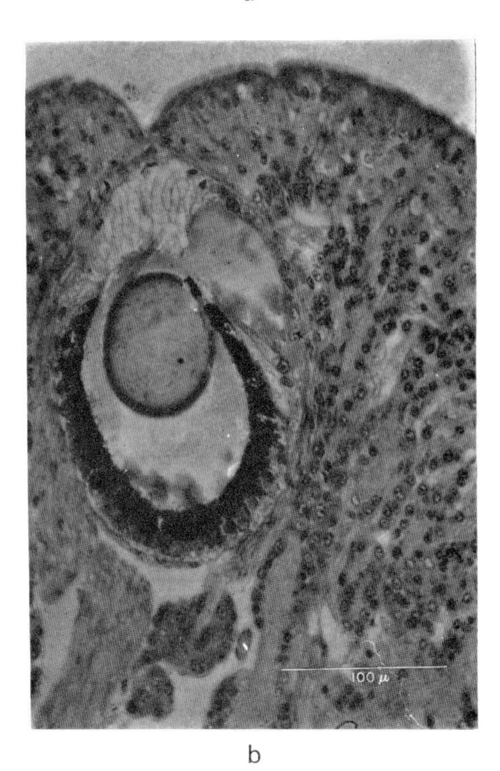

b

FIG. 8. (a) Electronmicrograph of the basal portion of two adjacent retinal cells showing cell contents. The three pointers in the top right hand corner show the cell membranes of a photoreceptor, of a pigment cell and of a pigment granule. Permanganate fixed. *G*—Golgi complex; *M*—mitochondria; *N*—nucleus of photoreceptor; *PG*—pigment granule; *V*—vesicles. (b) Photomicrograph of a nearly median section of the eye. For explanation see Fig. 1, p. 99.

retinal cell has a few rootlets inserted in the eye capsule. In addition each photoreceptor has a nerve fibre from its lower end which contributes to the optic nerve. The pigment cells and photoreceptors interlock throughout their upper thirds. The pigment cells bear short microvilli on their upper surfaces. The brushes of microtubules of adjacent photoreceptors overlap.

The core of each photoreceptor contains numerous large mitochondria and multivesicular bodies are plentiful near the base. There are several Golgi bodies near the nucleus.

The lens and cornea converge light to a focus on the retina but the receptors are few in number and are 20 μ apart at the back of the eye and two adjacent rods subtend an angle of 15° at the centre of the lens so that the visual acuity is poor. Moreover, the overlapping of the microtubules must produce blurring of the retinal image. The eye is adapted to detect changes in light intensity only and to operate at night.

A ventral pocket, lined by receptor cells, from the ventral side of the eye and termed the accessory retina, may be an infra-red receptor.

Acknowledgements

Our grateful thanks are due to Mr J. B. Kirkham for much good advice and for considerable help in the making of the electron micrographs at the E.M. Unit, Queen Mary College. We also wish to thank Mr Tovée for making many of the sections used for light microscopy. Much of the work was done at the Laboratory of the Marine Biological Association, Plymouth, and grateful acknowledgement is made of a grant from the Browne Fund of the Royal Society to G. E. Newell.

References

Barber, V. C. and Land, M. F. (1967). Eye of the cockle, *Cardium edule*: Anatomical and physiological investigations. *Experientia* **23**: 677–678.

Barber, V. C., Evans, E. and Land, M. F. (1967). The fine structure of the eye of the mollusc *Pecten maximus*. *Z. Zellforsch. mikrosk. Anat.* **76**: 295–312.

Callahan, P. S. (1965a). Intermediate and far infrared sensing of nocturnal insects. Part I. Evidences for a far infrared (FIR) electromagnetic theory of communication and sensing in moths and its relationship to the limiting biosphere of the corn earworm. *Ann. ent. Soc. Am.* **58**: 727–745.

Callahan, P. S. (1965b). Intermediate and far infrared sensing of nocturnal insects. Part 2. The compound eye of the corn earworm, *Heliothis zea*, and other moths as a mosaic optic-electromagnetic thermal radiometer. *Ann. ent. Soc. Am.* **58**: 746–756.

Chétail, M. (1963). Étude de la régéneration du tentacule oculaire chez un arionidae (*Arion rufus* L.) et un limacidae (*Agriolimax agrestis* L.) *Archs. Anat. microsc. Morph. exp.* **52** (suppl. 1): 129–203.

Eakin, R. M. (1963). Lines of evolution of photoreceptors. In *General physiology of cell specialization*: 393–425. Mazia, D. and Tyler, A. (eds.). New York: McGraw-Hill.

Eakin, R. M., Westfall, J. A. and Dennis, M. J. (1967). Fine structure of the eye of a nudibranch mollusc, *Hermissenda crassicornis*. *J. Cell Sci.* **2**: 349–358.

Evans, F. (1961). Responses to disturbance of the periwinkle, *Littorina punctata* (Gmelin) on a shore in Ghana. *Proc. zool. Soc. Lond.* **137**: 393–402.

Hesse, R. (1902). Ueber die Retina des Gastropodenauges. *Verh. dt. zool. Ges.* **12**: 121–125.

Krasne, F. B. and Lawrence, P. A. (1966). Structure of the photoreceptors in the compound eyespots of *Branchiomma vesiculosum*. *J. Cell Sci.* **1**: 239–248.

Land, M. F. (1968). Functional aspects of the optical and retinal organization of the molluscan eye. *Symp. zool. Soc. Lond.* No. **23**: 75–96.

Lane, N. J. J. (1962). Neurosecretory cells in the optic tentacles of certain pulmonates. *Q. Jl. microsc. Sci.* **103**: 211–224.

Macrae, E. K. (1963). Observations on the fine structure of photoreceptor cells in the planarian, *Digesia tigrina*. *J. Ultrastruct. Res.* **10**: 334–349.

Mattiessen, H. F. L. (1880). Untersuchungen über den Aplanatismus und die Perioscopie der Krystalllinsen in den Augen der Fische. Pflügers. *Arch. ges. Physiol.* **21**: 287–307.

Newell, G. E. (1965). The eye of *Littorina littorea*. *Proc. zool. Soc. Lond.* **144**: 75–86.

Newell, P. F. (1966). The nocturnal behaviour of slugs. *Med. biol. Illust.* **16**: 146–159.

Newell, P. F. (1968). The measurement of light and temperature as factors controlling the surface activity of the slug, *Agriolimox reticulatus* (Müller). In *The measurement of environmental factors in terrestrial ecology*. Wadsworth, R. M. (ed.). *Symp. Proc. Br. Ecol. Soc.* **1967**: 141–146.

Röhlich, P. and Török, L. J. (1963). Die Fienstruktur des Auges der Wienbergschnecke (*Helix pomatia* L.). *Z. Zellforsch. mikrosk. Anat.* **60**: 348–368.

Schwalbach, G., Lickfield, K. G. and Hahn, M. (1963). Der mikromorpholigische Aufbau des Linsenauges der Wienbergschnecke (*Helix pomatia* L.). *Protoplasma* **56**: 242–273.

Smith, G. (1906). The eyes of certain pulmonate gasteropods, with special reference to the neurofibrillae of *Limax maximus*. *Bull. Mus. comp. Zool. Harv.* **48**: 231–287.

Wolken, J. J. (1962). Photoreceptor structures. *Int. Rev. Cytol.* **11**: 195–218.

Yung, E. (1911). Anatomie et malformations du grand tentacule de l'escargot. *Revue suisse Zool.* **19**: 339–382.

NOTE ADDED IN PROOF

The retinal cells of many animals show traces of ciliary structures, either basal bodies, or axial filaments, or both. (Krasne and Lawrence, 1966; Eakin, Westfall and Dennis, 1967; Barber, Evans and Land, 1967; Barber and Land, 1967.) No evidence of any ciliary structure was found in the slug eye and in this respect it is similar to the eye of *Helix pomatia* described by Röhlich and Török (1963) and Schwalbach, Lickfield and Hahn (1963), and that of *Helix aspersa* (Eakin 1963).

Symp. zool. Soc. Lond. (1968) No. 23, 113–133.

THE PHOTORECEPTORS OF ARTHROPOD EYES

JEROME J. WOLKEN

Biophysical Research Laboratory, Carnegie-Mellon University, Pittsburgh, Pennsylvania, U.S.A.

SYNOPSIS

The compound eye was investigated in several species of insects and crustacea with respect to the structure of its ommatidium, the geometry of the retinula cells that form its rhabdom, and the *fine structure* of its rhabdomeres. It was found that the rhabdoms formed by the retinula cells are either of an *open* type or a *fused* type and that the rhabdomeres are paired except for one asymmetric rhabdomere. The rhabdomeres are analogous to the retinal rod outer segments of vertebrates and are packed tubules of the order of 500 Å in diameter. These tubules are formed of double membranes (lamellae) of the order of 50 Å in thickness, whose molecular structure is best visualized as *uni membranes*. The lamellae (microvilli) of the rhabdomeres are in contact with the cytoplasm of the retinula cells. Chemical analyses indicate that these organisms contain vitamin A. The photosensitive visual pigment therefore shows that these eyes also possess as the prosthetic group, retinene$_1$, in a concentration of 10^7 molecules per rhabdomere, which is similar to *rhodopsin*, the visual pigment in the retinal rods of vertebrate eyes.

INTRODUCTION

The compound eyes of arthropods are of considerable interest in elucidating the visual system of animals. Compound eyes are image-forming eyes which are particularly efficient in detecting movements in their total visual field. In addition many of the arthropods exhibit orientation relative to the direction of vibration of polarized light, which suggests the existence of a polarized light analyser within the eye. If we are to understand how the compound eye functions in response to a light stimulus it becomes important to study its structure and chemistry at the molecular level. Examples selected from studies in our laboratory of insect and crustacean eyes will be explored in an attempt to see how these compound eyes are structured for vision.*

STRUCTURE

The compound eye consists of eye facets, ommatidia, which vary in number from only a few, as in some species of ants, to more than 2000 in the dragonfly eye. Each ommatidium is a complete eye in the

* "Cajal, the gifted Spanish neurologist, gave special study to the retina and its nerves—lines to the brain. He turned to the insect eye thinking the nerve lines there 'in relative simplicity' might display schematically, and therefore more readably, some general plan which Nature adopts when furnishing animal kind with sight." (Sherrington, 1964.)

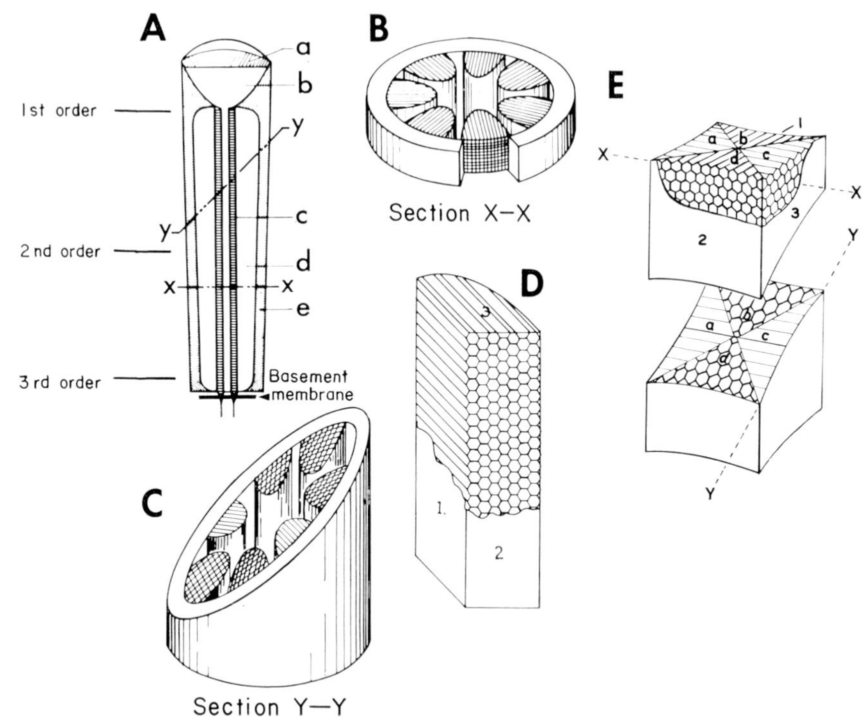

FIG. 1. Diagram of insect ommatidium. A. Longitudinal section: a, corneal lens; b, crystalline cone; c, rhabdomere; d, retinula cells; e, pigment sheath; cross section X–X; oblique section Y–Y. B. Open type rhabdom, three-dimensional cross section X–X. C. Open type rhabdom, three-dimensional oblique section Y–Y. D. Rhabdomere, three-dimensional section. E. Fused type rhabdom, three-dimensional view, cross section X–X and oblique section Y–Y.

sense that it has a corneal lens, a crystalline cone, and retinula cells (Fig. 1A). A differentiated structure of the retinula cell is the rhabdomere which is analogous in function to the retinal rod outer segment of the vertebrate eye. The association of rhabdomeres forms the "light-trapping" rhabdom within the ommatidium (Fig. 1B, C, E).

Exner (1891) described two anatomically distinct types of compound eye structures, the *apposition* and *superposition* eyes. In Exner's model, apposition eyes are those in which the rhabdomeres that form the rhabdom lie directly beneath or against the crystalline cone, and in which an inverted image is formed at the level of the rhabdom (Fig. 1A). Each ommatidium is entirely sheathed by a double layer of pigment cells. Only light striking the lens within about 10° of the perpendicular reaches the rhabdomeres. Light striking the lens at a more oblique angle may be reflected by the lens or absorbed by the pigment sheath. Light falling upon the lens of an ommatidium can only reach the rhabdomeres of that ommatidium. This type of compound eye structure was believed to be characteristic of diurnal insects.

Superposition eyes are those in which the receptor cells lie some distance away from the crystalline cone. The extent of the pigment sheath depends upon the degree of dark-adaptation of the eye. In bright light the pigment extends the full length of the ommatidium, as in the apposition eye. During dark-adaptation, however, the pigment granules migrate to the surface of the eye and are drawn up between the crystalline cones, leaving a light-permeable, non-refractile membrane between them. This migration of the pigment granules has analogies in function to the iris of the vertebrate eye and depends upon the light intensity. At high levels of illumination, the isolation of each ommatidium is nearly perfect, each rhabdom receiving only that light which enters its own ommatidium in a nearly axial direction. At low levels of illumination, the pigment granules are retracted, allowing convergence of light from neighbouring ommatidia and consequent brightening of the image. Thus, light striking the surface of the eye more obliquely is not absorbed by the pigment sheath but passes through to strike the rhabdomeres. In addition, light rays from several ommatidia can be brought to focus upon the rhabdomeres of a single ommatidium, increasing the intensity of the image formed. The superposition eye was believed to be characteristic of nocturnal species, and the superposition mechanism was thought to be important for its increased light-gathering power. However, superposition eyes have been found in both diurnal and nocturnal species.

Optics

Exner (1891) also suggested that the crystalline cones of superposition eyes had lens cylinder properties, the greatest index of refraction being at the axis of the lens cylinder, with concentric rings of decreasing indices of refraction as one proceeded to the periphery of the crystalline cone. Such a system would allow for an erect image on the receptor layer and increased light-gathering power, as the light from a point source entering the lenses of several ommatidia could be focused at a single spot on the retina.

Burtt and Catton (1962) performed a series of experiments in insect vision, using striped patterns and measuring the electro-retinograms to determine the limit of resolution of the insect eye. They found a degree of resolution higher than had been indicated from behavioural studies, and higher than the theoretical limits placed on the eye by diffraction. To account for this resolution Rogers (1962) proposed a theory of insect vision which involved complex diffraction images formed by the interaction of light from several ommatidia. According to this theory, the insect eye acted as a diffraction grating and produced several orders of images. The first and second order images lie within the receptor regions of the insect ommatidium (for the rhabdomeres are of the order of 100 μ in length) and the third image is still within the photoreceptor area (Fig. 1A). The optical images of a striped pattern formed by the corneal lens and crystalline cones of a grasshopper at various

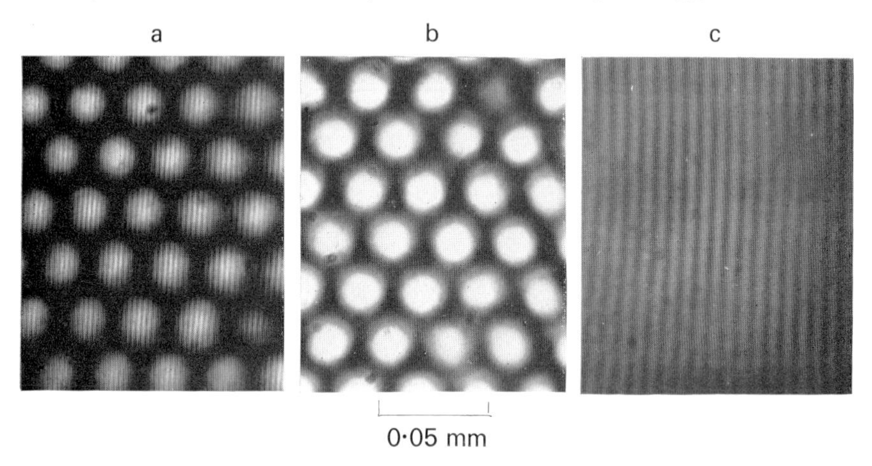

FIG. 2. a, First; b, second; c, third order images of a striped pattern photographed at different distances through corneal lens and crystalline cone of the grasshopper.

distances from it were photographed through a microscope and are seen in Fig. 2. The formation of these images depends on light passing between ommatidia so that the fields of adjacent ommatidia overlap. As a light moves across the visual field at any time, some of the individual retinula cells in an ommatidium "see" the light and some do not, while some retinula cells of adjacent ommatidia also "see" the light and some do not (Kuiper, 1962). One of the difficulties with Rogers' (1962) theory is an explanation for the function of the screening pigment granules which are present between the ommatidia and hence prevent the passage of light between ommatidia.

Electron microscopy of photoreceptors

Let us now examine the structure of the insect ommatidium and the fine structure of the rhabdom and the rhabdomeres to see if there is a structural basis for compound eye vision.

The compound eye of *Drosophila melanogaster*, for example, is composed of over 700 ommatidia. Each ommatidium consists of a corneal lens, a crystalline cone, seven retinula cells, and a sheath of pigment cells that extends over its entire length (Fig. 3a). The ommatidium is about 17 μ in diameter and from 70 to 125 μ in length. In cross-section it is observed that the seven retinula cells are radially arranged, each having a medial portion extending toward the centre of the ommatidium and terminating in a dense circular rhabdomere

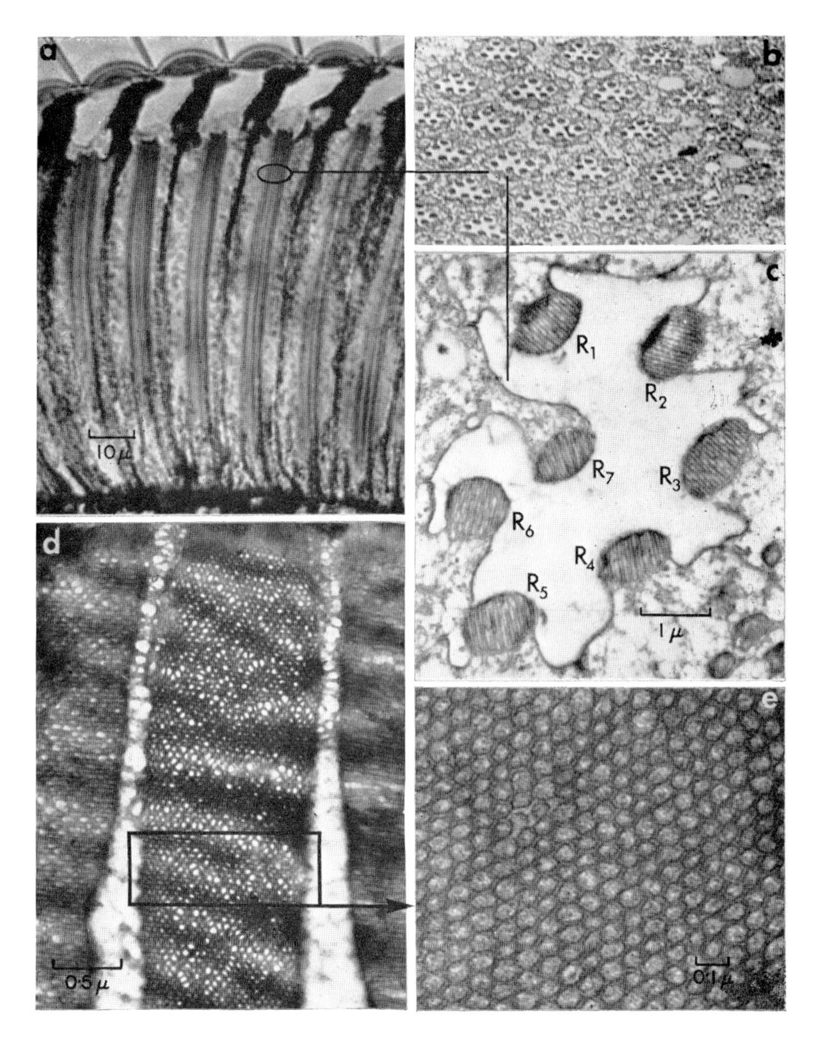

FIG. 3. *Drosophila melanogaster* ommatidium: a, Light microphotograph of a longitudinal section through several ommatidia (compare to Fig. 1A); b, cross section X–X through rhabdom; c, cross section of rhabdom illustrating the orientation and structure of the rhabdomeres; electron micrograph; d, longitudinal section of the distal end of the ommatidium, showing three adjacent rhabdomeres; electron micrograph; e, packing of tubules, note as in Fig. 1D and 1E.

(Fig. 3b, c). Each rhabdomere is distinct with respect to the retinula cells, with a finely differentiated line of attachment between them. The rhabdom consists of seven individual rhabdomeres, R_1—R_7, one of which is asymmetrical, which are situated in a relatively clear fluid cavity (Wolken, 1957). The rhabdomeres average 1·2 μ in diameter and are more than 60 μ in length. A definite fine structure of lamellae, separated by less dense interspaces, is found within each of the rhabdomeres. These dense laminations originate at the line of attachment and terminate in a scalloped border on the medial side of the rhabdomere. The lamellar structure is observed in all cross-sections of the rhabdomeres, while the tubular structure is seen in some oblique and longitudinal sections (Fig. 3c, d, e). A single structure of tubules (Fig. 1D) produces these two different geometric structures in thin section, depending upon the orientation of the individual rhabdomere with respect to the plane of cutting (Wolken, Capenos, and Turano, 1957). Each of the rhabdomere tubules is 500 Å in diameter and has a double membrane, with a wall thickness of the order of 50 Å.

In this study it was of interest to compare the ommatidium of the *Drosophila* eye to that of the cockroach. The cockroach is one of the more primitive of the unspecialized insects and the eye structure of two species of cockroaches, *Periplaneta americana* and *Blaberus giganteus*, were also examined by electron microscopy (Wolken and Gupta, 1961).

In these two species, the ommatidium is formed by seven retinula cells. The inner side of each retinula cell is differentiated to form a rhabdomere. All seven rhabdomeres are in close proximity and form a closed type rhabdom (as in Fig. 1E). Aggregates of intracellular pigment granules, which do not seem to be affected by dark-adaptation, surround the rhabdoms and extend the whole length of the retinula cells. The rhabdomeres which make up the rhabdom exhibit a regular pattern of organization; one of the rhabdomeres is asymmetrical, the others are arranged in pairs. In all rhabdoms, the main axes of the paired rhabdomeres have the same orientation with respect to the asymmetrical rhabdomere (Wolken and Gupta, 1961).

The cockroach rhabdomere is of the order of 2 μ in diameter and 100 μ in length. It is, like that of *Drosophila*, a single geometrical structure of tightly packed tubules. Each tubule is about 500 Å in diameter with walls of the order of 50 Å in thickness and there are approximately 80 000 tubules in a single rhabdomere (Wolken, 1966).

The structure of the compound eye of the fresh-water crustacean *Leptodora kindtii* is also of interest. *Leptodora* is a relatively large organism (up to 18 mm in length), nearly all transparent, with one median spherical eye and with various types of connecting neurons (Gerschler, 1911, 1912; Scharrer, 1964). The entire eye of *Leptodora kindtii* is contained within the transparent exoskeleton at the anterior end of the organism (Fig. 4a). The eye is free to move and can rotate

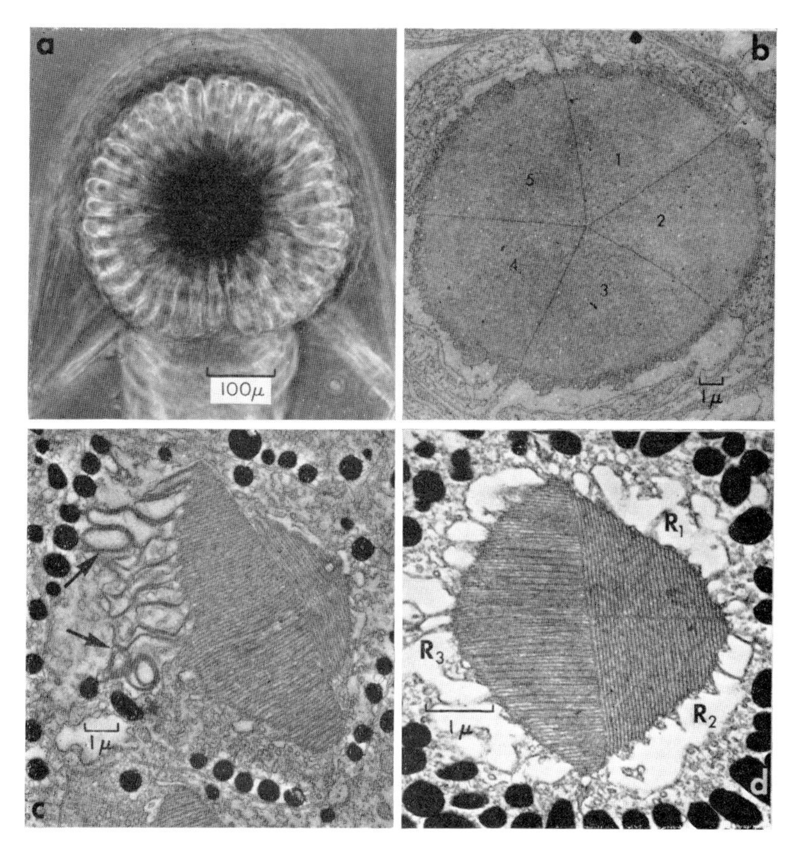

FIG. 4. *Leptodora kindtii*. a, eye × 90; b, crystalline cone; c, rhabdom and microvilli in contact with cytoplasm; d, rhabdom, fused rhabdomeres R_1, R_2, R_3.

10° in either direction. A small area in the back of the eye is for accommodation of the optic processes and the brain lies in close proximity to the eye structures.

The compound eye of *Leptodora* consists of approximately 500 ommatidia that are radially arranged as shown in Fig. 4a. The interstitial material between the surface of the eye sphere and the curved external chitinous wall probably serves as a common lens for all the ommatidia. The ommatidia are large conical structures 180 μ in length and of a diameter ranging from 30 μ at the outer portion to just a few microns at the base; they consist of pigment cells, a crystalline cone, and retinula cells which contain many vesicles and numerous mitochondria. The crystalline cone consists of five crystalline cone cells that are arranged in pie-shaped segments (Fig. 4b). Although the crystalline cone continues proximally to the surface of the rhabdom as in the apposition-type eye, observations of pigment migration indicate that, under certain conditions of dark adaptation, "crossing" among adjacent crystalline cones could result in the formation of a superposition image. As the crystalline cones continue upward, the space between them increases and is taken up with pigment cells.

The rhabdoms are affixed directly to the ends of the crystalline cones. The four radially arranged retinula cells that form the rhabdom show only three rhabdomeres (Fig. 4c, d). One of the rhabdomeres (R_3) is large in comparison to the other two (R_1 and R_2) and appears to be two fused rhabdomeres (Wolken and Gallik, 1965). The presence of four retinula cells that form only three fused rhabdomeres has been observed for the dragonfly, *Anisoptera* (Goldsmith and Philpott, 1957; Naka, 1960). The rhabdom of the fresh-water *Daphnia* is also similarly structured and the rhabdomere is also formed of tightly packed tubules (as in Fig. 3e). The rhabdomere fine structure here too is that of tightly packed microtubules (microvilli) with an outside diameter of 500 Å and a wall thickness of the order of 50 Å. The tubules of the small rhabdomeres R_1 and R_2 are arranged at 90° with respect to the large rhabdomere R_3. The ends of the rhabdomere tubules appear continuous with the cytoplasm (Fig. 4c). This has also been observed for the rhabdoms of other arthropods and Lasanky (1967) has attached considerable importance to this for visual excitation.

There are two geometric arrangements for the rhabdomeres that form the rhabdom. One, a "closed" or "fused" type, is found in such insects as the cockroach, honeybee, grasshopper, locust, moth, dragonfly, and butterfly, where the greater portion of the mesial border and the entire inner margin of the retinula cells are modified to form wedge-shaped rhabdomeres in close proximity with one another around a narrow axial cavity. The other is an "open" type, characteristic of the fruitfly, *Drosophila*, and the housefly, *Musca*, in which the rhabdomeres project through a neck-like portion of their retinula cells, extending into a comparatively large axial cavity. The structure of the ommatidium, the rhabdom, and the rhabdomeres for both the "open" type and "closed" type are schematized in Fig. 1B, C and E, and in the electron micrographs (Figs 3c and 4c, d).

Polarized light

One of the interesting aspects of studying animals with compound eyes is the question of how they are able to analyse the direction of vibration of polarized light, using this as a light compass for navigation. Karl von Frisch (1950, 1953) suggested that the answer may be in the arrangement and structure of the retinula cells that form the rhabdom. He constructed a model to show that the rhabdom consists of eight triangular polarizing elements, each transmitting a quantity of light proportional to the degree of polarization. The fine structure found in the individual rhabdomeres (Figs 1, 3, 4) may indicate that the rhabdomeres are the built-in analysers of polarized light. The rhabdomere, as we have described and illustrated, is the only structural unit found in the compound eye with a parallel rather than a radial symmetry in cross-section. The radial arrangement in the rhabdom of such radially asymmetrical units does suggest a relation to the analysis of polarized light in the insect eye. The analysis by Waterman (1966) on the eye structure and on the orientation of animals should be examined.

It has also been proposed that reflection patterns from the environment resolve polarized light into patterns of graded light intensity. According to this view, the compound eye merely discriminates intensity and is not a direct analyser.

Of the insects studied in our laboratory, ants (*Tapinoma sessile*, *Solenopsis saevissimae*), housefly (*Musca domestica*), and firefly (*Photurus pennsylvanicus*), each had a compass reaction $\pm 45°$ with respect to the plane of polarization of the incident light beam. All oriented at 0° and 90° (Marak and Wolken, 1965). The use of a white background, which reflects normally incident polarized light in a circular pattern, and a black background, which reflects elliptically, led to a consistent difference in the response curve. With the white background, there was more orientation at 45° and less at 0° and 90°. This response difference is consistent with the hypothesis that there are more cues for simple phototactic responses on the black background, which reflects more light which is perpendicular rather than parallel to the plane of polarization.

Houseflies and fireflies appear to orient to polarized light only within a narrow range of intensities, which is consistent with a model that includes two photoreceptors with different absorption vectors. If these two receptors are oriented at 90° with respect to each other and if the long axis of the visual pigment molecule is oriented in a single plane but free to vibrate in that plane, the receptors will absorb polarized light in a ratio of two to one when the electric vector is parallel to the absorption vector of one of the photoreceptors.

The reflectance pattern of non-polarized light is circular, while the reflectance pattern of polarized light is elliptical, with the long axis perpendicular to the plane of polarization. In normal daylight, the long axis of the ellipse will point to the solar azimuth. This indicates that insect navigation can be accomplished extra-ocularly with little difficulty. However, our understanding of the polarized light analysis, whether intra- or extra-ocular, is at present incomplete.

VISUAL PIGMENTS

Chemical analysis of all the eye pigments for the invertebrates has not yet been accomplished. In the absence of chemical isolation of the eye pigments, the response of the animal to light of different wavelengths and intensities, the action spectrum is a good indicator of the pigments or pigment system absorbing that light.

Fingerman (1952) and Fingerman and Brown (1952, 1953) have demonstrated that *Drosophila* possesses colour vision at high light intensities, but that at low light intensities there is a "Purkinje shift" from photopic to scotopic vision, similar to the shift in vertebrates from cone to rod vision, which suggests two types of receptor cells or visual pigments in the eye of *Drosophila*. For the three eye-colour mutants of *Drosophila* investigated (Fig. 5) an effectiveness or action spectrum was found which had a maximum near 508 nm (Wolken

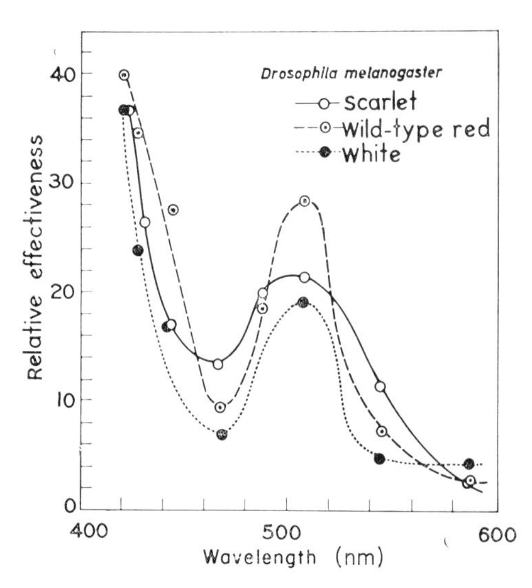

FIG. 5. Effectiveness or *action spectrum* for *Drosophila melanogaster*, three eye colour mutants.

Mellon and Contis, 1957). A similar action spectrum with a response peak at 505 nm was obtained for the ant, *Tapinoma sessile* (Marak and Wolken, 1965). In all insects investigated, fruitfly, ant, housefly, bee, major response peaks were also found in the ultraviolet near 365 nm.

To isolate and identify the visual pigment in these insects, it was necessary to extract the pigment; this is not an easy task, since thousands of insects must be collected and decapitated, and the heads fractionated. In one series of extractions, 2000 to 4000 heads from the housefly, *Musca domestica*, were ground with 0·05 M phosphate buffer, pH 7·0. Each mixture was frozen, thawed and centrifuged at 12 000 rev/min. The supernatant fraction was poured onto a prepared column and the fluid allowed to drain through. The column was calcium phosphate mixed with celite, a procedure described for the purification of cattle rhodopsin (Bowness, 1959). Adsorption proceeded best when the solution was diluted to give a 0·025 M phosphate buffer concentration.

After the heads were extracted and centrifuged, the supernatant was a dark red-brown colour. When it was placed on the column, a light-yellow fluid drained through. The presence of other pigments became apparent on eluting with 0·2 M phosphate buffer; a pigment with a major absorption peak at about 545 nm and a shoulder at about 515 nm, and another pigment with an absorption peak at about 408 nm were evident in the effluent fractions. The last to come off the column with 0·2 M phosphate buffer was yellow and photosensitive

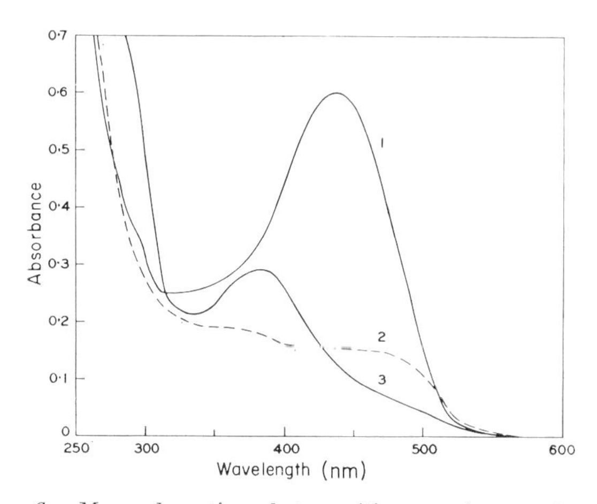

FIG. 6. Housefly, *Musca domestica*, photosensitive eye pigment. Spectrum 1, eluate P with 0·2 M phosphate buffer, pH 6·5; spectrum 2, after bleaching with white light, 1600 foot-candles for 2 h at 12°C; spectrum 3, eluate P mixed with 0·3 ml 2 N KOH and allowed to remain for 30 min in the dark.

and at pH 6·5 had an absorption peak at 437 nm (Fig. 6 spectrum 1). The fraction containing the 437 nm pigment showed no trace of the absorption maxima of the other pigments eluted from the column with phosphate buffer (Bowness and Wolken, 1959). After all the photosensitive pigment had been eluted, a dark red-brown material, which

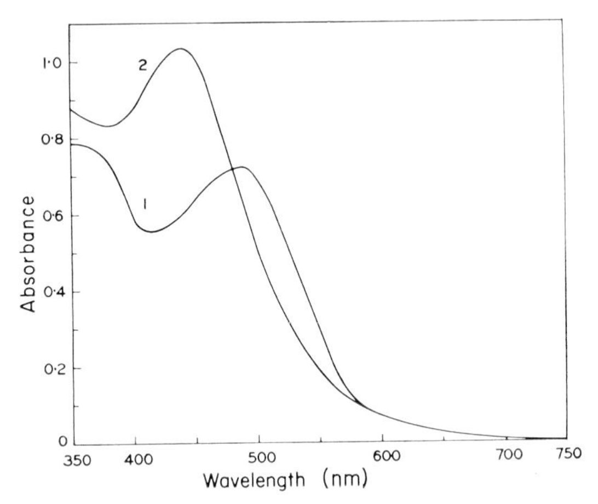

FIG. 7. Housefly, *Musca domestica*, pigment extract. Spectrum 1, extract made with 1 M acetate–acetic acid buffer, pH 4·8 (from top 2 cm of the extruded calcium phosphate and celite column, the loaded column having been eluted with 15 ml of phosphate buffer at pH 7·0 and 50 ml to pH 6·5; spectrum 2, extract 1 allowed to remain in the dark for 2 h.

could be eluted with 2 N potassium hydroxide, remained at the top of the column (Fig. 6 curve 3). If the pigment was allowed to remain on the column it turned red, and immediately after elution with 1 M acetate–acetic acid buffer, pH 4·8, showed an absorption peak at 490 nm (Fig. 7 spectrum 1) that changed to 440 nm on standing (Fig. 7 spectrum 2).

Goldsmith (1958) has also shown that a photosensitive pigment is present in phosphate buffer extracts of honeybee heads. Partial purification of the extracts indicated that a retinene$_1$-protein complex was present with an absorption maximum at 440 nm. The absorption maximum of 437 nm of the photosensitive housefly pigment at pH 6·5 lies very close to that of the honeybee pigment.

The absorption spectrum of the light-yellow pigment which drained through the column with 0·025 M phosphate buffer appears to be that of melanin. The red pigment which required 2 N potassium hydroxide or 1 M acetate–acetic acid buffer at pH 4·8 for elution from the column exhibited a shift in absorption maximum from about 490 to 440 nm in changing from alkaline to acid conditions (Fig. 7). This is similar to the shift shown by rhodommatin, a red pigment obtained from insects by Butenandt, Schiedt and Biekert (1954), which possibly is related to the pteridines (Forrest and Mitchell, 1954). One of the pigments obtained from *Drosophila* by Wald and Allen (1946) had an absorption maximum at 436 nm, and was not indicated to be photosensitive; however its absorption spectrum resembles the 440 nm form of the pigment shown in Fig. 7 curve 2.

There are interesting similarities in spectroscopic properties between the insect pigment and the visual pigments which have been found in many animals (Dartnall, 1957; Wald, 1959). Consider the pH indicator properties shown by the light-sensitive housefly pigment and by its bleached products. On bleaching in the light at pH 6·5, a solution with plateaus of absorption at 440 to 460 nm and 350 to 360 nm is produced. Addition of a strong acid to this material gives a plateau or a peak at 470 to 475 nm; in alkaline solution there is only a plateau at 360 nm. Absorption maxima in these wavelength regions are given by the retinylideneamines and indicator yellow under similar conditions of pH (Ball, Collins, Dalvi and Morton, 1949; Collins, 1954), although the 440 nm form of retinylideneamines is not stable except at pH 1 (Morton and Pitt, 1955). An absorption maximum at 380 nm is produced at pH 12 with the housefly pigment. An absorption peak at this wavelength is obtained with squid metarhodopsin at pH 9·9, and from cattle metarhodopsin at pH 13 (Hubbard and Kropf, 1959).

There are a number of observations which indicate that protein may be a part of the photosensitive housefly pigment. First, the ultraviolet absorption spectrum of the pigment, either bleached or unbleached, exhibits a peak at 290 nm in alkaline solution. Most proteins show a peak at 290 nm in alkaline solutions, though in neutral or acid solution the spectral peak shifts to 275 to 280 nm (Beaven and Holiday, 1952). Second, the heat bleaching of the pigment at 100°C gave a coagular precipitate, which when dissolved in 0·2 N sodium hydroxide showed a peak at about 290 nm. Third, a precipitate containing about 10·5% nitrogen was obtained from the light-sensitive pigment solution upon addition of sulphosalicylic acid.

The amount of retinene$_1$ present in the housefly heads was estimated by the method of Ball *et al.* (1949) in the antimony trichloride reaction, which gave an estimate of 0·3 μg of retinene/g fresh weight of heads. From the average number of 24 500 rhabdomeres per eye (Fernández-Morán, 1958) this corresponds to $2{\cdot}7 \times 10^7$ molecules of retinene$_1$ per rhabdomere (Wolken, Bowness and Scheer, 1960).

The presence of retinene$_1$ in the extract indicates that vitamin A_1 should also be present since in the visual cycle retinene$_1$ is reduced to vitamin A_1.* Extracts with an absorption band at about 328 nm, the principal absorption maximum for vitamin A_1, were obtained after bleaching the 437 nm housefly pigment, adding ethanol, evaporating to dryness, and extracting with chloroform. Antimony trichloride reactions with several fractions gave maxima near 620 and 696 nm, the maxima for vitamins A_1 and A_2; in some cases, these were found in the same fraction. The identification of vitamin A_2 is particularly crucial, since it is so far known to occur only in fresh-water fish and in some amphibians, and this would be the first instance of its presence outside these vertebrate groups.

To see if the cockroach (*Blatta orientalis*) eye also contains retinene$_1$, one group of 1000 cockroach heads was ground with 40 ml of 45% sucrose in pH 9·0 M/15 phosphate buffer to make a thin paste. The paste was then transferred to centrifuge tubes, layered over with phosphate buffer at pH 7·0 and centrifuged in a Spinco model L ultracentrifuge at 100 000 *g* for 20 min. The supernatant, containing most of the rhabdomeres, was drawn off, and the sediment was reground with 20 ml of sucrose solution and fractionated again as already described. The supernatants were combined, diluted with 1000 ml of buffer solution, and centrifuged in a chilled rotor (0°C) at 25 000 *g* for 40 min. The sediment was taken up in 100 ml of fresh buffer and left in the dark at 6°C for 15 h. The suspension was then centrifuged and the residue was taken up in 10 ml of distilled water and lyophilized. The dry powder was extracted twice in the dark with 300 ml of petroleum ether and dried again by evaporation. The defatted dry powder was then extracted at room temperature (25°C) for 3 h with 5 ml of 1·8% digitonin in pH 7·0 M/15 phosphate buffer. The absorption spectrum of the cockroach eye fraction has its major peak near 500 nm (Fig. 8). The bleached spectrum also shows a shift in the maximum from 500 nm to near 375 nm, which is indicative in the photochemistry of the release of retinene from the opsin of the rhodopsin complex.

Another group of cockroach heads was ground in a mortar and pestle with five times its weight of anhydrous sodium sulfate. The mixture was extracted repeatedly with petroleum ether in the dark at 10°C. The residue was then extracted in the light at room temperature, 25°C, with 250 ml of acetone containing 6% ethanol and 4% water. The acetone extract was evaporated to dryness under reduced pressure, taken up in 10 ml of petroleum ether, and placed on a column of

* In the new international chemical nomenclature, vitamin A is *retinol*; vitamin A aldehyde, or retinene, is *retinal*.

alumina which had been wetted with water to 5% of its dry weight. The column was eluted first with 100 ml of petroleum ether and then with mixtures of acetone in petroleum ether varying from 4 to 40% (v/v). The effluents from the columns were evaporated to dryness and dissolved in 1 ml of chloroform. The chromatographic fractions eluted from the alumina columns in acetone showed absorption spectra indicative of retinene$_1$. The fraction eluted with 17·5% acetone in

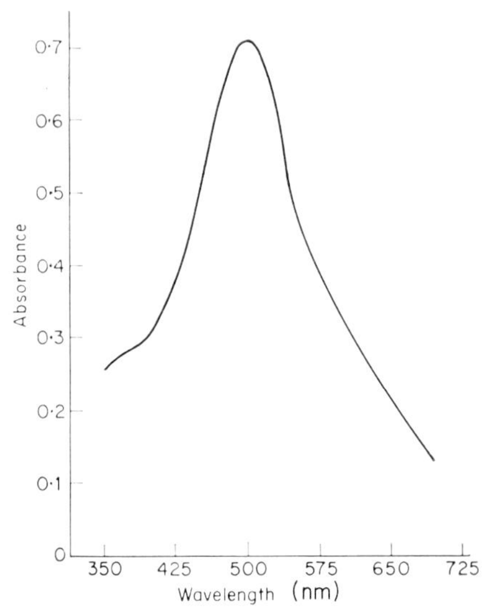

Fig. 8. Absorption spectrum of the cockroach, *Blatta orientalis*, rhabdomere fraction extracted with 1·8% aqueous digitonin, pH 7·0 M/15 phosphate buffer.

petroleum ether, when reacted with antimony trichloride, gave an absorption maximum at 664 nm which is characteristic of retinene$_1$. The amount of retinene per gram of cockroach heads (fresh tissue weight) was found to be 0·08 μg. On the basis of 28 000 rhabdomeres per eye (Wolken and Gupta, 1961), the amount of retinene per rhabdomere was calculated to be $4{\cdot}3 \times 10^7$ molecules.

The cockroach pigment extract differs in absorption maximum and in solubility from the pigments extracted from the housefly and the honeybee (Goldsmith, 1958) which are soluble in phosphate buffer and show maxima at 437 and 440 nm (Bowness and Wolken, 1959).

The spectral characteristics of the visual pigment of the cockroach are comparable to rhodopsin (a retinene$_1$–protein complex) extracted from marine invertebrate eyes; for example: the cephalopod mollusc *Loligo* (Bliss, 1948; St. George and Wald, 1949; Hubbard and St.

George, 1958); *Octopus* and *Sepia* (Brown and Brown, 1958); the crustacean *Homarus* (Wald and Hubbard, 1957); three species of euphausiids (Kampa, 1955); and the arachnid *Limulus* (Hubbard and Wald, 1960). Experimental evidence indicates that insect eyes possess vitamin A and a retinene complex for their visual pigments (Goldsmith and Warner, 1964).

Electrophysiological studies of the cockroach eye (Walther, 1958; Walther and Dodt, 1959) showed a maximum sensitivity peak at 507 nm. Studies of the ocellus of the cockroach *Periplaneta americana* indicated a single photoreceptor with a maximum also at about 500 nm (Goldsmith and Ruck, 1958). In addition, Walther (1958) suggested that the cockroach eye contains two types of photoreceptor, the lower half of the eye containing one type with the absorption maximum

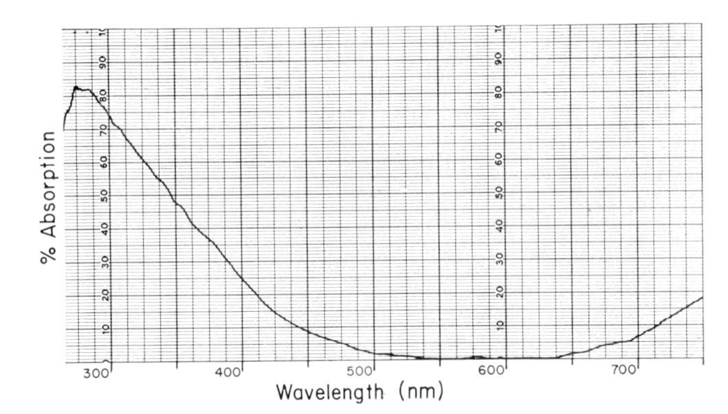

FIG. 9. Absorption spectrum, corneal lens of *Drosophila melanogaster*.

at 507 nm, and the upper half containing both types, one at 507 nm and another with maximum sensitivity between 314 and 369 nm. Electrophysiological studies of the eyes of honeybees show that they, too, have a complicated system which may involve several photoreceptors with maximum sensitivites in different regions of the spectrum (Goldsmith, 1961). Since only one photosensitive pigment complex was isolated from the eye of the cockroach (Wolken and Scheer, 1963), it is likely that the structure and absorption characteristics of the lens may be important. The lens is known to play a significant role in determining the sensitivity of the eye to light in the near ultraviolet region (see for example the corneal lens spectrum of *Drosophila*, Fig. 9). The absorption characteristics of the lens and the visual pigment could account for the dual sensitivity of the cockroach eye (Dodt and Walther, 1958; Buriain and Ziv, 1959).

DISCUSSION AND SUMMARY

Although much research remains before we can completely understand how compound eyes are structured for vision, progress has been made in unravelling the geometry of the rhabdom and the molecular organization of the rhabdomeres in a variety of arthropod eyes. All the arthropod rhabdomeres structurally are packed tubules of the order

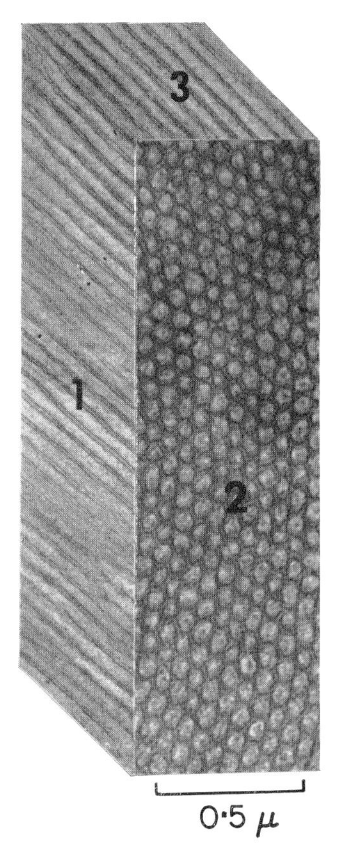

FIG. 10. A reconstructed three-dimensional structure of a section of a rhabdomere to indicate how the rhabdomere tubules are packed; electron micrograph.

of 500 Å in diameter with double walled membranes of the order of 50 Å, similar to the cell *unit membranes*. The microtubules or membranes of the rhabdomeres are in communication with the cytoplasm of the retinula cells that form the rhabdom. A reconstituted three-dimensional section of a rhabdomere showing how the microtubules are arranged is illustrated in Fig. 10. It is also of interest to compare the invertebrate rhabdomere microtubules to that of the vertebrate retinal rod outer segment sacs (Fig. 11) and to note the many similarities in microstructure.

A relationship may exist between the structural arrangements of the open and closed type rhabdoms and their visual physiology. The action potentials and the electrical responses of several arthropod eyes, as investigated by Autrum (1958), indicate that there are also two physiological types; a "slow" type eye characterized by a negative monophasic potential which is dependent upon the state of dark-adaptation, and a "fast" type eye in which the electro-retinogram (ERG) is diphasic, the magnitude and the form of the potential being independent of the state of dark-adaptation. It is of interest, then, that except for the adult dragonflies, all the arthropods having a

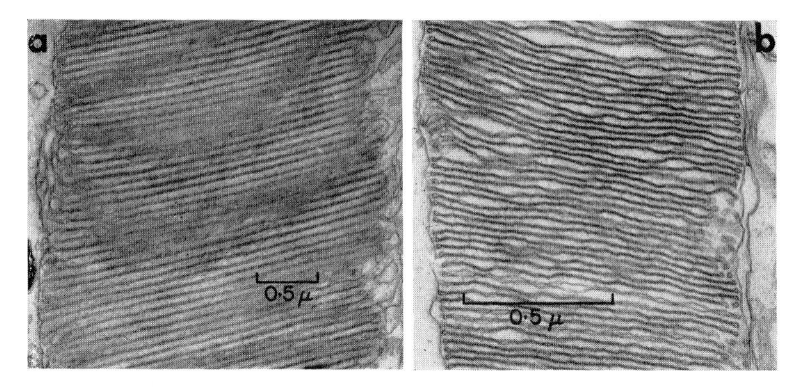

FIG. 11. a, Rhabdomere section of *Daphnia pulex*; compare with b, section of a cattle retinal rod outer segment.

"closed" arrangement for their rhabdoms possess a "slow" type electrical response. All the dipterous and hymenopterous insects that have an "open" type arrangement possess the "fast" type electrical response. In the "fast" type eye, the rhabdom occupies only a small part of the volume. In the "slow" type eye, characteristic of nocturnal insects which have a light-gathering problem, the rhabdom takes up a much larger part of the volume; for example, the volume of the rhabdom of the cockroach is about five times the volume of the *Drosophila* rhabdom (Wolken, 1966).

The arthropod eyes use the chemistry of the carotenoids, for they possess vitamin A and a retinene-complex for their visual pigment chemistry, as do all animals that possess eyes. The concentration for the visual pigment in insects studied was found to be of the order of 10^7

molecules, similar to the number of rhodopsin molecules found for all vertebrate retinal rods. It is questionable whether the light sensitive phosphate buffer extracts of the bee, 440 nm, and the housefly, 437 nm, pigments are rhodopsins, since they also contain photosensitive pigments with maximum absorption near 510 nm. The differences in solubility and spectral properties of invertebrate visual pigments may indicate that the proteins of the visual pigment complex differ from those of the vertebrate *opsins*.

Microspectrophotometry now being applied to obtain the absorption characteristics of the lens, screening pigment granules, rhabdoms and rhabdomeres is important (Langer and Thorell, 1966). This information together with the optics and electrophysiology is necessary if we are to understand more completely how the compound eye of arthropods functions in vision.

Acknowledgement

This research was supported in part by grants from the Pennsylvania Lions Sight Conservation and Eye Research Foundation and the National Aeronautics and Space Administration (NGR–39–002–011).

References

Autrum, H. (1958). Electrophysiological analysis of the visual systems in insects. *Expl Cell Res.* Suppl. **5**: 426–439.

Ball, S., Collins, F. D., Dalvi, P. D. and Morton, R. A. (1949). Studies in Vitamin A. II. Reactions of retinene$_1$ with amino compounds. *Biochem. J.* **45**: 304–307.

Beaven, G. H. and Holiday, E. R. (1952). Ultraviolet absorption spectra of proteins and amino acids. *Adv. Protein Chem.* **7**: 319–386.

Bliss, A. F. (1948). The absorption spectra of visual purple of the squid and its bleaching products. *J. biol. Chem.* **176**: 563–569.

Bowness, J. M. (1959). Preparation of rhodopsin using columns containing calcium triphosphate. *Biochim. biophys. Acta* **31**: 305–310.

Bowness, J. M. and Wolken, J. J. (1959). A light-sensitive yellow pigment from the housefly. *J. gen. Physiol.* **42**: 779–792.

Brown, P. K. and Brown, P. A. (1958). Visual pigments of the octopus and cuttlefish. *Nature, Lond.* **182**: 1288–1290.

Buriain, H. M. and Ziv, B. (1959). Electric response of the phakic and aphakic human eye to stimulation with near ultraviolet. *Archs Ophthal.* **61**: 347–350.

Burtt, E. T. and Catton, W. T. (1962). A diffraction theory of insect vision. I. An experimental investigation of visual acuity and image formation in the compound eyes of three species of insects. *Proc. R. Soc.* (B) **157**: 53–82.

Butenandt, A., Schiedt, V., and Biekert, E. (1954). Uber ommochrome. III.

Mitteilung synthese des zanthommatins. *Justus Liebigs Annln Chem.* No. 588: 106–116.
Collins, F. D. (1954). Chemistry of vision. *Biol. Rev.* **29**: 453–477.
Dartnall, H. J. A. (1957). *The visual pigments.* New York: John Wiley and Sons.
Dodt, E. and Walther, J. B. (1958). Fluorescence of the crystalline lens and electroretinographic sensitivity determinations. *Nature, Lond.* **181**: 286–287.
Exner, S. (1891). *Die Physiologie der Facettierten Augen von Krebsen und Insekton.* Liepzig.
Fernández-Morán, H. (1958). Fine structure of the light receptors in the eyes of insects. *Expl Cell Res.* Suppl. **5**: 586–644.
Fingerman, M. (1952). The role of the eye pigments of *Drosophila melanogaster* in photic orientation. *J. exp. Zool.* **120**: 131–164.
Fingerman, M. and Brown, F. A. (1952). A "Purkinje shift" in insect vision. *Science, N.Y.* **116**: 171–172.
Fingerman, M. and Brown, F. A. (1953). Color discrimination and physiological duplicity of *Drosophila* vision. *Physiol. Zöol.* **26**: 59–67.
Forrest, H. S. and Mitchell, H. K. (1954). Pteridines from *Drosophila*. I. Isolation of a yellow pigment. *J. Am. chem. Soc.* **76**: 5656–5658.
Frisch, K. von (1950). *Bees: their vision, chemical senses and language.* Ithaca, New York: Cornell University Press.
Frisch, K. von (1953). *The dancing bees.* New York: Harcourt, Brace, and Co.
Gerschler, M. W. (1911). Monographie der *Leptodora kindtii* (Focke). 1. *Arch. Hydrobiol. Planktonk.* **6**: 415–466.
Gerschler, M. W. (1912). Monographie der *Leptodora kindtii* (Focke). 2. *Arch. Hydrobiol. Planktonk.* **7**: 63–118.
Goldsmith, T. H. (1958). On the visual system of the bee (*Apis mellifera*). *Ann. N.Y. Acad. Sci.* **74**: 223–229.
Goldsmith, T. H. (1961). The physiological basis of wavelength discrimination in the eye of the honeybee. In *Sensory communications*: 357. Rosenblith, W. A. (ed.). New York: John Wiley and Sons, Inc.
Goldsmith, T. H. and Philpott, D. E. (1957). The microstructure of the compound eye of insects. *J. biophys. biochem. Cytol.* **3**: 429–440.
Goldsmith, T. H. and Ruck, P. R. (1958). The spectral sensitivities of the dorsal ocelli of cockroaches and honeybees. *J. gen. Physiol.* **41**: 1171–1185.
Goldsmith, T. H. and Warner, L. T. (1964). Vitamin A in the vision of insects. *J. gen. Physiol.* **47**: 433–441.
Hubbard, R. and Kropf, A. (1959). Molecular aspects of visual excitation. *Ann. N.Y. Acad. Sci.* **81**: 388–398.
Hubbard, R. and St. George, R. C. C. (1958). The rhodopsin system of the squid. *J. gen. Physiol.* **41**: 501–528.
Hubbard, R. and Wald, G. (1960). Visual pigment of the horseshoe crab, *Limulus polyphemus*. *Nature, Lond.* **186**: 212–215.
Kampa, E. M. (1955). Euphausiopsin. A new photosensitive pigment from the eyes of euphausiid crustaceans. *Nature, Lond.* **175**: 996–998.
Kuiper, J. W. (1962). The optics of the compound eye. In *Biological receptor mechanisms*: 58–71. Beament, J. W. L. (ed.). New York: Academic Press.
Langer, H. and Thorell, B. (1966). Microspectrophotometric assay of visual pigments in single rhabdomeres of the insect eye. In *Functional organization of the compound eye*: 145–149. Bernhard, G. G. (ed.). New York: Pergamon Press.

Lasanky, A. (1967). Cell functions in ommatidia of *Limulus*. *J. Cell Biol.* **33**: 365–383.

Marak, G. E. and Wolken, J. J. (1965). An action spectrum of the fire ant (*Solenopsis saevissima*). *Nature, Lond.* **205**: 1328–1329.

Morton, R. A. and Pitt, G. A. J. (1955). Studies on rhodopsin. 9. pH and the hydrolysis of indicator yellow. *Biochem. J.* **59**: 128–134.

Naka, K. I. (1960). Recording of retinal action potentials from single cells in the insect compound eye. *J. gen. Physiol.* **44**: 571–584.

Rogers, G. L. (1962). A diffraction theory of insect vision. II. Theory and experiments with a simple model eye. *Proc. R. Soc.* (B) **157**: 83–98.

St. George, R. C. C. and Wald, G. (1949). The photosensitive pigment of the squid retina. *Biol. Bull. mar. biol. Lab., Woods Hole* **97**: 248.

Scharrer, E. (1964). A specialized trophospongium in large neurons of *Leptodora* (Crustacea). *Z. Zellforsch. mikrosk. Anat.* **61**: 803–812.

Sherrington, Sir C. (1964). *Man on his nature.* New York: Mentor Books.

Wald, G. (1959). The photoreceptor process in vision. In *Handbook of physiology.* I: 671–692. *Neurophysiology.* Field, J. (ed.). Washington, D.C.: American Physiological Society.

Wald, G. and Allen, G. (1946). Fractionation of the eye pigments of *Drosophila melanogaster*. *J. gen. Physiol.* **30**: 41–46.

Wald, G. and Hubbard, R. (1957). Visual pigment of a decapod crustacean: the lobster. *Nature, Lond.* **180**: 278–280.

Walther, J. B. (1958). Changes induced in spectral sensitivity and form of retinal action potential of the cockroach eye by selective adaptation. *J. Insect Physiol.* **2**: 142–151.

Walther, J. B. and Dodt, E. (1959). Die spektralsensitivitat von insekten komplexaugen in ultraviolet bis 290 mu. *Z. Naturf.* **14B**: 273–278.

Waterman, T. H. (1966). Systems analysis and the visual orientation of animals. *Am. Scient.* **54**: 15–45.

Wolken, J. J. (1957). A comparative study of photoreceptors. *Trans. N.Y. Acad. Sci.* **19**: 315–327.

Wolken, J. J. (1966). *Vision: biophysics and biochemistry of the retinal photoreceptors.* Springfield, Illinois: Charles C. Thomas.

Wolken, J. J., Bowness, J. M. and Scheer, I. J. (1960). The visual complex of the insect: retinene in the housefly. *Biochem. biophys. Acta* **43**: 531–537.

Wolken, J. J., Capenos, J. and Turano, A. (1957). Photoreceptor structures. III. *Drosophila melanogaster*. *J. biophys. biochem. Cytol.* **3**: 441–448.

Wolken, J. J. and Gallik, G. J. (1965). The compound eye of a crustacean: *Leptodora kindtii*. *J. Cell Biol.* **26**: 968–973.

Wolken, J. J. and Gupta, P. D. (1961). Photoreceptor structures of the retinal cells of the cockroach eye. IV. *Periplaneta americana* and *Blaberus giganteus*. *J. biophys. biochem. Cytol.* **9**: 720 721.

Wolken, J. J., Mellon, A. D. and Contis, G. (1957). Photoreceptor structures. II. *Drosophila melanogaster*. *J. exp. Zool.* **134**: 383–410.

Wolken, J. J. and Scheer, I. J. (1963). An eye pigment of the cockroach. *Expl Eye Res.* **2**: 182–188.

Symp. zool. Soc. Lond. (1968) No. 23, 135–163.

ORGANIZATION OF THE LOCUST RETINA

S. R. SHAW

Gatty Marine Laboratory and Department of Natural History, University of St. Andrews, Scotland

SYNOPSIS

This article reviews recent experimental evidence bearing upon the organization of the locust eye, obtained with electrical recordings from single cells in the retinula and first synaptic regions.

Simultaneous recordings from pairs of receptors show that these can be maximally sensitive to different planes of polarization of light. Since such cells share identical receptive fields, all the optical reflections are therefore also shared. The differing polarized light responses therefore cannot be caused by reflection effects but must be the result of dichroic absorption by the visual pigment molecules. This finding is in accord with recent spectrophotometric measurements, and in addition suggests that the retinulas in each ommatidium are electrically independent of one another. This is confirmed when currents passed into one cell of an ommatidium via the recording electrode produce only small derivative potentials in neighbouring cells. Single-photon "bumps" in one cell are not detectable in the other. No lateral-inhibitory type of interaction has been seen in the locust eye, where the receptors do not produce spike potentials.

Receptors may be arranged into three groups having strongly overlapping spectral sensitivities, although a more continuous gradation between extreme types is probably more appropriate. No cell has a response spectrum like that of a single rhodopsin pigment, but all the response curves could be matched with a mixture of two such primary pigments in varying amounts. If interaction between receptors can be excluded, it is likely that there are two pigments in each cell in variable proportions, both pigments contributing to the membrane response. If this is so, the molecules of each apparently have the same orientation in the rhabdom.

In the final section the intracellular responses of one type of cell in the first synaptic region (lamina) are interpreted in the light of the known receptor properties. Large hyperpolarizations (HPs) appear in these cells when the receptors are producing only millivolt-sized depolarizations. Both the anatomy and angular sensitivity measurements show that only cells of one ommatidium converge on each HP site, and there is evidence of at least two presynaptic contributors in some HP recordings. Allowing for this convergence, the efficiency of photon capture measured from the HP response is of the same order as in the receptors when only a few photons per second are incident on each cell. This shows that the transmission efficiency of the system is high even at very low intensity. The latency difference between the receptor and lamina responses is about 5 msec, which is less than that for known monosynaptic inhibitory synapses. The above measurements and marking experiments suggest that the HPs are probably from second-order fibres in the lamina. If this is the case, a presynaptic depolarization produces hyperpolarizations in the lamina cells, which suggests that the first synapse is chemically mediated. The receptor depolarizing afterpotential does not produce HPs in the lamina, and a possible explanation for this is offered. The polarized-light responses of the lamina cells are poor by comparison with the receptors, which possibly means that receptors with unlike polarization responses converge on to each site. The implications of the results for the transmission mechanisms operating between receptors and lamina, and lamina and medulla are discussed.

A survey of some of the earlier work on locust retinula cells undertaken in his laboratory was given by Horridge (1965) and to this work and to a paper by Horridge and Barnard (1965) the reader is referred for the detailed anatomy of the receptors. It is here intended to extend the survey to cover more recent developments in this field, involving work by Drs R. Bennett and J. Tunstall besides myself. All the projects have concerned the compound eye of *Locusta migratoria*, studied with stimulating and intracellular recording techniques described in detail elsewhere (Shaw, 1967; Bennett, Tunstall and Horridge, 1967).

RECEPTOR CELL POLARIZED-LIGHT SENSITIVITY

Reflection effects

Some preliminary recordings from single retinula cells illuminated on their visual axes had shown them to give receptor potentials of different amplitudes upon rotation of the plane of polarization of a light beam, in common with receptors in the Dipteran eye (Kuwabara and Naka, 1959; Burkhardt and Wendler, 1960; Autrum and von Zwehl, 1962). One explanation of this is that preferential reflection of some planes of vibration from a surface inclined at an angle to the light beam causes intensity changes at the receptors (see Goldsmith, 1964). This is not an attractive explanation because some receptors are as much as nine times more sensitive to one plane of light polarization than to the plane at right angles (Shaw, 1966). In addition, if analysis were external, cells in the same part of the eye would not be expected to have different peaks of sensitivity whereas they often do show such differences (Autrum and von Zwehl, 1962). In order to rigorously exclude this explanation a few simultaneous recordings were taken from cells in the same ommatidium in a locust eye (Shaw, 1967). The evidence for the location of the recordings comes from the identity of the angular responses to the closest detail (Fig. 1a, b), whereas the recordings must come from separate cells because the polarized light responses measured on the visual axis show two maxima some 60° apart (Fig. 1c). Because both cells have identical angular sensitivities and are viewing the same source, any artefactual reflections in the optical path will be the same for each cell. The large angle between the polarization maxima cannot therefore be explained by reflection effects, and so this theory must be discarded for the locust eye at least.

It seems to be a fair extension of the few results to say that probably all the cells in any one ommatidium share the same directional sensitivity. In passing it may be noted that this conclusion allows the exclusion of Burtt and Catton's (1962) diffraction theory of insect vision, which required the retinulas in one ommatidium of the locust to have different fields of view. The experimental basis for their theory is itself no longer acceptable however, because edge effects and spatial non-uniformities in the striped patterns used give rise to artificially high measurements of acuity, which cannot be repeated when these defects are removed (Palka, 1965; McCann and MacGinitie, 1965).

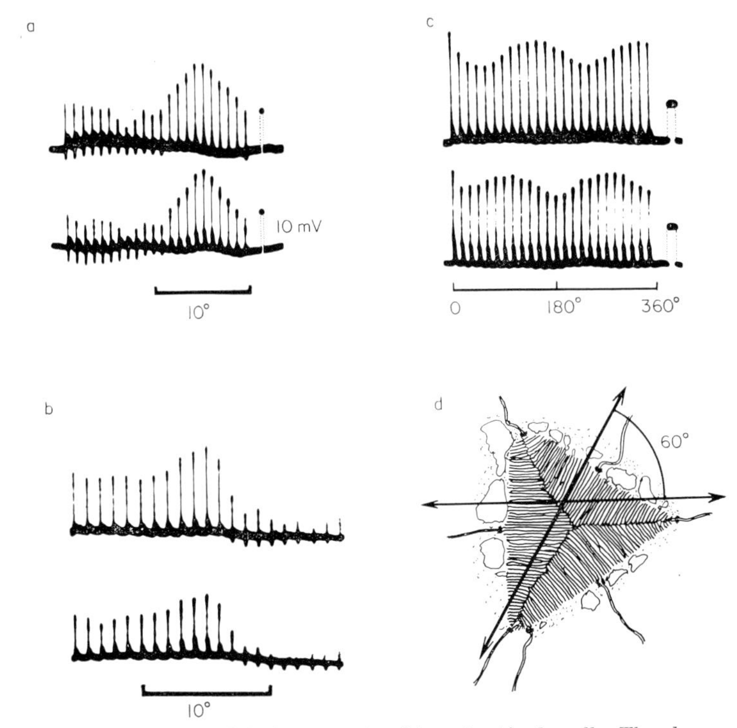

FIG. 1. Receptor potentials from a pair of locust retinula cells. The slow potential responses to a repetitively flashed light appear spike-like because of the slow oscilloscope sweep speed. In between flashes, (a) the light source is moved around the eye in small steps in the horizontal and (b) in the vertical plane, and (c) a polaroid is rotated in the beam in 15° steps on the visual axis. The angular responses are identical for each cell (a, b) but their polarization maxima are 60° apart (c). A possible basis for this is shown in (d) in the regular arrangement of microtubules seen in some electronmicrographs. After Shaw (1967).

Absorption effects

The alternative explanation of polarized light sensitivity is that it is not a reflection but a selective absorption phenomenon. The photopigment molecules in the vertebrate eye are known to be dichroic absorbers regularly arranged in the receptor outer segment membranes (Wald, Brown and Gibbons, 1963). Several writers have speculated that if there were a similar regular arrangement in the parallel tubules that comprise the rhabdomeres of arthropod eyes, each retinula cell would function as a polarized light analyser (Fig. 1d). This view has recently received direct support from the microspectrophotometric measurements on the fly eye by Langer and Thorell (1966), and is also supported by the demonstration that cells in the crab eye fall into two groups which are maximally sensitive to light polarized in line with the two

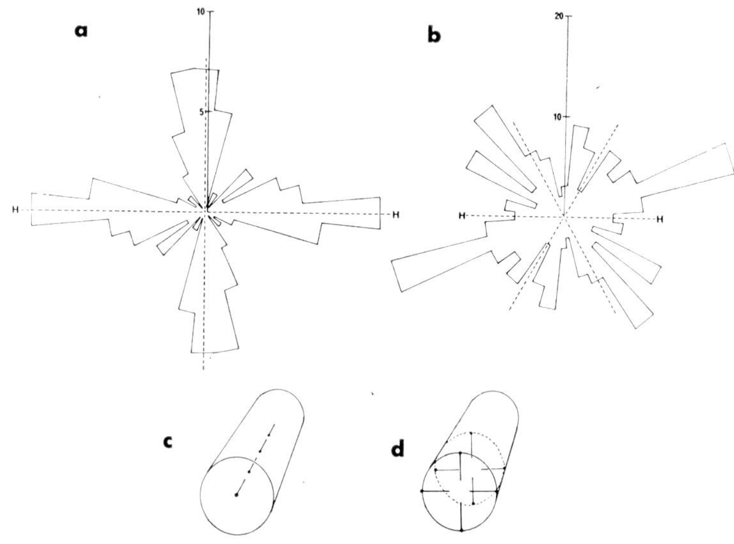

FIG. 2a. Polar diagram of the numbers of cells (plotted outwards) which respond maximally to each particular plane of polarization (plotted circularly), for 58 photoreceptors in the crab eye. Results were combined by reference to a fixed axis through the eyestalk, relative to which the two orientations of the rhabdom tubules are known (Horridge, 1967) and so can be superimposed (dotted lines; H—H is the set held horizontally). b, A similar plot with data from four locust eyes, where there are often three sets of microtubules (see Fig. 1d). There is no clear grouping along the axes or across them as might be expected if the dichroic axis lay regularly along (c), or across (d), the tubules, but this may be caused by variability across the eye or distortion by the microelectrode.

sets of rhabdom tubules (Fig. 2a; Shaw, 1966). It would be anticipated from the vertebrate results (Wald *et al.*, 1963) that the dichroic axes of the photopigment molecules would lie along the tubules and not across, but Fig. 2a does not allow this distinction to be made. However, in some locust eyes there appear to be three sets of tubules forming triangular rhabdoms which are regularly repeated across the eye thus providing an opportunity to test pigment orientation (Fig. 1d; G. A. Horridge, personal communication). The results of an experiment similar to that on the crab eye are shown in Fig. 2b, with the approximate alignment of tubule axes shown as dotted lines. If the pigment is arranged axially (Fig. 2c) the results should group around the dotted lines, and if radially (Fig. 2d), on the bisectors of these lines. Although the radial arrangement appears to some extent to be favoured there is too much scatter to allow firm conclusions. Indeed, from first principles, the axial arrangement would seem more likely since all the molecules would then be fixed at right angles to the light path and so capture light most efficiently, whereas the radial scheme would be even less efficient than a random arrangement of molecules (Wald *et al.*, 1963). Guilio (1963) has provided electroretinogram (ERG) evidence of an axial arrangement in the fly eye.

Unless the variability in Fig. 2b results from cellular distortion by the microelectrode, any analysing system summing responses across the locust eye will encounter so much scatter that it is difficult to see how any reliable polarized light analysis could be performed. There is no behavioural evidence that locusts can perform such discriminations, as there is for other animals (Jander and Waterman, 1960).

SPECTRAL SENSITIVITY

In common with the eyes of the honeybee and fly (Autrum, 1965; Burkhardt, 1962, 1964) different receptors in the locust eye have been shown to respond over different spectral ranges (Bennett *et al.*, 1967). The results were classified rather arbitrarily into three groups but more likely represent a continuous gradation between extreme types (Fig. 3a). The striking result is that none of the cells encountered had spectral absorptions similar to that of either retinene$_1$ or retinene$_2$ (Dartnall, 1953; Munz and Schwanzara, 1967), although the available evidence suggests that retinene can be extracted from invertebrate eyes (see Goldsmith, 1964).

Absorption by accessory screening pigments seems an unlikely explanation of the unusual spectral sensitivity curves since the same results were obtained whether on axial or off-axial source was used. Direct measurement of screening pigment absorption in *Limulus* and *Musca* shows this to be neutral or red transmitting (Wassermann, 1967; Strother, 1966), adding further weight to this argument.

Another explanation is that there are two visual pigments present in each cell in differing amounts, so that the spectral response is a combination of their individual absorption spectra (Fig. 3b). This situation apparently exists in the fly retinula cell where absorption at one wavelength can be increased by selective bleaching at another, suggesting that two pigments are present, one of which is a derivative of the other

(Langer and Thorell, 1966). In vertebrate rods one pigment derivative has its dichroic axis at right angles to its precursors (Denton, 1959; Wald *et al.*, 1963), so it seemed worthwhile to test locust retinula cells through the visible spectrum with polarized light to see if two separate pigment orientations occurred (Fig. 4). The negative results obtained from all cells so far does not of course imply conversely that only one pigment was present.

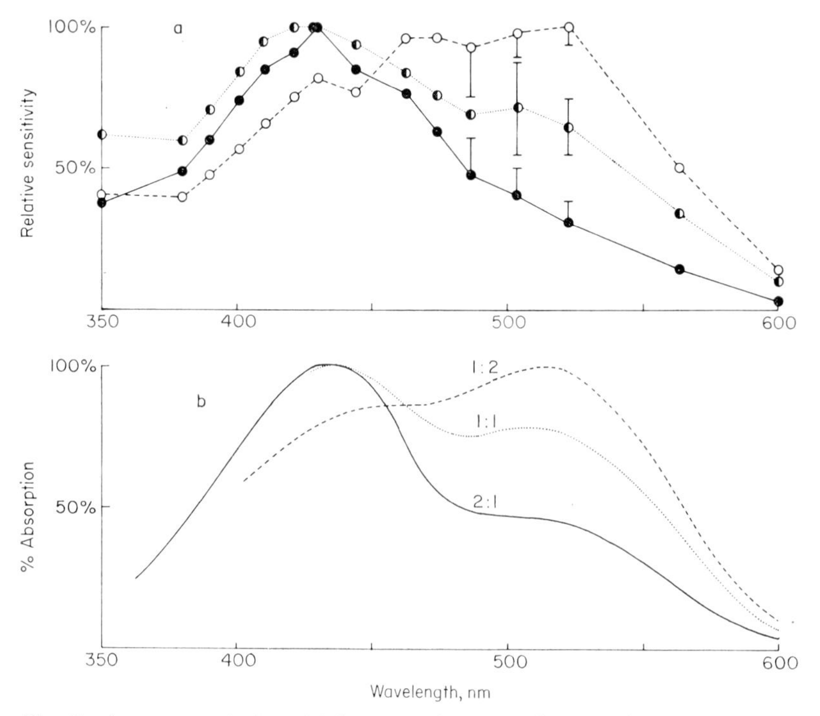

FIG. 3a. Average spectral sensitivity curves for twenty locust retinula cells previously separated into three groups. b. For comparison, the absorption curves that would result from mixing two retinene$_1$ pigments with peak absorptions at 430 and 515 nm in the ratios shown. (After Bennett, Tunstall and Horridge, 1967.)

Burkhardt (1962), recording similarly from single cells, also found two peaks in each receptor's spectral sensitivity curve. This suggests that if two pigments are present, both take part in producing the membrane response. Burkhardt tested this possibility by adapting cells to green light and plotting the subsequent recovery of sensitivity in the dark, using either a green or a UV test light. Since recovery rate did not depend on the test colour he concluded that this demonstrated only one pigment was present. This does not necessarily follow, since the recovery effect after light adaptation probably involves not so much photopigment regeneration as recovery of membrane processes (membrane permeability, local ion concentrations—see p. 158). In a few experiments here the attempt has been made to control the membrane processes by using a steady monochromatic light for bleaching with a test flash superimposed, but this has been hampered by the lack of intensity of the present test source. A twenty-fold reduction of

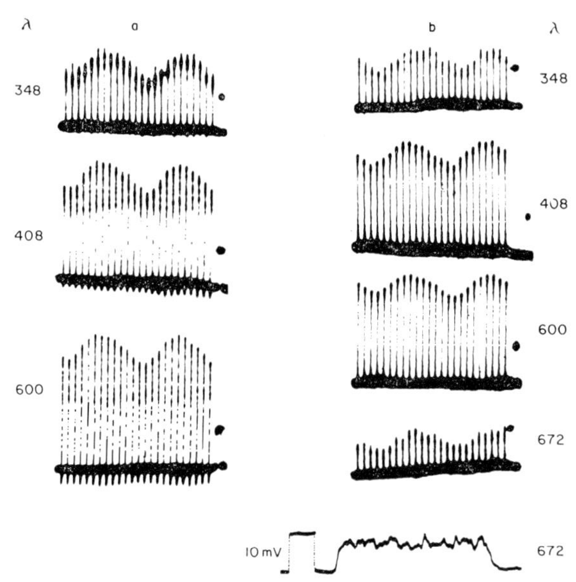

FIG. 4. Records from locust visual cells, a and b, to a 360° rotation of polarization plane at different peak wavelengths (λ) isolated by narrow-band interference filters. There is no detectable shift of the plane which gives peak responses, nor does the relative sensitivity to polarized light change. The differences in amplitude modulation seen in records from the same cell arise because the amplitude–light intensity relationship is non-linear, and this can be corrected for subsequently. At the extremes of each cell's visual spectrum, responses to the lamp are small and rather variable because of the irregular summation of individual quantal potentials (b, bottom record).

receptor sensitivity appears to cause no pronounced shift in the peak wavelength response but in the absence of objective evidence of the extent of pigment bleaching it is difficult to evaluate this result.

RECEPTOR INTERACTION

Effects of an off-axial flash

Another interpretation of the aberrant spectral sensitivity curves is that neighbouring receptors with single visual pigments are in some way interacting. Tunstall discovered that a bright flash delivered several degrees off the visual axis produced a small hyperpolarization in some retinula cells, but that in other cells this effect only occurred when the flash coincided with a response to an axial light. This was investigated since it initially seemed likely that it represented some kind of lateral inhibitory system. Fig. 5a shows the development of the effect with increasing brightness of the axial light. The more the receptor itself

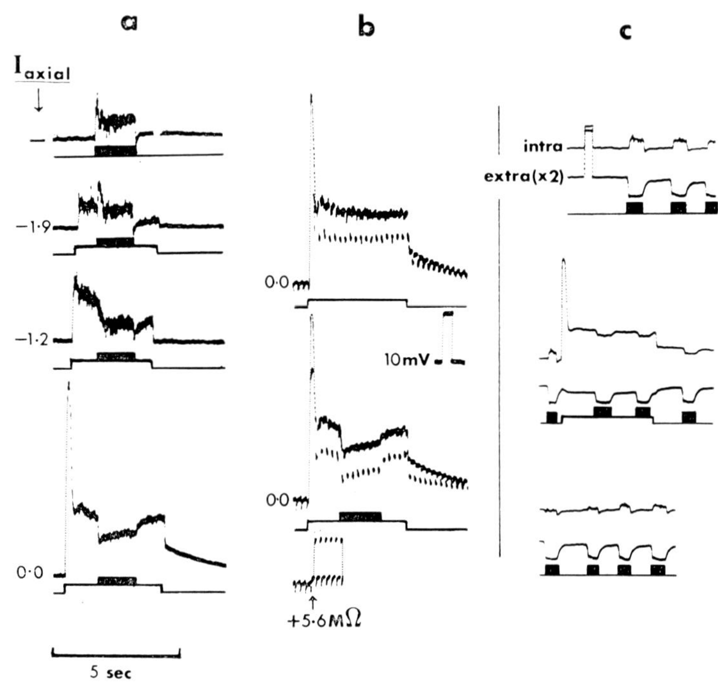

FIG. 5a. Response of a cell to a bright flash delivered 10° off-axis, shown under each record as a black bar. From top to bottom, this flash is superimposed on a progressively brighter axial flash, also monitored and of log intensity given at the left. The off-axial light alone depolarizes the receptor, but hyperpolarizes with a bright axial flash. b, The cell input resistance is measured as the unbalance in a Wheatstone bridge during current pulses down the microelectrode. There is no detectable extra resistance change during the off-axial flash (bottom record) beyond that recorded without it (top). Notice the persistent reduction of membrane resistance underlying the afterpotential. c, Records from two electrodes, one inside a cell and the other a few microns away during tests similar to (a). The points of interest are that the hyperpolarization resembles the extracellular ERG, and reverts to a depolarization as membrane resistance recovers and repolarization proceeds.

responds to the axial flash the greater the "inhibition" becomes, in contrast to the situation in the *Limulus* eye. Indeed in that eye the inhibition acts at the spike locus and not on the slow potential itself (Hartline, Ratliff and Miller, 1961). If the hyperpolarization represents the usual type of inhibitory effect it would be expected to act upon the ion conductances generating the retinula slow potential, which are measured by the bridge method in Fig. 5b. Thus if during the effect these ion fluxes were partially suppressed or if permeability to some other species of ion came to dominate, an easily measurable increase or decrease in membrane resistance would be respectively expected (Purple and Dodge, 1965), but none such can be measured here (Fig. 5b).

In order to clarify the nature of the effect, recordings were taken with an extracellular electrode separated by a few microns from an intracellular one (Fig. 5c). A large off-axial source is flashed before, during and after a bright flash delivered on the penetrated cell's axis. Significantly, the intracellular hyperpolarization has a time course and size similar to that of the locally recorded ERG, and persists after the axial flash goes off. Its reversion to a depolarization is correlated with the persistent lowered membrane resistance that follows a large response (e.g. Fig. 5b). The reversion is not connected with changes in the adjacent ommatidia since these are exposed to constant flashes throughout, apart from the small axial flash which is at the edge of their visual fields.

Since the hyperpolarization has none of the expected characteristics of an inhibitory effect, the most likely explanation of its presence is that it represents the electrotonic invasion of the cell by local currents from adjacent receptors, when the recording site is not insulated from the exterior by a high membrane resistance. The potential record results from the summation of local intra- and extracellular changes relative to a remote indifferent electrode. Where membrane resistance is lowered by damage caused by an electrode, a negative response to an off-axial light alone would be anticipated, since the extracellularly-arising potential would then be large enough to swamp the small response of the cell itself. In all cases the extracellular change would equal or exceed the intracellular resultant, so that the nett polarization of the membrane would be zero or even a small depolarization. The effect is therefore not inhibitory.

Direct measurements of receptor interaction

In some of the paired recordings from retinula cells, currents were passed into one cell through one microelectrode and potentials recorded from the other. In cells where the angular sensitivities were different, no coupling could be detected apart from the expected capacity currents (Fig. 6a). In pairs of cells from the same ommatidium a small amount of coupling is revealed by this test (Fig. 6b, c). An estimate of the coupling ratio in the dark (voltage in one cell relative to the voltage developed in the other) in the case shown is that it is poorer than 7 : 1 (Shaw, 1967).

If it were possible to optically stimulate one cell in an ommatidium in isolation of its neighbours, the coupling would not necessarily be the same as that detected in the above test, since the e.m.f. in the two cases would not arise across the same structures. In particular, a selective increase in conductance of one cell membrane would not necessarily produce a potential change in the other cells of the group: the size and polarity of any effect would depend on the species and rate of ion flux, and the relative occlusion of the extracellular space. It is worth noting that at low light intensities there is no measurable

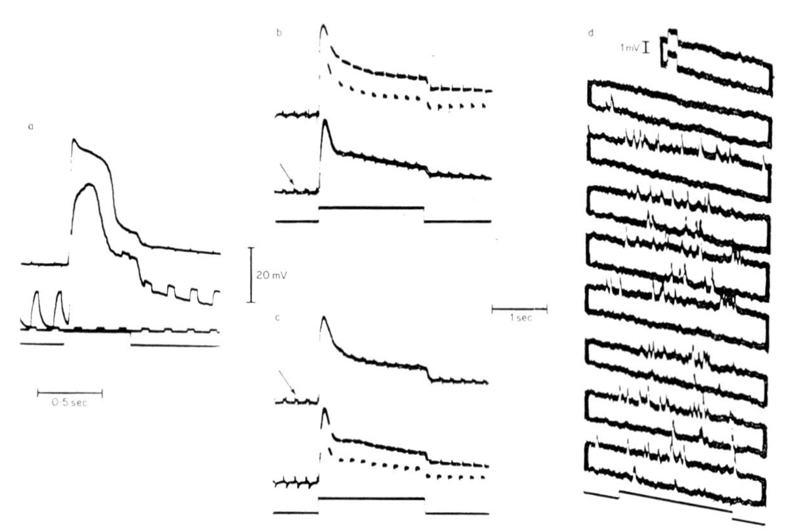

Fig. 6a. Simultaneous records from a pair of receptors of slightly different angular sensitivity. Hyperpolarizing current pulses applied to the bottom cell produce responses of about 15 mV which decrease during illumination. Lack of deflection in the other cell shows the two are not coupled. (b, c, d), The same pair of cells as in Fig. 1, in the same ommatidium. Positive pulses applied to either cell (b, c) produce small deflections in the other (arrows). The coupling ratio is poorer than 7 : 1. (d) At low light intensities individual photon captures or bumps are not coincident in the two cells. Simultaneous sweeps have been bracketed together.

coincidence of the small potentials ("bumps") which probably result from single photon absorptions (Fig. 6d; see Scholes, 1965), and that the retinula cells at high intensities appear to act independently when tested with polarized light (Fig. 1c).

Coupling between cells in one ommatidium is strong in *Limulus* when tested by polarization via the microelectrode (Smith, Baumann and Fuortes, 1965; Borsellino, Fuortes and Smith, 1965), but there are some indications that the individual cells do not in all circumstances act as a concerted group. The individual receptors have distinct angular sensitivities (Ratliff, 1965; Stieve, 1965), and spikes originating in the eccentric cell sometimes produce only small potentials in the retinula cell bodies (Behrens and Wulff, 1965). Destruction of one retinula cell does not lead to depolarization of the others (Behrens and Wulff, 1967) although this could be due to junctional sealing which occurs in some conditions in other connected cells (Loewenstein, 1966). However, since there is usually only one conducting axon per ommatidium in *Tachypleus* (Kikuchi and Ueki, 1965) and *Limulus* (Behrens and Wulff, 1965; but see Borsellino *et al.*, 1965), the cells are probably functionally coupled anyway in having only one output.

Retinula cell spikes

Scholes (1965) noticed that occasionally a spike discharge could be recorded in the retinula layer. This is so in a few percent of penetrations

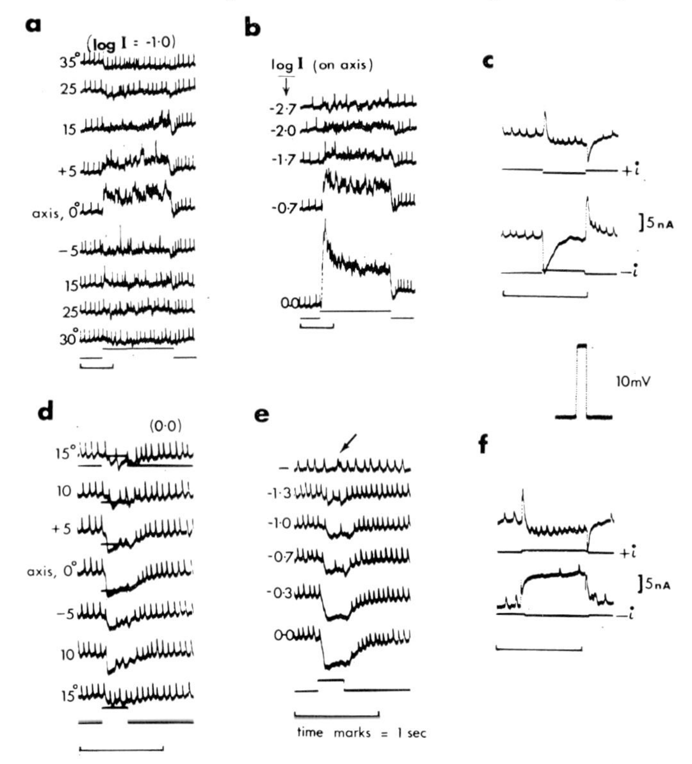

FIG. 7. Records from two sites (a–c, d–f) in the retinula cell layer which give regular, attenuated spike-like discharges. a, d, A constant flash is delivered at different angles in the horizontal plane relative to the visual axis, and at every position inhibits the discharge. Note the post-inhibitory rebound effect. b, e, Intensity of the flash is varied at the visual axis. Arrow denotes apparent summation of "spikes" in (e). c, f, Depolarization via the microelectrode increases and hyperpolarization decreases the frequency in both cases. In the upper set of records, however, illumination decreases frequency but causes a depolarization.

even where the electrode is located just beneath the cornea and could not possibly be in the first synaptic region. Since the data shown by Scholes resembles the result of the lateral inhibitory process in *Limulus*, it warrants closer inspection. In Fig. 7a and d, the angular sensitivity of two sites in the distal retinula region is shown, and it can be seen that there is no part of the receptive field of either cell where the light flash increases spike frequency, in contrast to the *Limulus* eye. In fact the inhibitory effect has a lower threshold on the slow potential axis (Fig. 7b, e). The depolarizing discharge is indeed inhibited and not just shunted out by a conductance increase between the recording site and the response origin, as can be judged by the "off"-discharge and the presence of a few responses during inhibition on some records. Whether the depolarizations are attenuated spikes or post-synaptic potentials (PSPs) is not clear. Their shape in the upper records of Fig. 7 suggests spikes but the occasional summation (Fig. 7e, arrow) perhaps suggests PSPs.

The main anomaly is that whilst "spike" frequency is always increased during imposed depolarization and decreased during hyperpolarization (Fig. 7c, f), light, which depolarizes the receptors, inhibits the response. Occasionally the discharge can be recorded on top of a negative potential (Fig. 7d, e) but normally occurs along with a depolarization which is sometimes smaller and more variable than the usual retinula potentials. These differences seem to imply that the spike phenomenon and retinula potential are not closely linked and probably arise in different structures. Possibly the spikes originate in the rudimentary retinula cells as Scholes (1965) suggests, but it seems more likely that they are electrotonically conducted back up the receptor

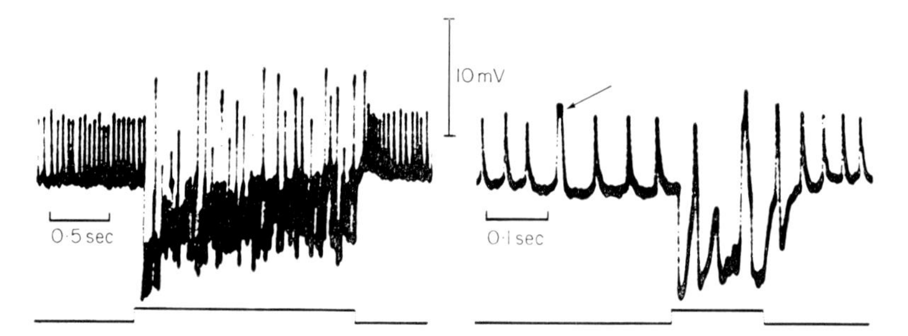

FIG. 8. Two records from a cell in the lamina which gave depolarizations similar to those in the previous figure. Illumination produces large negative-going potentials which reduce the frequency of the depolarizing effect. Depolarizations occasionally summate (arrow), and are of increased amplitude where they occur during the flash, presumably when they fall during the repolarization phase of an IPSP. Note the post-inhibitory rebound.

axons, since similar but sometimes larger potentials can be recorded occasionally in the lamina synapse (Fig. 8). These depolarizations summate and can be of increased amplitude during the hyperpolarization caused by light, and so behave similarly to the PSPs recorded from locust muscle (Usherwood and Grundfest, 1964).

The results described in this section suggest that interactive effects amongst receptors are small and do not resemble a lateral inhibitory process. Similar conclusions have been drawn for the fly eye (Washizu, Burkhardt and Streck, 1964) and for the locust eye from studies on ventral cord units (Palka, 1967).

POSTSYNAPTIC POTENTIALS FROM THE LAMINA

The preceding account, together with the reviews of Scholes (1965) and Horridge (1965) gives some idea of our current knowledge of the factors affecting retinula cell responses in the locust. With this in mind it has been possible to turn attention to the way information is transmitted to the second stage of the visual system in the locust. Some of the results obtained with intracellular electrodes in the lamina synapse will be summarized here, and a fuller account will appear elsewhere (S. R. Shaw, in preparation).

Anatomy of the lamina region

Electron microscope data on the insect lamina synapse is so far confined to the fly. The lamina is composed of compact synaptic complexes (cartridges) each containing six retinula endings and two second-order fibres, along with a few small ancillary fibres (Trujillo-Cenóz, 1965). It is preceded by a precise local chiasma of the retinula axons so that each cartridge receives axons from six separate ommatidia (Braitenberg, 1967; Trujillo-Cenóz and Melamed, 1966). The function of this arrangement is described in Kirschfeld (1967, 1968). A similar anatomical arrangement is found in the lobster although the reported crossover of axons should be treated with reserve (Hámori and Horridge, 1966). It has been suggested for both animals that the synapses between retinula axons and ganglion cells are electrically-conducting because of the reduced synaptic gap, but they have been more convincingly interpreted as chemically-mediated on different morphological grounds (Osborne, 1967).

The locust lamina has been scrutinized carefully with the light microscope in thirteen serial silver-stained preparations, and in none of these has any convincing instance of axon crossover been seen, although it is quite obvious in similar preparations of dipteran material

(Burtt and Catton, 1962; Pedler and Goodland, 1965; Braitenberg, 1967). It is difficult to follow the courses of many axons over long distances since they are rather irregular and the fibres bunch together at one stage (Fig. 9), as also found by Burtt and Catton (1962) who

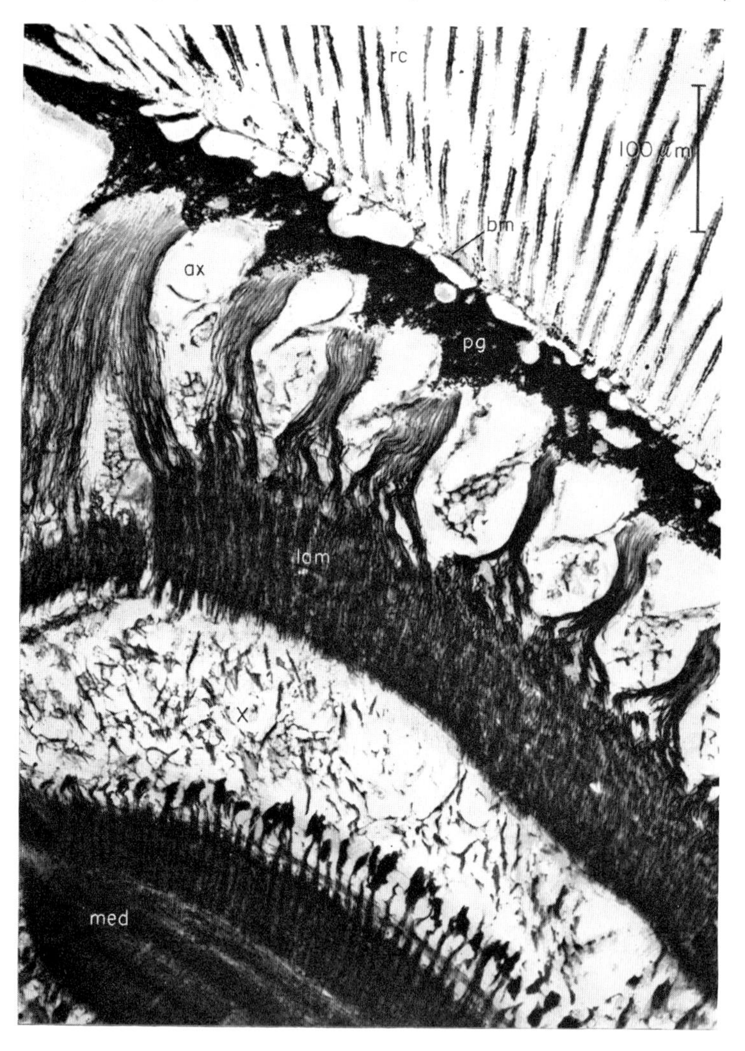

Fig. 9. Silver-stained section through the locust lamina transverse to the body axis. Axons (*ax*) of the retinula cells (*rc*) penetrate the basement membrane (*bm*) in regularly-spaced ommatidial bundles and bunch together before diverging again just outside the lamina synapse (*lam*). There are no chiasmata in the retinula axons such as occur in the Diptera. The first chiasma (*X*) is in the horizontal plane between the lamina and medulla (*med*) and so the connecting fibres appear in oblique section. Numerous pigment granules (*pg*) are clustered beneath the basement membrane.

likewise failed to demonstrate chiasmata. Since bundles of eight axons from single ommatidia leave the basement membrane (Horridge, 1965) and similar bundles enter the lamina in silver stained preparations, it therefore seems likely that the bundles are the same, so that only cells of one ommatidium contribute to each lamina cartridge. The structure of the cartridge itself in the locust is unclear and must await an electron microscope study.

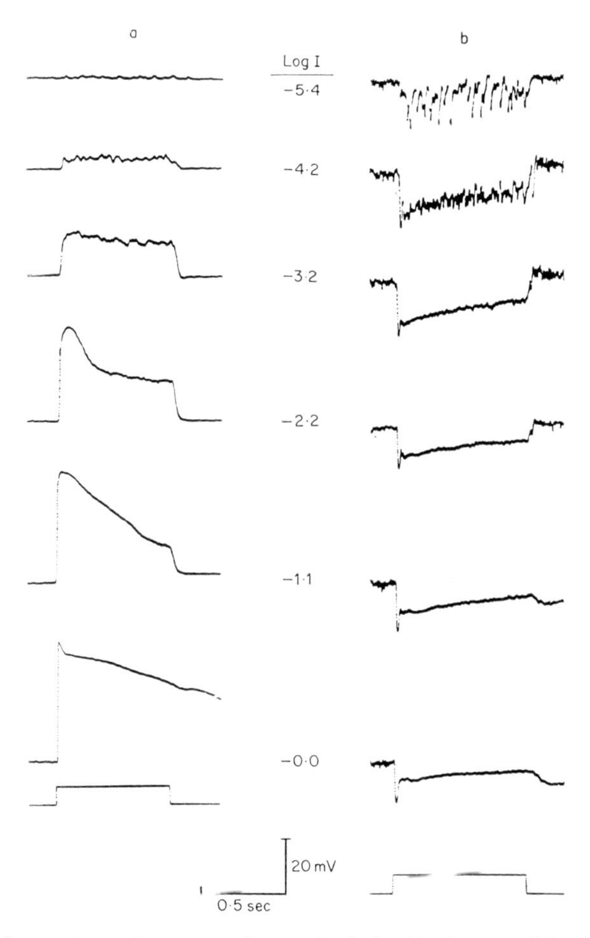

Fig. 10. Comparison of responses from a typical retinula, a, and lamina cell, b, to an axial light flash of increasing intensity from top to bottom of the record. The cell in the lamina gives a strong hyperpolarizing response even when the average retinula is producing depolarizations of only a millivolt (top records). At higher intensities the lamina response appears reduced, possibly as a result of the superposition of a depolarization at the recording site (see text).

Response localization

Since the locust retina is curved and pigment-covered, eight marking experiments have been conducted to confirm definitely where the responses to be described were being recorded. In seven cases the electrode tips were located in the lamina, and in one case in the fibres leaving that synapse, but the fibres are too small to reveal which sort of cell was penetrated.

Hyperpolarizing potentials from lamina cells

Different sorts of activity have been obtained from the lamina region, but the most frequently encountered type is characterized by intracellularly recorded hyperpolarizations of up to 15 mV amplitude in response to a dim light, and usually occurs without superimposed depolarizations like those in Fig. 8. The typical form of response to a light flash is shown in Fig. 10 with a retinula cell response for comparison, over the same 5·4 log-units range of light intensity. The large hyperpolarizations summate together at higher light intensities, and with bright flashes the waveform often consists of an initial negative

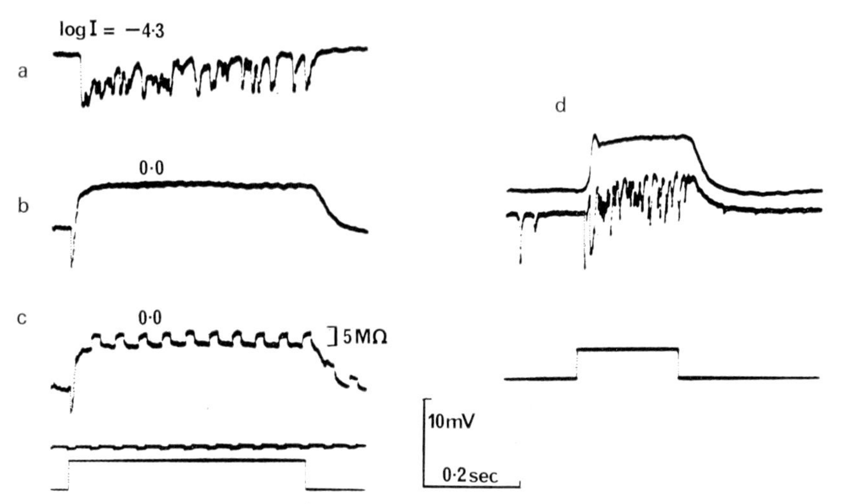

FIG. 11. Response of a lamina cell to a dim flash, a, and to one 4·3 log units brighter, b. The brighter flash produces an initial negative peak followed by a slow positive potential. c, A resistance drop of about 4 MΩ can be measured during the bright flash, indicating that at least part of the potential record represents an active membrane phenomenon. Current pulses through the microelectrode and flash duration are monitored on the two lower beams. d, Intracellular response of another cell (middle trace) to a bright flash, with the extracellular response from a glass-coated metal electrode (tip diameter 15 μm) located about 20 μm away also in the lamina (top trace). The intracellular record appears to consist of hyperpolarizations as in (a) combined with the local extracellular response.

peak followed by a slow depolarization (e.g. Fig. 11b). This depolarization is similar to that recorded extracellularly with large electrodes (Fig. 11d) so perhaps again represents the addition of an extracellular component to the intracellular response. Since the nature of some of the positive potentials recorded in the lamina is not at present clear, they will be excluded from this account.

It is, however, clear that at least a part of the response in Fig. 11b is due to an active membrane phenomenon, since a membrane resistance decrease of about 4 MΩ can be detected (Fig. 11c). An analogous experiment (Fig. 12) shows that the reversal point for the hyperpolarizing potentials (hereafter HPs) in that case lies about 25 mV negative to resting potential, as is similarly found for inhibitory post-synaptic

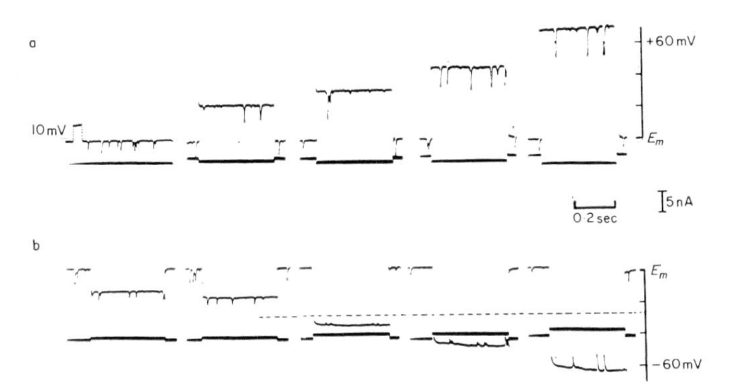

FIG. 12. A potassium acetate-filled double barrel microelectrode depolarizes a lamina cell, a, or hyperpolarizes it, b, during dim background illumination. Response height is altered during polarization and the reversal potential appears to be about 25 mV negative to resting potential (*Em*). This is an overestimate since there is some resistive coupling between the electrode and the location of the conductance change, although it is unusually small here. Polarization does not alter response frequency.

potentials in other systems (Bullock and Horridge, 1965). Most cells, however, need to be hyperpolarized more than 100 mV before reversal is effected, which probably indicates that the origin of the response is remote from the recording site in these instances.

Absolute sensitivity and receptor adaptation

The most striking fact arising from the comparison between receptors and HP cells is that for very dim flashes a brisk response occurs in the lamina when the receptors are experiencing depolarizations of only a millivolt or less (Fig. 10, log I $\equiv -5{\cdot}4$). The most obvious comment on this is that probably the receptor response is not being recorded at full amplitude because the microelectrode tip is not exactly at the site of origin. But in many hundreds of penetrations, the maximum amplitude of individual receptor bumps is seldom more than 1 mV, those in Fig. 16 being exceptionally large. Moreover, even if the bump amplitudes really do represent larger depolarizations of some inaccessible parts of the receptor, the fraction of this which is recorded in the cell body will be the important one from the point of view of potential spread down the cell axon. The conclusion seems inescapable that potentials of the order of 1 mV or less in the cell bodies cause an effect in the lamina which results in the large negative potentials. Kirschfeld (1965) reached a similar conclusion concerning receptor axon transmission from a combination of his work and that of Fermi and Reichardt (1963).

To estimate the transmission efficiency of the axons, a measure is first needed of the light-capturing efficiency of the receptors. Scholes (1964, 1965) showed statistically that a single "event" and not the temporal coincidence of two or more events was sufficient to produce a single receptor bump, and the most reasonable assumption was that each event was a photon absorption. But Scholes estimated that only 0·1% of the incident quanta were actually absorbed, so this figure has been checked for a few cells in the present study, where the light-intensity calibration is simplified since it does not depend on angular sensitivity. The most sensitive cell measured had a capturing efficiency of about 8% (one photon in twelve) for a long dim flash delivered on axis at the optimum wavelength peak. Other cells ranged up to about ten-fold less sensitive, so that the values lie in the same range as was found for the fly eye by Kirschfeld (1965).

A similar determination for a typical HP locus is shown in Fig. 13a, where the intensity/frequency curve is plotted over a limited range. If only one of the common blue-sensitive type of retinula cell (p. 139, Bennett *et al.*, 1967) were supplying the HP cell, the absolute calibration on the linear scale would apply and the combined efficiency of photon capture and response transmission would be 60% if based on the lowest measurement point. This value would of course be artificially high if input convergence of several retinulas on to one HP cell existed, so this is investigated in Fig. 13b where the angular sensitivity of the cell is plotted. This function is narrow in common with most cells, the half-width value of 4·5° being well within the 7° reported for the average dark-adapted retinula cell (Tunstall and Horridge, 1967). Since the angular sensitivity of the individual retinula is the same as that of its neighbour in that ommatidium (p. 137) and different from that of the neighbouring ommatidia, between one and seven cells could therefore be contributing to the response, the estimated overall efficiency being reduced proportionately. But it would still be of the same order as the efficiency of the receptors themselves, which shows directly that a substantial fraction of the bumps (single photon captures) do in fact get through the system to cause an equivalent response in the lamina HP cells.

Figure 13 shows a typical result where the response frequency is non-linear with intensity at very low light flux levels. The data fit a

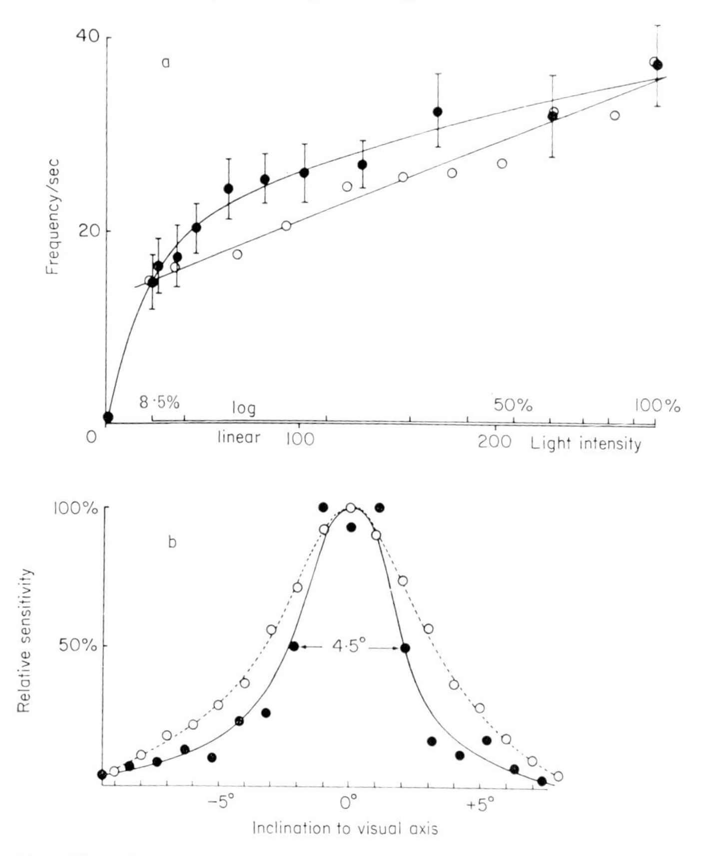

FIG. 13 a. Mean frequency of hyperpolarizing potentials from a lamina cell to long dim flashes on the visual axis, at intensities shown on the abscissa. The response fits a logarithmic relation (open circles, upper abscissa) better than a linear one (filled circles, S.D. values for $N = 6$). The calibration on the lower abscissa is of numbers of photons per second incident on the retinulas directly in line of the flash, expressed as if these were delivered at the spectral peak of a blue sensitive cell. At the lowest measurement point, about 25 photons per retinula would cause about 15 potentials in the lamina cell. b, Input convergence on to the above cell is from no more than the seven cells of one ommatidium, since the angular sensitivity (filled circles) of the cell, taking account of the non-linearity in (a), is well within the average for the dark-adapted retinula cell (open circles; from Tunstall and Horridge, 1967).

logarithmic relation somewhat better (Fig. 13a, open circles). Possibly part of the levelling-off is due to the difficulty of counting peaks where extensive summation is occurring at higher frequencies, but this would not affect most readings. The effect might be attributed to some synaptic property such as the progressive depletion of a transmitter substance, or to a non-linear relationship between voltage and the amount of transmitter released. Alternatively, the cause may be that some receptor-pigment desensitization occurs even where only a few light quanta per second are being captured. It has been shown that locust retinula cells have a linear intensity/bump-frequency relation at low frequencies of response and also at higher frequencies by extension of data obtained with very short flashes (Scholes, 1965). It therefore seems most likely that the non-linearity arises at some stage in the transmission process.

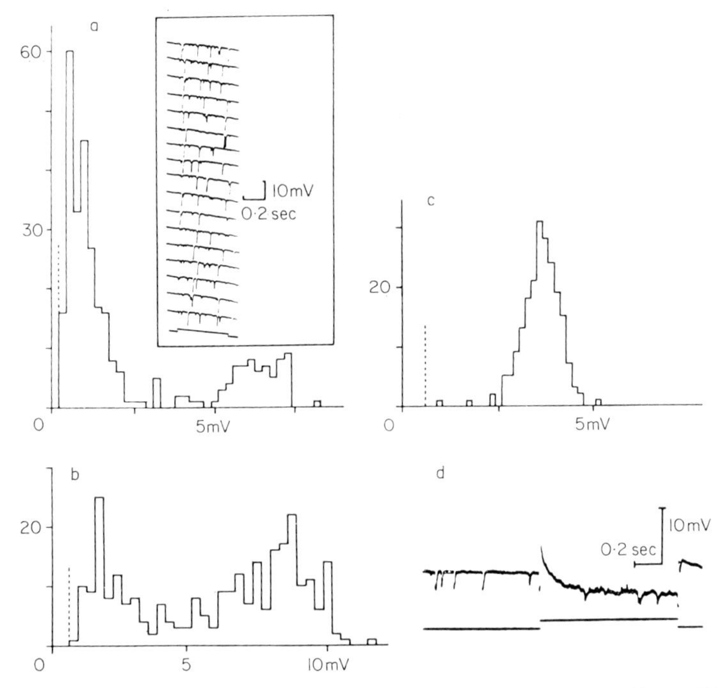

FIG. 14 (a–c). Amplitude histograms from three lamina cells, providing evidence of nput convergence (Abscissa : amplitude; ordinate : frequency). Amplitude variability is normally great as in (b), but occasionally cells show unimodal (c) or bimodal (a) amplitude distributions. Inset in (a) is a specimen record. (d) During dim illumination, hyperpolarization of one cell, partly balanced out by the bridge circuit, reverses some potentials but not others. This is explained if there is one synapse near the recording site and another distant from it.

Evidence for input convergence

Occasionally the individual potential amplitude variation from a HP cell reveals a clear amplitude division (Fig. 14a) or is strikingly invariant (Fig. 14b) but most frequently considerable scatter occurs (Fig. 14c). This could reasonably be due to the occurrence of two, one and several different synaptic inputs respectively, on to the different cells above. The clear division in Fig. 14a could be a result of one synapse releasing less transmitter on average, or of its being located further from the recording site. Some support for the second alternative was provided in one experiment where upon hyperpolarization of the cell to a certain level, some HPs were reversed in polarity and others not (Fig. 14d). This might be anticipated if the two synapses were spatially separated so that electrotonic decay of the polarizing current were more severe at one than the other. The inference that several presynaptic elements are contributing in most cases could be tested with extended data, but unfortunately the lifetime of individual units is usually very short, with a rapid mean response amplitude decline. Additional causes of amplitude variation can be also imagined: there is for example a continuous variation of bump amplitude in the receptors (Scholes, 1964; Kirschfeld, 1965).

Latencies of receptor and lamina responses

In the lamina synapse of the fly and lobster the main neural elements are the retinula axon endings and the pairs of postsynaptic axons. Since large negative potentials would not be expected to occur in the retinula terminals as a result of small depolarizations in the cell bodies, and since there is evidence of limited input convergence, it is tempting to place the HP origin in the second-order axons. This is at least consistent with the anatomical picture (p. 147) and also with the latency determinations below. If correct it suggests that the lamina synapse is a chemically mediated one.

If transmission down the retinula axons is by electrotonic spread, transmission time should occupy only a small fraction of the determined latency in Fig. 15. The measurements are arbitrary in the sense that there exists no basis for determining what potential level in the receptors must be exceeded before the lamina cells start to hyperpolarize. Latencies in Fig. 15b are measured to a potential level of 1 mV in the receptors and to half peak-amplitude in the HP cells. Mean latency is roughly 5 msec for the two highest intensities, with scatter increasing progressively at lower flash intensities. This figure is well below that of 15–30 msec for the only two invertebrate inhibitory synapses well-documented (Bullock and Horridge, 1965, p. 184), and so is consistent with the proposition that there is only one synapse between the receptors and the HP response.

There appears to be no direct information from other synapses on the minimum depolarization necessary to release effective quantities of transmitter, though in the squid giant fibre synapse the minimum

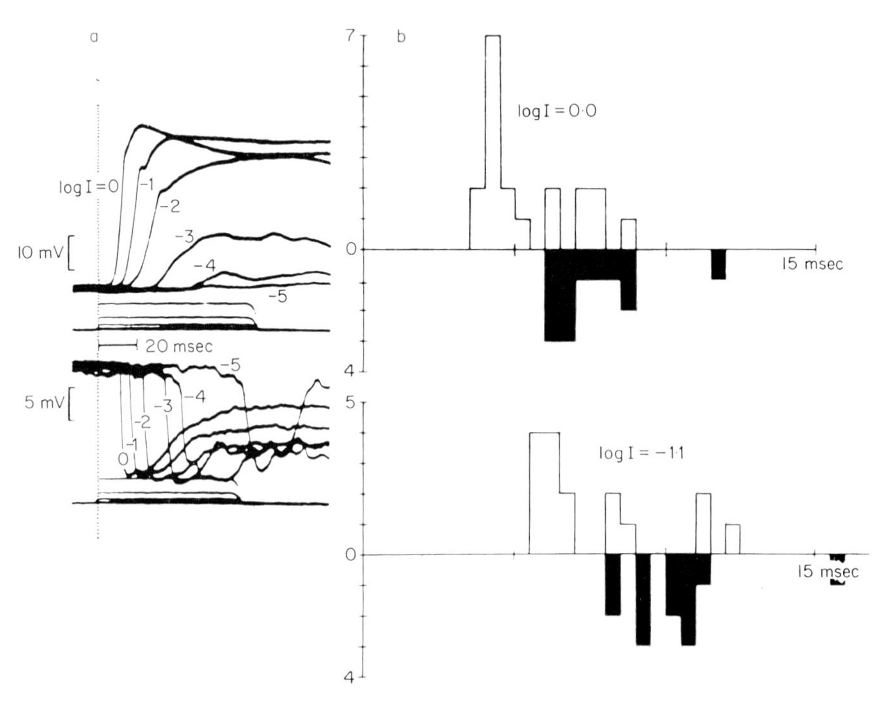

FIG. 15 a. Comparison of retinula cell latency (top) with lamina cell latency (bottom) over the same light intensity range as in Fig. 10. The actual delay between responses depends on which points on the waveforms are chosen for measurement. b, Latency histograms for several retinulas (plotted upwards) and lamina cells (plotted downwards, black) at the two highest intensities in (a). "Latency" is measured to a 1 mV criterion in the receptors and to half peak amplitude in the lamina cells. The mean difference is about 5 msec.

figure so far is about 25 mV (Bloedel, Gage, Llinas and Quastel, 1966; Katz and Miledi, 1966; Kusano, Livengood and Werman, 1967). This may not be a good comparison for in the squid the prefibre is normally invaded by the spike, the post fibre is very large, and the synaptic effect is anyway excitatory.

Depolarizing responses with a slower time-course but similar latency of first appearance to the receptor potential have been recorded from the lamina. These possibly represent the activity of the receptor terminals, but their nature is equivocal at present.

The receptor after-potential

If a receptor is depolarized by a strong flash, the depolarization persists after the light is switched off, and gradually disappears over

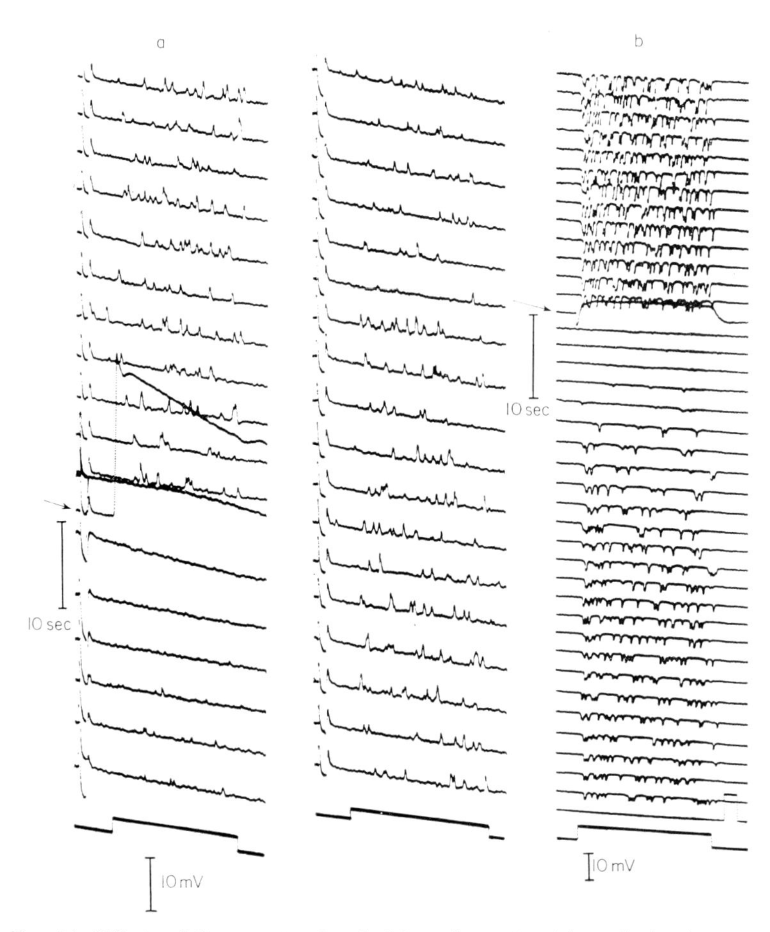

FIG. 16. Effects of the receptor depolarizing after-potential on the lamina response. In (a), records from a retinula cell read down the first column then down the second. On each sweep, a dim light flashes which evokes large bumps in the receptor. At the arrow, the flash is made 4 log units brighter for one sweep and produces the usual receptor response with a persistent after-potential. Cell resistance, monitored as bridge unbalance at the left, slowly falls by about 37 MΩ and then gradually recovers. Bumps are of reduced amplitude during this period. (b) An analogous experiment on a lamina cell shows that the after-potential in the receptors does not produce hyperpolarizations in these cells as might be expected (see text).

about 20 sec with the brightest flash used here. This is correlated with a persistent reduction of membrane resistance (e.g. Figs. 5, 6, 16a). It might be expected on the lines of the preceding argument that this depolarization would propagate out of the receptors and stimulate a dark discharge of HPs in the lamina. This is not so, nor can HPs be evoked by a dim flash that was previously adequate (Fig. 16b). The cause of the latter effect is clear from Fig. 16a, namely that receptor bumps are also absent during this period.

The nature of the residual conductance and depolarization is at present not understood, in common with the rest of the receptor waveform. In other visual cells sodium ion appears to be implicated at least in the initial depolarization to light (Kikuchi, Naito and Tanaka, 1962; Stieve, 1965, Benolken and Russell, 1967; Eguchi, 1965) and there is evidence that potassium controls resting potential (Kikuchi *et al.*, 1962; Yeandle, 1967). The repolarization phase might then be interpreted as persistent sodium permeability, which would account for its sign (depolarization), and to some extent for reduction of bump amplitude if the voltage recorded is a simple non-linear function of a particular variable membrane resistance (Naka and Rushton, 1966). But bumpiness is usually still seen during the plateau phase of the receptor potential when depolarization is even greater, so the observed complete abolition of bumps would not be anticipated, as neither would the sharp repolarization of about 5 mV usually seen at light-off. Increased potassium permeability therefore seems a more likely explanation, since this would reduce bump amplitude and is consistent with the hyperpolarizing after-potential that occurs when a receptor is illuminated after first having been depolarized by a background light (Naka and Kishida, 1965). The persistent depolarization discussed above is probably due to accumulated extracellular potassium (Frankenhaeuser and Hodgkin, 1956). It appears that this depolarization should spread to the lamina terminals, and the afterpotential present in the lamina positive potentials suggests that this occurs. The lack of corresponding response from the HP cells may therefore reflect a parallel desensitization of them by the strong flash.

Lamina polarized-light responses

Most of the cells in the lamina that were tested showed no differential response to plane of polarization, but this might well have been due to the limited sampling time available before the responses deteriorated. A positive result is shown in Fig. 17, where the troughs would be deeper on a true sensitivity plot.

The importance of this result is that if input convergence on to one

HP locus is of cells of the same ommatidium, then since these will have several different polarized light responses (p. 135), addition of inputs would largely cancel any differential response. The situation could be saved if there were selective convergence of cells with similar responses on to two or more HP cells per cartridge, or if some retinulas by-passed

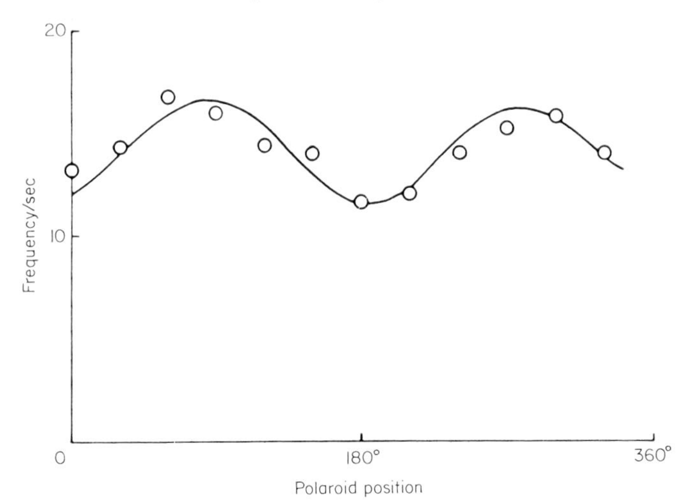

FIG. 17. Hyperpolarizing-potential frequency in a lamina cell in response to rotation of the plane of polarization of light.

the lamina (Braitenberg, 1967). No lamina responses have yet been found with a polarized-light sensitivity as large as that of the best locust receptors (about 7 : 1), and the behavioural discrimination of polarization plane by the locust has yet to be convincingly shown.

CONCLUSIONS

It appears at this stage in the investigation that the physiological results outlined above are at least not in conflict with the limited anatomical picture, but nevertheless some distinct problems remain. Not only can many invertebrates distinguish the plane of light polarization, but some can discriminate colours as well (Burkhardt, 1964). There is good evidence that some wavelength discrimination can take place within one ommatidium in *Calliphora* (Langer and Thorell, 1966) but it appears that in other species the different pigments are confined to different patches of the eye (Bennett, 1967). If the former situation holds in the locust eye the problem of unselective input convergence discussed in the previous section would be magnified. But there is as yet no evidence bearing on this, and the behavioural evidence for colour vision in Orthoptera is slender (Mazokhin-Porshniakov, 1964).

A major problem concerns the ultimate effect of the hyperpolarizing potentials in the presumed second-order cells. Only rarely have these been seen to inhibit any kind of excitatory or spike discharge (Fig. 8), and it is just possible that this is merely a trivial result of microelectrode damage to small cells. It is fairly certain that lack of a spike response does not result from inappropriateness of visual input, since it cannot be evoked by other stimuli, such as movement. How then does the lamina response reach the next relay station in the medulla, some 400 μm away along the lamina cell axons? If millivolt-sized bumps can produce a response after passive transmission down the 200 μm-long receptor axons, there seems to be no good reason why the same should not apply to the ganglion cell axons where the initial response is often 15 mV. Certainly, no impulse activity was detected peripheral to the medulla with metal-filled extracellular electrodes (Horridge, Scholes, Shaw and Tunstall, 1965).

Acknowledgement

This work was carried out in the laboratory of Dr. G. A. Horridge, to whom I am indebted for his continued support and encouragement. I should also like to thank him and several of his former associates for invaluable discussions: particularly Drs John Scholes, Ruth Bennett and John Tunstall. The work was partly financed by an SRC grant.

References

Autrum, H. (1965). The physiological basis of colour vision in honeybees. In *Ciba Foundation Symposium on colour vision*: 286–300. de Reuck, A. V. S. and Knight, J. (eds.). London: Churchill.

Autrum, H. and von Zwehl, V. (1962). Die Sehzellen der Insekten als Analysatoren für polarisiertes Licht. *Z. vergl. Physiol.* **46**: 1–17.

Behrens, M. E. and Wulff, V. J. (1965). Light initiated responses of retinula and eccentric cells in *Limulus* lateral eye. *J. gen. Physiol.* **48**: 1081–1093.

Behrens, M. E. and Wulff, V. J. (1967). Functional autonomy in the lateral eye of the horseshoe crab, *Limulus polyphemus*. *Vision Res.* **7**: 191–196.

Bennett, R. (1967). Spectral sensitivity studies on the whirligig beetle *Dineutes ciliatus*. *J. Insect Physiol.* **13**: 621–633.

Bennett, R., Tunstall, J. and Horridge, G. A. (1967). Spectral sensitivity of single retinula cells of the locust. *Z. vergl. Physiol.* **55**: 195–206.

Benolken, R. M. and Russell, C. J. (1967). Tetrodotoxin blocks a graded sensory response in the eye of *Limulus*. *Science, N.Y.* **155**: 1576.

Bloedel, J., Gage, P. W., Llinas, R. and Quastel, D. M. J. (1966). Transmitter release at the squid giant synapse in the presence of tetrodotoxin. *Nature, Lond.* **212**: 49–51.

Borsellino, A., Fuortes, M. G. F. and Smith, T. G. (1965). Visual responses in *Limulus*. *Cold Spring Harb. Symp. quant. Biol.* **30**: 429–443.

Braitenberg, V. (1967). Patterns of projections in the visual system of the fly. I. Retina–lamina projections. *Expl Brain Res.* **3**: 271–298.

Bullock, T. H. and Horridge, G. A. (1965). *Structure and function in the nervous systems of invertebrates.* San Francisco: Freeman.

Burkhardt, D. (1962). Spectral sensitivity and other response characteristics of single visual cells in the arthropod eye. *Symp. Soc. exp. Biol.* **16**: 86–109.

Burkhardt, D. (1964). Colour discrimination in insects. *Adv. Insect Physiol.* **2**: 131–173.

Burkhardt, D. and Wendler, L. (1960). Ein direkter Beweis für die Fähigkeit einzelner Sehzellen des Insektauges, die Schwingungrichtung polarisiertes Lichtes zu analysieren. *Z. vergl. Physiol.* **43**: 687–692.

Burtt, E. T. and Catton, W. T. (1962). A diffraction theory of insect vision. I. *Proc. R. Soc.* (B) **157**: 53–82.

Dartnall, H. J. A. (1953). The interpretation of spectral sensitivity curves. *Br. med. Bull.* **9**: 24–30.

Denton, E. J. (1959). The contributions of the orientated photosensitive and other molecules to the absorption of the whole retina. *Proc. R. Soc.* (B) **150**: 78–94.

Eguchi, E. (1965). Rhabdom structure and receptor potentials in single crayfish retinula cells. *J. cell. comp. Physiol.* **66**: 411–429.

Fermi, G. and Reichardt, W. (1963). Optomotorische Reaktionem der Fliege *Musca domestica*. *Kybernetik* **2**: 15–28.

Frankenhaeuser, B. and Hodgkin, A. L. (1956). The after-effects of impulses in the giant nerve fibre of *Loligo*. *J. Physiol., Lond.* **131**: 341–376.

Goldsmith, T. H. (1964). The visual system of insects. In *The physiology of Insecta*: 397–462. Rockstein, M. (ed.). London: Academic Press.

Guilio, L. (1963). Elektroretinographische Beweisführung dichroitischer Eigenschaften des Komplexauges bei Zweiflüglern. *Z. vergl. Physiol.* **46**: 491–495.

Hámori, J. and Horridge, G. A. (1966). The lobster optic lamina. I–IV. *J. Cell Sci.* **1**: 249–280.

Hartline, H. K., Ratliff, F. and Miller, W. H. (1961). Inhibitory action in the retina and its significance for vision. In *Nervous inhibition*: 241–284. Florey, E. (ed.). London: Pergamon Press.

Horridge, G. A. (1965). The retina of the locust. In *The functional organisation of the compound eye*: 513–541. Bernhard, C. G. (ed.). London: Pergamon Press.

Horridge, G. A. (1967). Perception of polarization plane, colour and movement in two dimensions by the crab, *Carcinus*. *Z. vergl. Physiol.* **55**: 207–242.

Horridge, G. A. and Barnard, P. B. T. (1965). Movement of palisade in locust retinula cells when illuminated. *Q. Jl microsc. Sci.* **106**: 131–135.

Horridge, G. A., Scholes, J. H., Shaw, S. R. and Tunstall, J. (1965). Recordings from neurones in the locust brain and optic lobe. In *The physiology of the insect central nervous system*: 165–202. Treherne, J. E. and Beament, J. W. L. (eds.). London: Academic Press.

Jander, R. and Waterman, T. H. (1960). Sensory discrimination between polarized light and intensity patterns by arthropods. *J. cell. comp. Physiol.* **56**: 137–160.

Katz, B. and Miledi, R. (1966). Input–output relation of a single synapse. *Nature, Lond.* **212**: 1242–1245.

Kikuchi, R., Naito, K. and Tanaka, I. (1962). Effect of sodium and potassium ions on the electrical activity of single cells in the lateral eye of the horseshoe crab. *J. Physiol., Lond.* **161**: 319–343.

Kikuchi, R. and Ueki, K. (1965). Double-discharges recorded from single ommatidia of horseshoe crab, *Tachypleus tridentatus*. *Naturwissenschaften* **52**: 458–459.

Kirschfeld, K. (1965). Discrete and graded receptor potentials in the compound eye of the fly *Musca*. In *The functional organization of the compound eye*: 291–307. Bernhard, C. G. (ed.). London: Pergamon Press.

Kirschfeld, K. (1967). Die projektion der optischen Umwelt auf das Raster der Rhabdomere im Komplexauge von *Musca*. *Expl Brain Res.* **3**: 248–270.

Kirschfeld, K. (1968). The *Musca* compound eye. *Symp. zool. Soc. Lond.* No. 23: 165–166.

Kuwabara, M. and Naka, K-I. (1959). Response of a single retinula cell to polarized light. *Nature, Lond.* **184**: 455–456.

Kusano, K., Livengood, D. R. and Werman, R. (1967). Tetraethylammonium ions: effects of presynaptic injection on synaptic transmission. *Science, N.Y.* **155**: 1257–1259.

Langer, H. and Thorell, B. (1966). Microspectrophotometry of a single rhabdomere in the insect eye. *Expl Cell Res.* **41**: 673–676.

Loewenstein, W. R. (1966). Permeability of membrane junctions. *Ann. N.Y. Acad. Sci.* **137**: 441–472.

Mazokhin-Porshniakov, G. A. (1964). Methods of studying and the state of our knowledge on the colour-sight in insects. *Ent. Obozr.* **43**: 503–513. (English translation by *Ent. Soc. Am.*: 257–266.)

McCann, G. D. and MacGinitie, G. F. (1965). Optomotor response studies of insect vision. *Proc. R. Soc.* (B) **163**: 369–401.

Munz, F. W. and Schwanzara, S. A. (1967). A nomogram for retinene$_2$-based visual pigments. *Vision Res.* **7**: 111–120.

Naka, K-I. and Kishida, K. (1965). Retinal action potentials during light and dark adaptation. In *The functional organization of the compound eye*: 251–266. Bernhard, C. G. (ed.). London: Pergamon Press.

Naka, K-I. and Rushton, W. A. H. (1966). S-potentials from luminosity units in the retina of fish (Cyprinidae). *J. Physiol., Lond.* **185**: 587–599.

Osborne, M. P. (1967). The fine structure of neuromuscular junctions in the segmental muscles of the blowfly larva. *J. Insect Physiol.* **13**: 827–833.

Palka, J. (1965). Diffraction and visual acuity of insects. *Science, N.Y.* **149**: 551–553.

Palka, J. (1967). An inhibitory process influencing visual responses in a fibre of the ventral cord of locusts. *J. Insect Physiol.* **13**: 235–248.

Pedler, C. and Goodland, H. (1965). The compound eye and first optic ganglion of the fly. *Jl R. microsc. Soc.* **84**: 161–179.

Purple, R. L. and Dodge, F. A. (1965). Self-inhibition in the eye of *Limulus*. In *The functional organization of the compound eye*: 451–464. Bernhard, C. G. (ed.). London: Pergamon Press.

Ratliff, F. (1965). Selective adaptation of local regions of the rhabdom in an ommatidium of the compound eye of *Limulus*. In *The functional organization of the compound eye*: 187–191. Bernhard, C. G. (ed.). London: Pergamon.

Scholes, J. H. (1964). Discrete subthreshold potentials from the dimly lit insect eye. *Nature, Lond.* **202**: 572–573.

Scholes, J. H. (1965). Discontinuity of the excitation process in locust visual cells. *Cold Spring Harb. Symp. quant. Biol.* **30**: 517–527.

Shaw, S. R. (1966). Polarized light responses from crab retinula cells. *Nature, Lond.* **211**: 92–93.

Shaw, S. R. (1967). Simultaneous recording from two cells in locust retina. *Z. vergl. Physiol.* **55**: 183–194.

Smith, T. G., Baumann, F. and Fuortes, M. G. F. (1965). Electrical connections between visual cells in the ommatidium of *Limulus*. *Science, N.Y.* **147**: 1446–1448.

Stieve, H. (1965). Interpretation of the generator potential in terms of ionic processes. *Cold Spring Harb. Symp. quant. Biol.* **30**: 451–457.

Strother, G. K. (1966). Absorption of *Musca domestica* screening pigment. *J. gen. Physiol.* **49**: 1087–1088.

Trujillo-Cenóz, O. (1965). Some aspects of the structural organization of the intermediate retina in Dipterans. *J. Ultrastr. Res.* **13**: 1–34.

Trujillo-Cenóz, O. and Melamed, J. (1966). Compound eye of Dipterans: anatomical basis for integration–an electronmicroscope study. *J. Ultrastr Res.* **16**: 395–398.

Tunstall, J. and Horridge, G. A. (1967). Electrophysiological investigation of the optics of the locust retina. *Z. vergl. Physiol.* **55**: 167–182.

Usherwood, P. N. R. and Grundfest, H. (1964). Inhibitory postsynaptic potentials in grasshopper muscle. *Science, N.Y.* **143**: 817–818.

Wald, G., Brown, P. K. and Gibbons, I. R. (1963). The problem of visual excitation. *J. opt. Soc. Am.* **53**: 20–35.

Washizu, Y., Burkhardt, D. and Streck, P. (1964). Visual field of single retinula cells and interommatidial inclination in the compound eye of the blowfly *Calliphora erythrocephala*. *Z. vergl. Physiol.* **48**: 413–428.

Wassermann, G. S. (1967). Density spectrum of *Limulus* screening pigment. *J. gen. Physiol.* **50**: 1075.

Yeandle, S. (1967). Some properties of the components of the *Limulus* ommatidial potential. *Kybernetik* **3**: 250–254.

Symp. zool. Soc. Lond. (1968), No. 23, 165–166.

THE *MUSCA* COMPOUND EYE

KUNO KIRSCHFELD

Max-Planck-Institut für Biologie, Abt. Reichardt, Tübingen, Germany

SYNOPSIS

This paper is published in fullas: Kirschfeld, K. (1967). Die Projektion der optischen Umwelt auf das Raster der Rhabdomere im Komplexauge von *Musca*. *Expl Brain Res.* **3**: 248–270.

The lens and the crystalline cone of the *Musca* ommatidium act together as an inverting lens system. The distal endings of the rhabdomeres at the basis of the dioptric apparatus are separated and arranged in a typical asymmetric pattern. The optical axes of the individual rhabdomeres of one ommatidium are the geometric projections of the distal rhabdomere endings into the environment, inverted by 180° by the dioptric apparatus. The divergence angles between the optical axes of the rhabdomeres of one ommatidium create the situation in which seven rhabdomeres, each from a different ommatidium are looking at one point in the environment. These facts were established from sections of living eyes and confirmed by analysing the "pseudopupil" in the intact animal. A model with the same optical properties was demonstrated.

The pattern of the decussation of individual retinula cell axons between retina and lamina was predicted adopting the hypothesis that the fibres of retinula cells number one to six, whose rhabdomeres are looking at one point in the environment, project into a single "cartridge" in the lamina. These connexions are identical with the fibre distribution which has been analysed by Braitenberg (1967) in *Musca*, and, in another eye region of *Lucilia* by Trujillo-Cenóz and Melamed (1966). These results show that a one to one correspondence between a lattice of points in the environment and the lattice of the "cartridges" in the lamina exists.

The complicated optical projection between points in the environment and rhabdomeres on one hand, which is exactly compensated by the anatomical projection between retina and lamina on the other, is interpreted as a means of increasing the light gathering power of the eye by a factor of seven compared to the classical apposition eye. The *Musca* compound eye is regarded as a "neural superposition eye".

References

Braitenberg, V. (1967). Patterns of projection in the visual system of the fly. I. Retina–Lamina projections. *Expl Brain Res.* **3**: 271–298.

Trujillo-Cenóz, O. and Melamed, J. (1966). Compound eye of dipterans: Anatomical basis for integration. An electron microscope study. *J. Ultrastruct. Res.* **16**: 395–398.

Symp. zool. Soc. Lond. (1968) No. 23, 167–198.

THE FINE STRUCTURE AND FUNCTIONAL ORGANIZATION OF CHORDOTONAL ORGANS

P. E. HOWSE*

Department of Zoology, University College, Cardiff, Wales

SYNOPSIS

Chordotonal sensilla occur in the insect tympanal organ, subgenual organ, and Johnston's organ. They are also found in organs that span joints or body segments. It is proposed that these are called connective chordotonal organs: they are found in the trunk segments of insects, but more commonly in the appendages of both insects and Crustacea.

Much of the physiological evidence supports theories that the sensilla respond when they are flexed.

Descriptions are given of the fine structure of chordotonal sensilla from the subgenual organ and tibio-tarsal connective organ of the cockroach, and from the Johnston's organ and an antennal connective organ of the termite. These are compared with the locust tympanal organ sensillum, the sensillum from the crab leg, and the Johnston's organ sensillum of *Drosophila*. A striking feature of all these is the presence of some form of ciliary dilatation which may be restricted to the part of the cilium just proximal to the cap. It is suggested that the adequate stimulus may be stretch of the membrane of this dilatation by flexion at the scolopale-cap junction. An exception may be the Johnston's organ sensillum of the termite, which is comparable with campaniform and hair-plate sensilla, and where it is possible that movements of the cuticle can compress the terminal directly.

INTRODUCTION

The term "chordotonal organ" was first used by Graber (1882) for all sense organs that contained terminal nerve elements similar to the "auditory spikes" (*Hörstifte*) of Orthopteran hearing organs. Graber reserved the term "tympanal apparatus" for organs associated with a drum-like membrane and referred to those without a drum as "*atympanale stiftführende Organe*". Later authors maintained this distinction, referring to tympanal organs on the one hand and chordotonal organs on the other. Of the terms "*stiftführende*" and "*scolopofere*" used by Graber to describe chordotonal organs only the latter has come into the English language as scolopophorous (adj.), scolopale (noun) for the spikes themselves, and scoloparium or scolopidium (nouns) for the sensory unit (see also Whitear, 1962). The present terminology is so varied that it seems advisable not to restrict the meanings of the terms, but to reduce them. In this paper the term "chordotonal organ" will be used in the same wide sense that Graber originally used it, and the

* Present address: Department of Zoology, The University, Southampton.

sensory units will be referred to as "chordotonal sensilla" or "scolopidia". Reference to the classical review of Eggers (1928) shows that chordotonal organs occur in a wide variety of situations, and there seems no valid reason to split off tympanal organs as a separate group. Other kinds of chordotonal organs have received second names, e.g. Johnston's organ in the insect antenna, and the subgenual organ in the insect tibia.

Graber distinguished between *amphinematic* scolopidia in which the cap was drawn out into a terminal thread that inserted on to the cuticle, and *mononematic* scolopidia in which there was no such terminal thread. These terms may well lose their usefulness as more is found out about the fine structure of scolopidia.

There is, however, a common tendency to refer to chordotonal organs in which scolopidia are contained in a connective tissue strand as "stretch receptors". This is unfortunate in two senses, for firstly the implied function may not have been proved and secondly even if the organ responds to stretch the scolopidia may not (see for example Taylor, 1967). It is therefore proposed to refer to such chordotonal organs that span body segments as "connective".

Scolopidia have been found in all the insect orders including the Apterygota. They have been reported in Crustacea from the second antennal segment of *Caprella* (Wetzel, 1934), in the basal antennular segments of *Panulirus* (Laverack, 1964), in the basal flagellar segment of the hermit crab *Petrochirus* (Taylor, 1967) and in all joints of the walking leg of the crab *Carcinus* (Whitear, 1962). It is striking that chordotonal organs have not yet been found in Myriapoda or Chelicerata; the stretch receptors of *Limulus* joints are multipolar neurones (Barber, 1960). The fine structure of the locust tympanal scolopidium has been described by Gray (1960), that of crab leg scolopidia by Whitear (1962, 1965), and that of the Johnston's organ scolopidia of *Drosophila* by Uga and Kuwabara (1965). The present paper reports the fine structure of three further insect scolopidia and attempts to relate the fine structure of scolopidia to theories of the mode of action of chordotonal organs.

THE DISTRIBUTION OF CHORDOTONAL ORGANS*

Connective chordotonal organs of trunk segments

Scolopidia are common in the abdominal segments of insect larvae, and Eggers (1928) gives many references to their occurrence in the

* Where no direct reference is given it may be taken that further details and references will be found in the extensive review of Eggers (1928).

larvae of Diptera, Lepidoptera, Coleoptera and Hymenoptera. The organs generally contain only a small number of scolopidia enclosed in a ligament that spans the pleura horizontally, but may also be found in other planes within segments. Finlayson and Lowenstein (1958) report that the oblique receptor in the abdomen of *Aeschna juncea* and *Ephemera danica* contains scolopidia, and this organ would be stretched during inspiration and relaxed during expiration. Janet (1894) described a similar chordotonal organ in the ant *Myrmica rubra* running

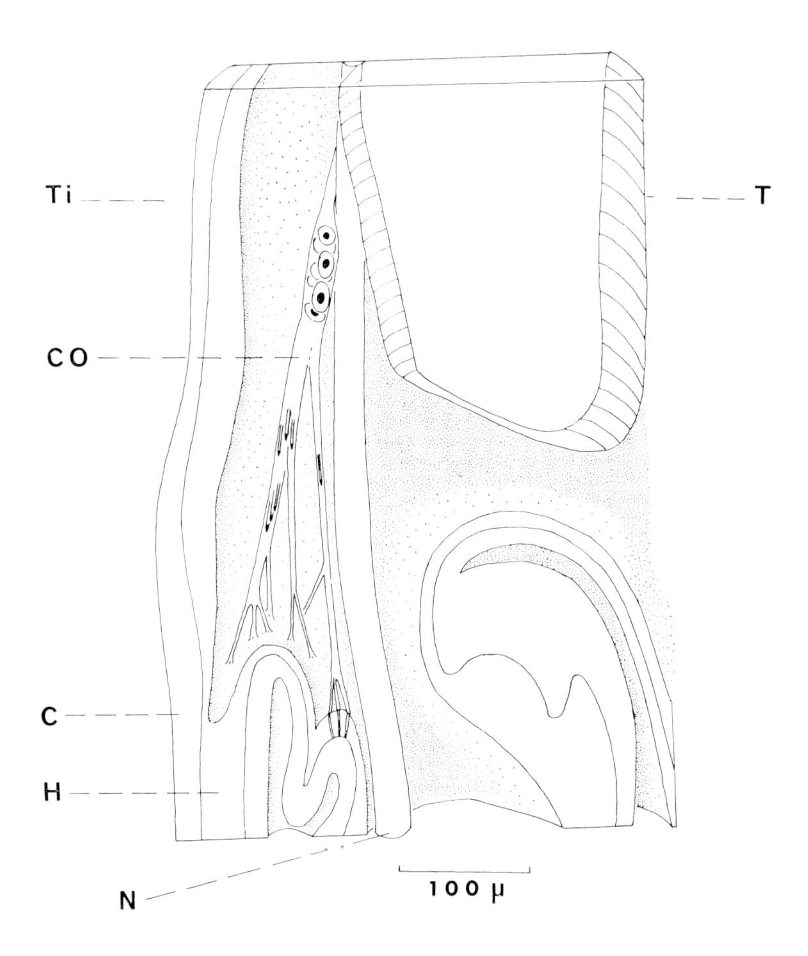

FIG. 1. Horizontal section through the mesothoracic leg of the cockroach, *Periplaneta americana*, showing the chordotonal organ (*CO*) which inserts on to the tibio-tarsal joint membrane. Reproduced with the kind permission of D. Young. *T*, trachea; *Ti*, tibia; *C*, cuticle; *H*, hypodermis; *N*, tarsal nerve.

between the prothoracic ganglion and the neck membrane. The same organ in *Drosophila* has up to 150 sensilla (Hertweck, 1931).

Connective chordotonal organs in legs

Chordotonal organs similar to those found in trunk segments have been recorded from the femur, tibia and tarsus of the insect leg (Eggers, 1928; Debaisieux, 1935, 1938) and are probably of ubiquitous occurrence in those situations in insects. In the cockroach *Periplaneta americana* there is a chordotonal organ in the tibia, the terminal ligament of which inserts on the tibia-tarsal joint membrane (D. Young, personal communication). The ends of the ligament attach directly to the cuticle in a tent-like fashion (Fig. 1) so that different component strands will be stretched according to the direction of movement of the tarsus. About seven or eight scolopidia are present in the strand, as in a similar organ in the termite (Howse, 1965), but this organ in the honeybee contains about 50 scolopidia (Lukoschus, 1962). In the latter example the terminal filaments form a series graded in length.

In Crustacea, Barth (1934) described chordotonal organs in strands of connective tissue in the meropodites of the pereiopods of many decapods. Burke (1954) described a similar organ at the propopodite-dactylopodite joint of *Carcinus maenas*, and Alexandrowicz and Whitear (1957) at the coxopodite-basipodite joint of decapods and at the thoracico-coxal junction (Macrura only). In the CP and MC joints there are two chordotonal organs. CP1, which is a membrane-like organ, is attached to the propopodite productor tendon and will be stretched by production, while CP2, which is attached to the reductor tendon, will be stretched during joint reduction (Bush, 1965b).

Chordotonal organs of wings

Chordotonal organs in insects have been found in the wings (some times associated with a tracheal bladder or tympanum) associated with the wing articulation, and associated with halteres.

Where chordotonal organs occur in the wing they are usually near the base but often appear to have no connection with the wing articulation and run obliquely between the upper and lower wing membranes (Eggers, 1928). In the hind wing of *Dytiscus marginalis* they are found in the basal region of the subcostal vein and are inserted ventrally on thickened cuticle (Lehr, 1914) but in *Agrion puella* the chordotonal organ is enveloped by tracheae and the terminal ligaments insert on an oval membranous area of cuticle (Erhardt, 1916). Vogel (1912) investigated the wing chordotonal organs of Lepidoptera and found that they were generally contained in cavities formed from enlarged veins at the wing bases. In the family Satyridae a tympanic membrane is present at the base of the forewing. In *Maniola jurtina* three chordotonal organs insert on to thickenings of the drum. Swihart (1967) has recently recorded electrophysiologically from similar organs in two Nymphalids and found that they are sensitive to sound with optimal frequencies at 1200 Hz. Pringle (1957) has catalogued the occurrence of chordotonal organs in the wing base, where ante-alar, radial, medial and cubital organs may be present. All are apparently absent in Orthoptera, but more recently Gettrup (1962) has found a connective organ in the region of the subalar sclerite of the locust, suspended between the pleuron and the mesophragma. Wilson and Gettrup (1963) found this organ also in several species of Orthoptera, and showed it to be concerned with the regulation of wing beat frequency. When the wing lifts, the chordotonal organ fires one to three nerve impulses; if all four organs are cut the wingbeat frequency drops to one half. Two chordotonal organs that occur in the basal region of Dipteran halteres (Pflugstaedt, 1912) may have a similar function; it has been suggested that they function as "stimulatory organs" since Diptera fly for only a short time, if at all, when the halteres are removed (for references see Wigglesworth, 1965).

The subgenual organ

The insect subgenual organ is found in each leg in the proximal tibial region (Fig. 2a). It has been described as fan-shaped in Orthoptera (Schwabe, 1908) and Dictyoptera (Friedrich, 1929; Debaisieux, 1938), club-shaped in some Hymenoptera (Schön, 1911; Debaisieux, 1938), club-shaped or cup-shaped in termites (Richard, 1950; Howse, 1965), diffuse and apparently unattached distally in Lepidoptera (Debaisieux, 1935), while it is apparently absent in Coleoptera, Hemiptera, and Diptera. The scolopidia are sometimes divided into two groups, one orientated obliquely towards the leg wall and the other at right angles to it. Autrum and Schneider (1948) have shown that this organ is sensitive to vibrations of the substratum and has different threshold sensitivities according to its structure, the fan-shaped variety being the most sensitive. Howse (1964) has shown that the organ responds to changes in steady-state induced by vibrations.

The detailed structure of the subgenual organ of the cockroach *Periplaneta americana* has recently been investigated by P. E. Howse (unpublished). It consists of a leaf-like membrane which spans the dorsal tibial blood-space (Fig. 2). The attachment cells form the distal wall of the membrane inserting at one end on to the cuticle,

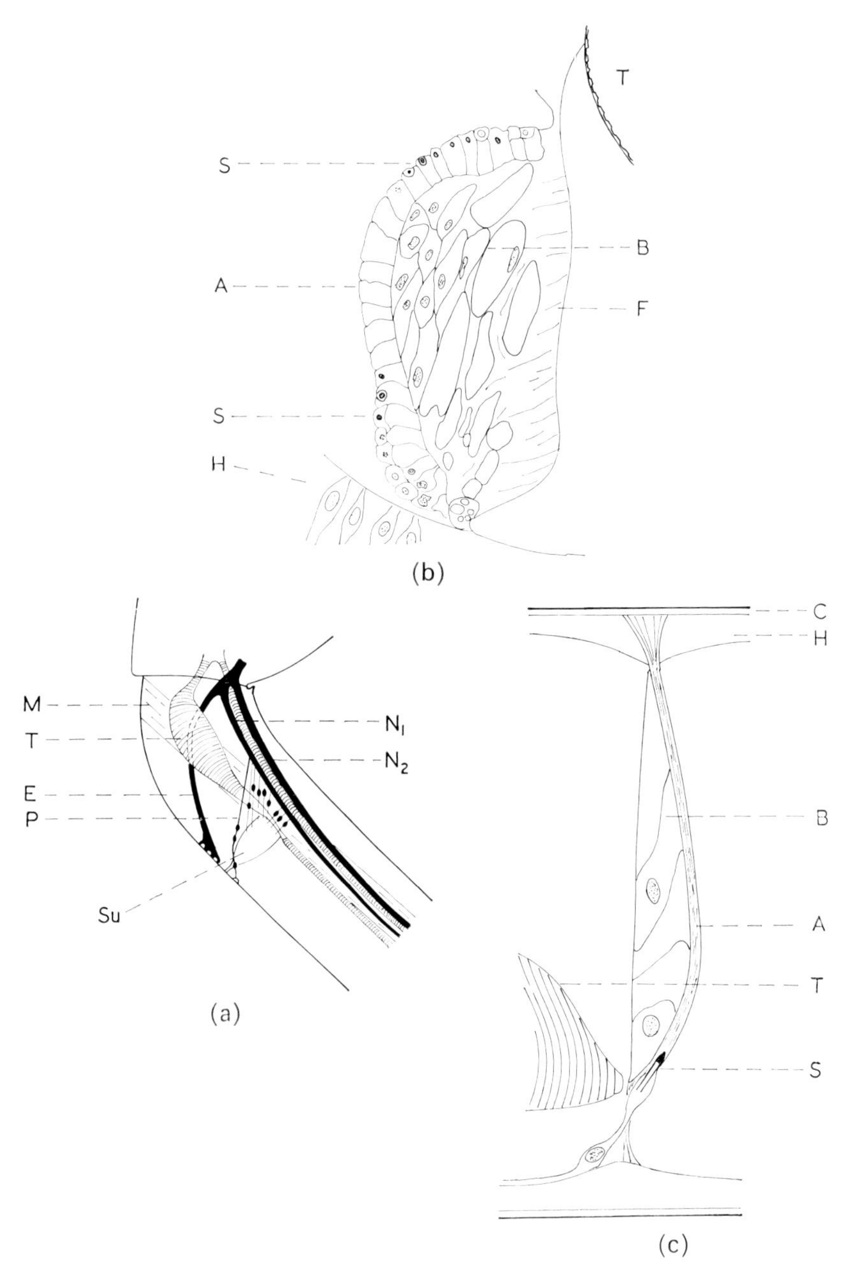

FIG. 2. (a) The innervation of the subgenual organ in the tibia of the termite, *Zootermopsis angusticollis*, reproduced with permission from Howse (1965). M = muscle, T = trachea, E = nerve to proximal group of campaniform sensilla, P = nerve to distal group of campaniform sensilla, Su = subgenual organ, N_1 and N_2 = tibial nerves. (b) The subgenual organ of the cockroach, *Periplaneta americana*, in a transverse section of the tibia. The scolopidia also appear in T.S. T, trachea; S, scolopidia; A, attachment cells; B, accessory cells; F, fibrous connective tissue; H, hypodermis. (c) A horizontal section through the cockroach tibia showing a scolopidium in L.S. in the subgenual organ. Key to letters in Figs 1 and 2(b).

while the other end of each attachment cell receives the scolopale units.*

The plaque of attachment cells is saucer-shaped, the hollow of the saucer being filled in with large vacuolated accessory cells in a matrix of fibrous connective tissue. The general form is very similar to that of the subgenual organs of the various Orthoptera described by Schwabe (1908). It is attached dorsally to the cuticle by a broad stalk formed from the attachment cells and ventrally it has a small attachment to the membrane connecting the two tracheal branches. The cell bodies of the bipolar nerve cells are attached by connective tissue to the hypodermis.

The tibial tympanal organ

This organ is found in the prothoracic legs of crickets and long-horned grasshoppers (Grylloidea and Tettigonoidea), and has been described in considerable detail by Schwabe (1908), some of whose fine illustrations are widely known. In *Acheta domesticus* the tympanal organ scolopidia fall into three groups. These are (1) a subgenual organ, (2) a "proximal tracheal organ" in which the scolopidia run along the enlarged anterior trachea in a distal direction and only the attachment cells (*Deckzellen*) turn towards the cuticle, and (3) a "true tracheal organ" in which the bipolar nerve cells are attached to the trachea but the scolopidia are orientated at right angles to the tracheal wall. Figure 3 shows the tympanal organ in transverse section. There is a small anterior and a large posterior tympanum between which the tracheae form an almost continuous air space. The attachment cells of the true tracheal organ are bound together by cellular processes and draw together to find a common point of insertion on the cuticle. The whole organ is enveloped in a tent-like fashion by a cuticular membrane (*Deckmembran*), but the scolopidia do not attach to this.

The tympanal organ of the tettigoniid *Decticus verrucivorus* is by comparison more extensively differentiated. The two tympana are equivalent in size but enclosed in cavities in the tibia that open by slits to the exterior. Between the subgenual organ and the tracheal organ is a small group of scolopidia forming an intermediate organ (*Zwischenorgan*), the attachment cells of which come together in a single mass which is joined by a strand of tissue to the cuticular covering membrane of the tracheal organ. The endings of the latter are arranged in a double row, the attachment cells are very short and join not with the cuticle but with the cuticular covering membrane. They

* The scolopidium minus its attachment cell is referred to here as the scolopale unit.

therefore all run approximately at right angles to the leg wall, and the row of cells is known as the "crista acustica". The scolopidia of the crista have unusually broad caps and also form a graded series in their dimensions, as reference to Table I will show.

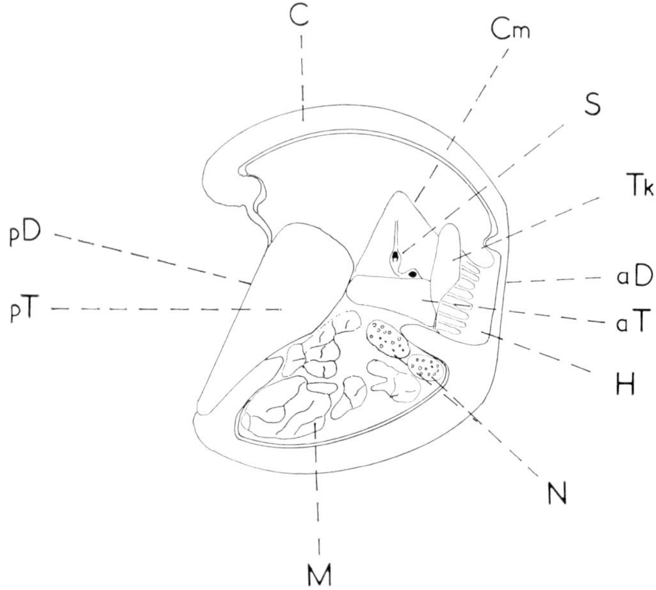

FIG. 3. Transverse section through the tibia of *Acheta* (=*Gryllus*) *domesticus* in the region of the two tympana (after Schwabe, 1908). *C*, cuticle; *Cm*, cuticular membrane; *S*, scolopidium of tracheal organ; *Tk*, cuticular tracheal body; *aD*, anterior tympanum; *aT*, anterior tympanal trachea; *H*, hypodermis; *N*, nerves; *M*, muscle; *pT*, posterior tympanal trachea; *pD*, posterior tympanum.

TABLE I
The dimensions of the scolopidia (scolopales + cap) from the crista acustica of Tettigonia (=Locusta) viridissima (*after Schwabe*, 1908)

	Length in μ	[Width of] cap in μ	Ratio
1st sensillum of proximal group	21	3·4	6·18
Last sensillum of proximal group	20	5·5	3·63
1st sensillum of "true" crista	19	5·5	3·46
5th sensillum of "true" crista	19	6·8	2·80
10th sensillum of "true" crista	19	7·5	2·53
15th sensillum of "true" crista	17·8	8·2	2·18
20th sensillum of "true" crista	17·8	6·8	2·62
25th sensillum of "true" crista	16	6·2	2·58
30th sensillum of "true" crista	16	6·2	2·58
35th and antepenultimate	17·8	4·1	4·34

It is hard to avoid the conclusion that the crista does not play some role in the analysis of sound. Subgenual organ "spikes" are much narrower at the cap and therefore more pencil-shaped. Schwabe gives the length of one of these as 23·3 μ and the width at the cap as 2·7 μ. Autrum (1941) found by electrophysiological methods that the subgenual organ in *Decticus* and *Tettigonia* responded to vibration, and the crista acustica exclusively to airborne sounds. Similar chordotonal organs are present in the meso- and meta-thoracic legs although these possess no tympana. Autrum found that the cristae here had an auditory sensitivity over a much lower range than the tympanal organs.

Thoracic tympanal organs

Tympanal organs are found on the meso- and metathorax in some Hemiptera, and on the metathorax in many Lepidoptera. In the aquatic bug *Corixa punctata* (= *Macrocorixa geoffroyi*) projections from the thoracic subcoxae form a cavity which is filled with air, and on the inner surface of this cavity is a tympanum from which a club-like projection arises (Hagemann, 1910). Inserted into the base of this club is a strand of tissue containing two scolopidia. Popham (1961) has produced evidence that this organ functions as a depth indicator. These insects both produce and react to sounds, and although it is unlikely that the club acts as a resonator an auditory function may still be possible in this and similar biscolopidial organs that have been found on the metathoracic and first abdominal segments by Larsen (1957). In *Notonecta*, *Naucoris* and *Plea* a mesothoracic tympanum is present but the club-shaped projection is lacking.

In Lepidoptera, tympanal organs are present on the metathorax in the Noctuoidea. A strand of tissue containing two scolopidia is inserted on to the tympanic membrane. This strand is surrounded by air spaces formed from enlarged tracheal sacs, and the tympanal tracheae are in apposition medially (Eggers, 1928) so that any vibration of the sensilla must be very lightly damped. The physiological responses of these organs and the role that the auditory sense plays in the behaviour of the moths have been extensively investigated by Roeder and his co-workers (for a recent summary see Roeder, 1966).

Abdominal tympanal organs

Tympanal organs are present on the first abdominal segment in short-horned grasshoppers (Acridoidea), and the anatomy of these organs has also been described in great detail by Schwabe (1908). Again, large air sacs, some of which meet medially, form an essential part of the organ. The chordotonal sensilla are contained in a large "ganglion" attached to thickened ridges or projections of the well-developed tympanic membrane. There are four of these structures; the folded body, the elevated process, the styliform body, and the pyriform vesicle. In *Locusta migratoria*, 60 to 80 scolopidia insert in separate groups with different orientations on to these thickenings of the drum (Gray, 1960).

In moths belonging to the families Pyralidae and Geometridae tympanal organs are found on the first or second abdominal segments, and in the Axiidae on the seventh abdominal segment. A tympanic cavity is present, opening to the exterior by a slit, and the tympanum is on the medial wall of this cavity. A cuticular bridge with a central bulge spans the tympanum internally, and the chordotonal organ is suspended between this bulge and the centre of the tympanic membrane (Eggers, 1928). The chordotonal organ is a strand with two bulbous swellings, one of which contains the bipolar nerve cells and the other the sensory endings, which are few in number. Remarkably, however, it is the swelling nearest to the tympanum that contains the nerve cell bodies so that the scolopidia in the more medial swelling point away from the drum.

The tympanal organ of cicadas lies ventro-laterally on the second abdominal segment. The scolopidia which may number up to 150 are attached to a cuticular process extending from the outer side of the tympanic membrane into a tympanic capsule (Vogel, 1923). The scolopidia are attached to the inner wall of this capsule, and again point more or less away from the tympanum. Pringle (1954) has found that this tympanal organ responds to the song of the species and is prevented from damage during stridulation by a muscle which contracts to bend the rim of the tympanum allowing the membrane to go slack.

A tympanal organ superficially reminiscent of that of the Acridoidea is found on the first abdominal segment of the winged form of the termite *Zootermopsis angusticollis* (Howse, 1963) but there is as yet no evidence for this being a sense organ.

Connective chordotonal organs of the antennae and mouthparts

The distribution of chordotonal organs in the antennal segments of insects has been reviewed by Eggers (1928). Chordotonal organs have also been reported from the mouthparts of some insects. In Hymenoptera there is an organ attached to the antennal base, and connective organs have been found in various insects in every segment up to and including the fourth. In *Hydropsyche longipennis*, Debauche (1935) has described a connective chordotonal organ in the pedicel which traverses the segment obliquely to insert on a cuticular projection at the base of the third segment. Debauche described this organ as having four scolopidia, one of which had one nerve process, and the others two processes each, one large and one small. There is one connective organ in the scape of the termite *Calotermes flavicollis* and one in the pedicel (Richard, 1957). The same applies to *Zootermopsis angusticollis*,

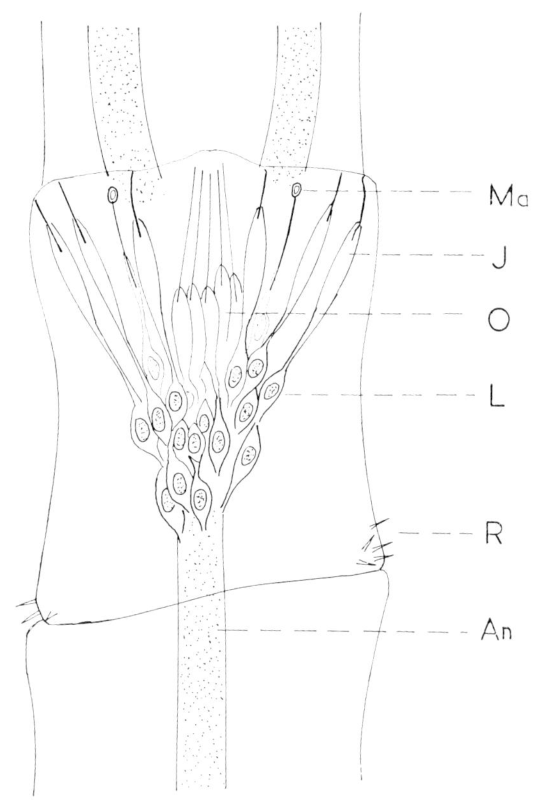

Fig. 4. The sensilla of the antennal pedicel of the termite, *Zootermopsis angusticollis*. *Ma*, campaniform sensillum; *J*, Johnston's organ sensillum; *O*, sensillum of connective chordotonal organ; *L*, bipolar sense cells; *R*, chaetic sensilla of hair plate; *An*, antennal nerve.

where the pedicelar organ runs obliquely to end on the distal mid-dorsal cuticle of that segment (Fig. 4).

Johnston's organ

In 1855 Johnston suggested that the swollen second antennal segment of *Culex* was a hearing organ, and this segment, which has been found to contain a cone or cylinder of scolopidia in all insects investigated, has borne his name ever since. Typically (Fig. 4) the sensory nerves join the antennal nerve at the base of the pedicel. The terminations of the scolopidia insert on to the intersegmental membrane between the pedicel and the base of the flagellum. Debauche (1935) found that the Johnston's organ scolopidia of *Hydropsyche* had two nerve processes. He referred to the scolopidia as *isodynal* when the processes were equal in size and *heterodynal* when they were not. He reserved the term *monodynal* for scolopidia with only one process.

In various Lepidoptera, Eggers (1924) found that the cuticle of the pedicelar-flagellar joint was considerably thickened and the terminal filaments of the chordotonal sensilla ran through pores in the thickened region.

In the whirligig beetle *Gyrinus* which lives on the water surface the club-shaped flagellum is held in the surface film, any disturbances of which are then apparently registered by Johnston's organ sensilla. De Wilde (1941) found that damaging the pedicel produced disturbances in the orientation of this insect.

Johnston's organ has been shown to register air currents in some Diptera, namely *Aëdes aegyptae* (Bässler, 1958), *Muscina stabilans* (Hollick, 1940) and *Calliphora erythrocephala* (Burkhardt and Schneider, 1957), and also in the honeybee, *Apis mellifera* (Heran, 1959). Responses can be recorded from Johnston's organ of the blow-fly in response to angular movements, but not to static displacements. Interference with the freedom of movement of the pedicelar-flagellar joint abolishes the control of wingbeat (Burkhardt and Gewecke, 1966).

Organs that respond to air flow must, to some extent at least, also be capable of registering airborne sound. Thus in *Calliphora* and *Apis* (Burkhardt and Gewecke, 1966) a sensitivity of Johnston's organ to sound of frequencies up to 500 Hz has been established. Roth (1948) showed that *Aëdes* could respond to sound of frequencies up to 800 Hz, the flight tone of the female serving to attract the male. The structure of the Johnston's organ of several culicids has been described in more recent times by Risler (1955) and appears exquisitely adapted for registering flagellar movements. The numerous scolopidia contained in the swollen pedicel insert on to plates forming the "foot" of the slender flagellum.

The chordotonal organ discovered in the basal segment of the antenna of the hermit crab *Petrochirus* by Taylor (1967) appears to resemble Johnston's organ. There are about 80 cells in this organ forming a complete ring. There are two bipolar nerve cells to each scolopidium and the terminal dendrites are embedded in the cuticle at the base of the next distal segment. The organ is capable of responding to water-borne vibrations of up to 1000 Hz.

THE ORGANIZATION OF CHORDOTONAL SENSILLA IN RELATION TO THEIR FUNCTION

Connective chordotonal organs

Information on the functioning of insect connective chordotonal organs is as yet scanty. Only in Crustacea have there been attempts to investigate the functioning of the organs at the cellular level. Wiersma and Boettiger (1959) showed that the PD organ of *Carcinus maenas* contained phasic movement fibres that were responsive unidirectionally to flexion or extension of the joint, and tonic position fibres that responded in the two extreme positions. Bush has extended this work to investigate the CB organ (Bush, 1965a) and the four MC and CP organs (Bush, 1965b). A summary of the characteristics of all the organs is in the last paper cited. Afferent fibres were always found responding to the release of the organs, but, on the whole, fewer fibres were found responding to stretch (none were found by Bush in MC2 and CP2).

Whitear (1962) proposed that the differential directional sensitivity of the chordotonal organs might lie in different properties of the unequal nerve processes of the scolopidia found in most of the organs, one cell (the ciliary cell) being responsive to relaxation of the organ and the other (the paraciliary cell) to stretch. However, Bush (1965a) was able to show that the CB organ responded to both stimulus parameters although it contains only isodynal scolopidia.

Since the evidence available generally supports the view that flexing of the dendrites is the adequate stimulus to the chordotonal sensillum, Mendelson (1963) suggested that a differing elasticity of components of the receptor strand would lead to displacement of the tip of the scolopidium from the long axis. This must be unlikely as the scolopidia lie parallel to the receptor strand (Whitear, 1962) and not across it as Mendelson assumed. Bush (1965a) put forward an alternative hypothesis in which he proposed that an elastic fibre with distal and proximal attachments to the scolopidium would tend to flex it one way on relaxation of the receptor strand, while a terminal fibre sharing only the distal attachment would tend to flex the sensillum the other way on contraction of the strand. Following a suggestion of Mendelson's he indicated that movement responses could result from the sensilla slipping in the matrix in a ratchet-like fashion. Certainly, it is apparent that movement and position fibres do not form two distinct groups. Hartman and Boettiger (1965) found movement cells in the PD organ that showed position responses to contractions of the strand in 25–100 μ increments.

It remains to be seen whether the fine structure of the flagellar chordotonal organ is similar to that of the leg organs, but Taylor (1967) found that its responses were inhibited by stretch but elicited by flexion or intrascolopidial compression of the dendrites. He held this to be the reason for the poor response to stretch in the leg organs and suggested that movement fibres responsive to receptor stretch would result in scolopidia that crossed the interface between the mainly fibrous connective tissue in the body of the strand and the amorphous peripheral connective tissue. Such a sensillum would be flexed during stretch and straightened during relaxation. Some findings of Hartman and Boettiger (1965) may prove to be relevant to this theory. They found relaxation-sensitive cells in the antero-lateral plane of the PD organ and extension-sensitive cells in the dorsal plane. They also found neurons of like sensitivity to be paired in scolopidia.

Insect tympanal organs

Pumphrey and Rawdon-Smith (1939) found that impulses recorded from the tympanal nerve of *Locusta migratoria* were synchronous with the amplitude modulation pattern of sounds but did not reflect the carrier frequency of the sounds. It is a general feature of insect songs that they show amplitude modulation to a marked degree, often resulting in pulse modulation. Busnel and his co-workers maintain that transients are the features of sounds to which tympanal organs are responsive (see, for example, Busnel, 1963; Busnel and Burkhardt, 1962). Transients in this particular instance are taken to mean rapid rises in the intensity of a sound from zero, or the reverse. Haskell (1956) has disputed the important role that the French workers ascribe to transients. In an electrophysiological study he found that only the modulation frequency was reflected in the response of grass-hopper tympanal organs to sound. Howse (1967) has recently put forward a theory of the mode of action of tympanal organs which is based on the similarity of their structure with that of the subgenual organ. Both have surfaces that transmit forced vibrations, i.e. the tympanum and the leg wall. Both have attachment cells locked together to form a single mass and scolopale units that are attached by their tips to the attachment cells. The membrane formed by the accessory and attachment cells of the subgenual organ is suspended across the leg between the cuticle and the tracheae and it appears that it can oscillate in response to a sudden displacement of the tibia, damping forces being supplied by the blood in the haemocoel. Earlier work (Howse, 1964) has confirmed that the subgenual organ is sensitive to acceleration and shown that its responses are phasic, occurring when a change in the steady state is induced by the onset or sudden change in the intensity of a vibration. These responses could be explained if a sudden displacement of the tibia set the attachment cell membrane into natural oscillation, and this in turn set the scolopale units into natural oscillation. These natural or transient oscillations would, of course, be very short-lived. Since the attachment cells form a single unit their natural frequency would be different from that of the scolopale units.

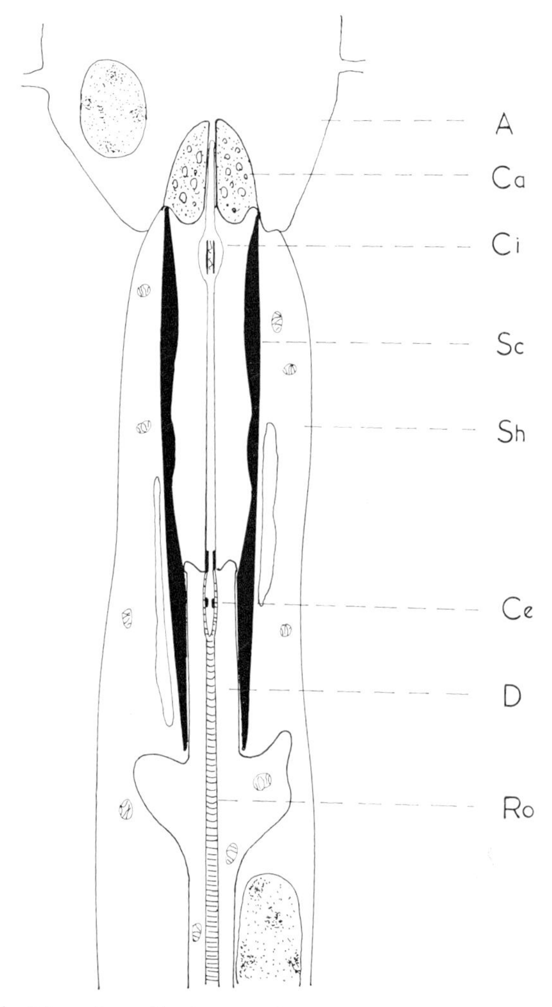

Fig. 5. Scolopidium from the tympanal organ of the locust, *Locusta migratoria*, redrawn from Gray (1960) in proportion to a phase-contrast photograph of an isolated sensillum. *A*, attachment cell; *Ca*, cap; *Ci*, ciliary dilatation; *Sc*, scolopale; *Sh*, scolopale sheath cell; *Ce*, centriole; *D*, tip of dendrite; *Ro*, root of ciliary process.

This would mean that there would be considerable flexion at their junctions with the attachment cells, which could result in excitation of the receptors. Under the action of a maintained vibration of constant characteristics the natural oscillations would die out exponentially in a very short time and all the cells of the organ would then be vibrating at the frequency of the maintained vibration.

The argument will not be set out in detail again here, but it has been suggested that the tympanal organ, because of the similarity of its responses and structure, could function in a very similar way. In the locust tympanal organ it is envisaged that a sudden displacement of the tympanic membrane would cause the attachment cells, which are interlinked by cytoplasmic processes, to vibrate *en masse* and the scolopale units would simultaneously move at a different frequency. It has been possible to see this effect under the microscope in parts of the organ mounted in Ringer's solution. On tapping the slide the attachment cells vibrated together and the scolopale units oscillated about their tips at a much lower frequency. It seems, then, that flexion of the scolopidium may be the adequate stimulus in auditory receptors.

If this explanation of the functioning of tympanal organs is correct it would explain features such as the responses to amplitude modulation or the "transients" of the French workers, the after discharge following a very brief stimulus and the similarity of the threshold curves for tympanal organs to resonance curves. The phasic response could easily be explained, and the tonic response could result from resonances set up in the organ which is very lightly damped in comparison with the subgenual organ. Michelsen (1966), however, has recently suggested that the phasic response in *Schistocerca* results from the synchronization of action potentials at the onset of a sound and ascribes the tonic response to lack of synchronization. He also found that sense cells from three of the four parts of the auditory ganglion differed in their sensitivity in a number of respects.

THE FINE STRUCTURE OF CHORDOTONAL SENSILLA

The locust tympanal organ sensillum

Gray (1960) described the fine structure of the auditory sensillum of *Locusta migratoria*, which is redrawn in Fig. 5 in proportion to a light-microscope photograph of an isolated sensillum. This was the first scolopidium to be examined under the electron microscope. The "spike" was found to consist of a cylinder of six to eight fibrous rods which at their bases enclose the tip of a dendrite and articulate distally with the cap, which is extracellular. The "axial filament" of earlier authors was found to consist of two structures; a ciliary root within the dendrite and a ciliary process that arises from its tip. The ciliary process has at its base a ciliary body and below this a centriole-like structure. The central filaments of the cilium are lacking, but there is an outer ring of nine double filaments in which one of each pair is solid

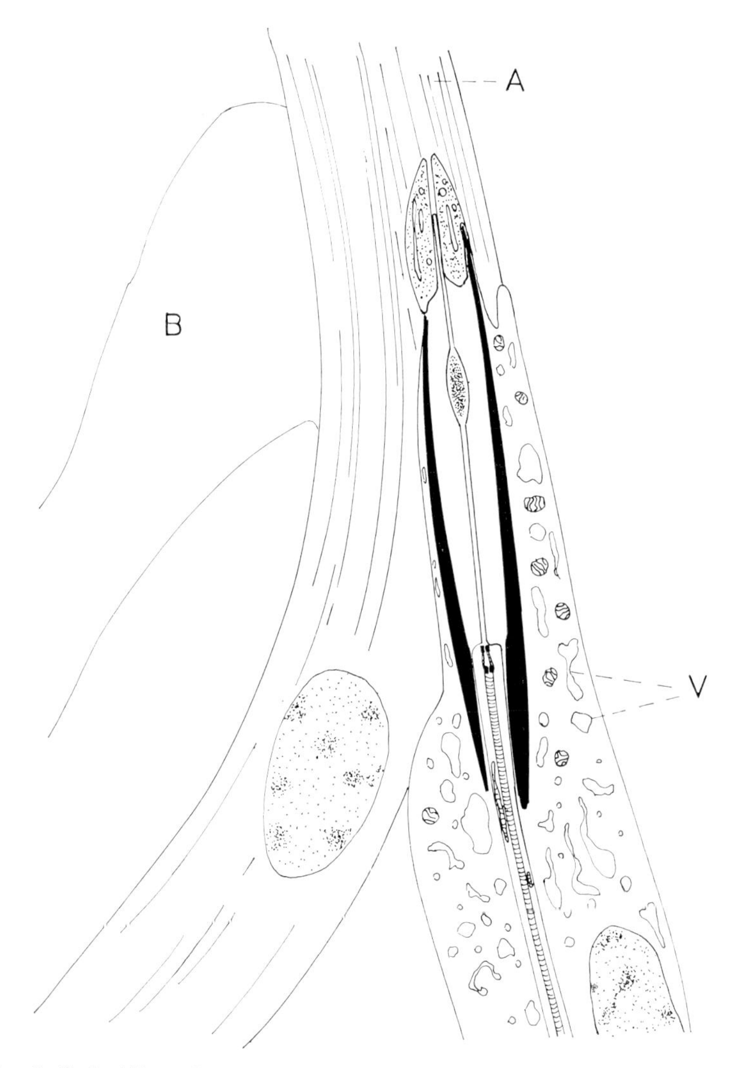

FIG. 6. Scolopidium from the subgenual organ of the cockroach, *Periplaneta americana*. *A*, attachment cell with microtubules; *B*, accessory cell; *V*, vacuoles in the scolopale sheath cell. Otherwise the scolopidium is comparable with that in Fig. 5.

and bears arms. Distally the cilium ends in a channel in the cap, but just proximal to the cap has a dilatation in which there are apparent expansions of some of the filaments and some microtubules are present. The cap is composed of amorphous moderately electron-dense material with many small cavities, and it is embedded in the attachment cell.

The subgenual organ sensillum

The cockroach subgenual organ sensillum (Figs. 6 and 15) is broadly similar in structure to the locust tympanal scolopidium. The attachment cell gains its fibrillar appearance from the dense packing of microtubules that it contains, many of which arise from the cell wall applied to the cap. The attachment cell runs between the hypodermal cells distally and here the microtubules associate together in bundles the end knobs of which form part of the attachment to the cuticle.

The cap is more pointed than in the locust tympanal scolopidium, but is similarly lacunar. These lacunae, however, are channels in the cap, mainly with a longitudinal orientation, and they contain processes of the scolopale sheath cell with microtubules and fibrous scolopale material. This is clear from Fig. 18 which also shows the tips of the scolopales which form a thin continuous ring at the base of the cap but fork to terminate in grooves in the cap. The scolopales thicken proximally and are completely separate where they enclose the tip of the dendrite (Fig. 11). The cilium again has nine solid filaments and nine hollow ones (Fig. 18). It has a marked dilatation proximal to the cap, which contains dense granular material (Figs 15 and 17). The cilium terminates in a cleft in the cap. There is clearly, as in the locust, very close contact between the two, and it is not difficult to believe that there is direct attachment.

The ciliary root divides into nine processes which enclose the centriole at the base of the cilium. At this point connecting filaments are clearly visible in some sections between the root processes and the points at which the inner edges of the scolopales are in contact with the nerve (Fig. 11). Traces of these connections are found at all levels between the root and the attachment areas of the dendrite and scolopale cell membranes. The scolopale sheath cell contains many mitochondria and numerous irregularly shaped vacuoles.

The crab connective organ sensillum

The crab chordotonal sensillum, the fine structure of which has been described by Whitear (1962), differs in a number of important respects from the sensillum just described (see Fig. 7). Fibrous scolopales are present but they form an incomplete cylinder proximally that appears roughly horse-shoe shaped in transverse section, but is a complete cylinder distally. The cap is represented by an extracellular "tube" which apparently has no firm articulation with the scolopales, and takes the form of a long drawn out cone. It is enveloped by two processes of the scolopale sheath cell, and terminates in the receptor strand.

The dendrites entering the base of the scolopale cylinder are anchored to the sheath cell by attachment plaques. There are two sensory cells to each scolopidium. In one kind, the ciliary cell, the tip of the dendrite has a ciliary segment with a structure similar to that of the chordotonal

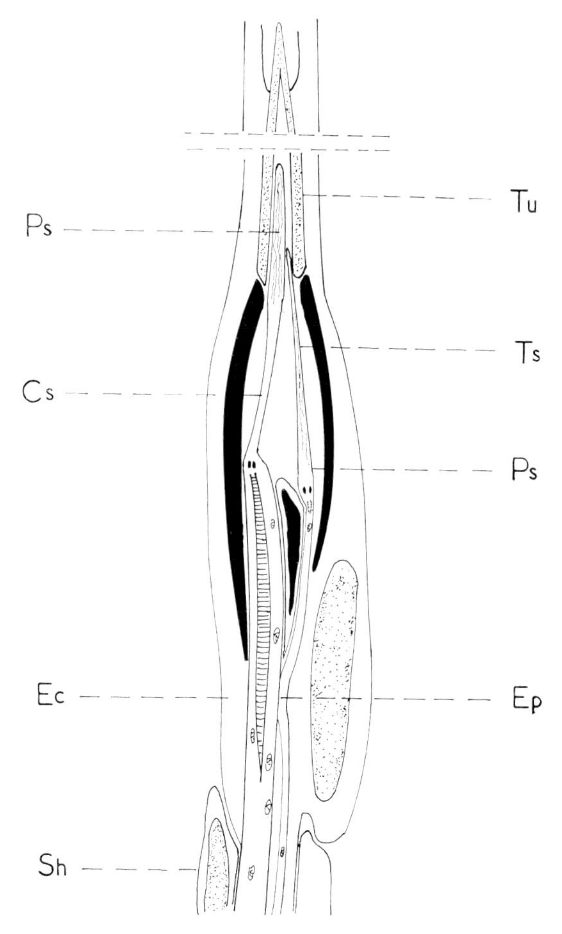

FIG. 7. Heterodynal scolopidium from the leg of *Carcinus* (after Whitear, 1962). This contains a ciliary sense cell on the left and a paraciliary cell on the right. *Ps*, paraciliary segment; *Cs*, ciliary segment; *Ec*, envelope cell; *Sh*, sheath cell; *Ep*, ephapse; *Ts*, terminal segment; *Tu*, tube.

cilia already described. This then dilates further distally to twice its diameter, forming a paraciliary segment. In the paraciliary cell there is no ciliary segment. As the distal process of both types of cell enters the

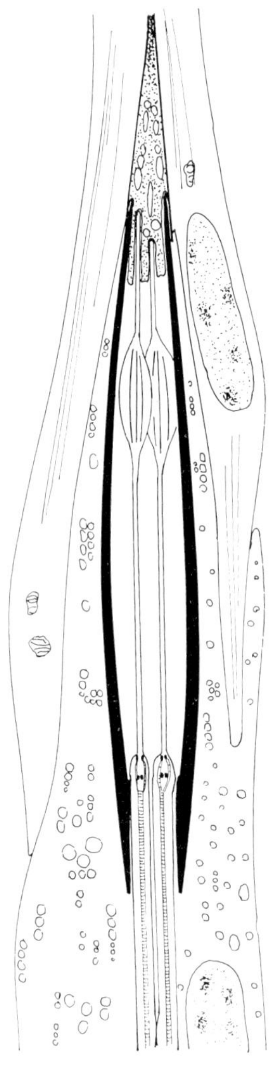

FIG. 8. Scolopidium from the connective chordotonal organ in the pedicel of the termite, *Zootermopsis angusticollis*. This is directly comparable with scolopidia in Figs. 5 and 6.

tube the paraciliary segment merges into a terminal segment which contains numerous microtubules. The terminal segments continue in the tube for some way and around their tips material resembling tangled connecting membranes is present: this material has been referred to as "glue".

Heterodynal scolopidia were commonly found, in which there was a ciliary cell and a paraciliary cell, but the CB organ contained only isodynal scolopidia with two ciliary cells. In all scolopidia an area was found near the base of the scolopales where the dendrite membranes were in apposition. This area has been called the ephapse.

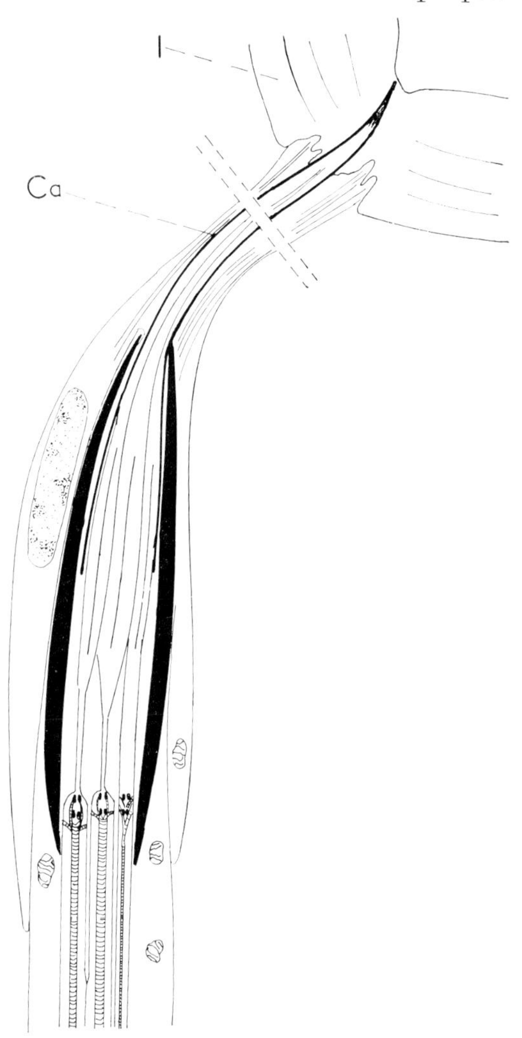

FIG. 9. Scolopidium from the Johnston's organ of the termite, *Zootermopsis angusti collis*. This contains three ciliary processes which run into the cap or tube (*Ca*) which ends in the intersegmental membrane (*I*) between the pedicel and flagellum.

The connective chordotonal organ of the pedicel

The scolopidia of this organ in the termite resemble the crab connective sensilla in the possession of two dendrites and correspondingly two ciliary processes, although in all the examples met the latter appeared similar in size and structure (Fig. 8). The scolopale sheath cell contains numerous large vacuoles. The scolopales form a fairly thick cylinder of material proximally, and they can not be said to form separate rods. The cilia have nine double hollow filaments. Distally they dilate enormously. They first become paraciliary, then bodies which are crescent-shaped in transverse section appear in association with the ciliary filaments. These bodies fuse to form an axial structure consisting of a central rod surrounded by six others (Fig. 16). The diameter of the cilium reaches about one micron in this region. The cilia assume their normal dimensions again when they enter separate channels in the cap (Fig. 13) where they are in intimate contact with the cap material. The ends of the scolopales with their associated cytoplasm are inserted into the cap (Fig. 13) which is long and pointed. Although it can not be said with complete certainty, the cap appears to continue as a solid electron-dense filament which runs to the cuticle and inserts into it a short way. This interpretation would be consistent with light-microscope observations (Howse, 1965).

The connective chordotonal organ of the tibia

D. Young (1967, personal communication) has found that the scolopidium of the cockroach tibio-tarsal organ is broadly similar in structure to that of the connective organ in the termite antenna. Two ciliary processes are present which dilate to a diameter of about half a micron before they enter the cap. The dilatation has a central core of fibrous or particulate material. The scolopidia are embedded in shell-like envelopes of connective tissue, and the organ is bounded by a membrane that contains small amounts of fibrous scolopale-like material at intervals.

The Johnston's organ

The scolopidia of Johnston's organ in the termite appear to be unique in two respects. Firstly, they contain three ciliary processes, and secondly, the cap is in the form of a tube which arises within the scolopales and passes right through the pedicelar-flagellar intersegmental membrane almost to the surface (Figs. 9, 19 and 20). All the cilia dilate distally, and in the dilatation is an axial structure which is here more in the form of a single rod (Fig. 19). The cilia continue into

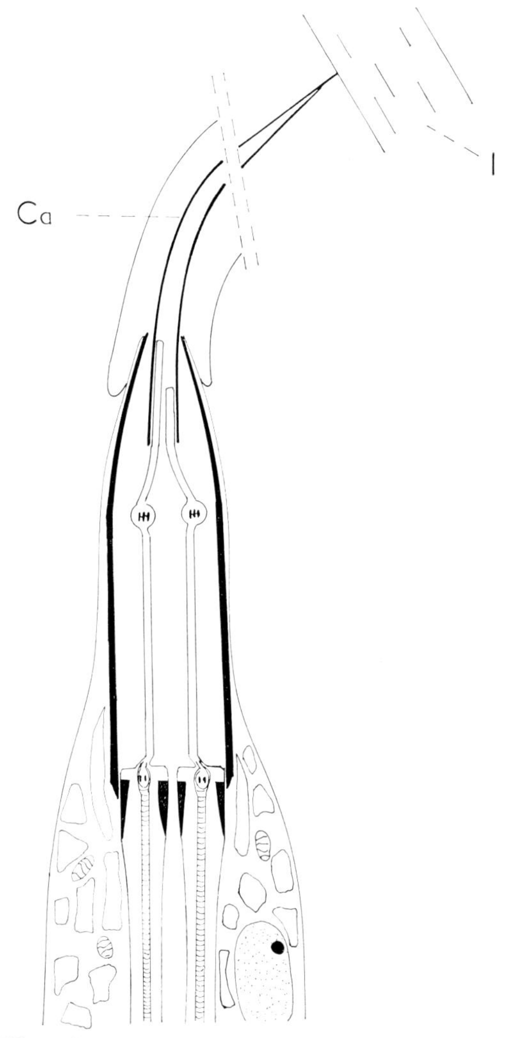

FIG. 10. Scolopidium from the Johnston's organ of *Drosophila melanogaster* (redrawn after Uga and Kuwabara, 1965). This contains two ciliary processes, but is otherwise directly comparable with Fig. 9.

the cap as terminal segments in which the outer ciliary filaments lose their identity and large numbers of microtubules are present. There are some indications that the cap may be divided at some levels into separate tubes each containing a terminal segment. There is some variation in the size of the ciliary processes; one is often smaller than the other two and may stem from a smaller nerve ending (Fig. 14). The smaller process may not have a ciliary segment. Ephapses have been found below the level of the scolopales. It has not been possible to determine whether all three processes are continued throughout the length of the cap.

The Johnston's organ scolopidium of *Drosophila melanogaster* described by Uga and Kuwabara (1965) is again in some respects quite different from those already described (Fig. 10). Two ciliary processes are present and before entering the cap together these each have a small dilatation which contains an electron-dense "sieve-like structure". The cilia terminate in the cap which is proximally less than half the diameter of the scolopale ring and has no articulation with the scolopale sheath cell. The cap continues as a hollow tubular structure to attach on to the intersegmental membrane between the pedicel and flagellum.

DISCUSSION

The fine structure of the Johnston's organ scolopidium in the termite is of particular interest in relation to that of the campaniform sensilla and hair-plate sensilla in the pedicel of the same insect. Both the latter sensilla are similar in fine structure to those described by Thurm (1964) from the honeybee, but they each have three dendrites with ciliary processes that enter a tubular cap that runs into the cuticle. Fibrous material is sometimes evident in the sheath cell wall around the ciliary processes. Pringle (1961) has suggested that the presence of an axial filament in sensory terminals relates insect chordotonal organs, campaniform sensilla and hair sensilla, among others, as a single class. Here now, in the antenna of the same insect, is a series of sensilla; the chaetic sensillum, campaniform sensillum, Johnston's

---

Fig. 11. T.S. through a cockroach subgenual organ scolopidium at the level of the centriole. Five scolopale rods (*sc*) surround the tip of the dendrite (*d*), and connections can be seen running between the scolopale–dendrite junctions and the nine root processes surrounding the centriole. Scale mark = 0·5 μ.

Fig. 12. T.S. through a scolopidium of the antennal connective organ of the termite at a level just above or below the centriole. A scolopale ring (*sc*) surrounds two dendrites (*d*) in which the divided root processes may be seen. Fibres may be seen running between these processes and scolopale–dendrite junctions. Scale mark = 0·5 μ.

Fig. 13. T.S. through the cap of the termite antennal connective organ scolopidium. A number of scolopale processes (*sc*) are inserted into the cap, in which the tip of one cilium may be seen. There are numerous microtubules in the surrounding attachment cell. Scale mark = 0·5 μ.

Fig. 14. T.S. through a Johnston's organ scolopidium of the termite. A scolopale ring (*sc*) surrounds three dendrites (*d*), one much smaller than the others, which are cut approximately at the level of the centriole. Scale mark = 1 μ.

Fig. 15. A subgenual organ scolopidium of the cockroach, in logitudinal section. This micrograph is directly comparable with Fig. 6. Scale mark = 5 μ.

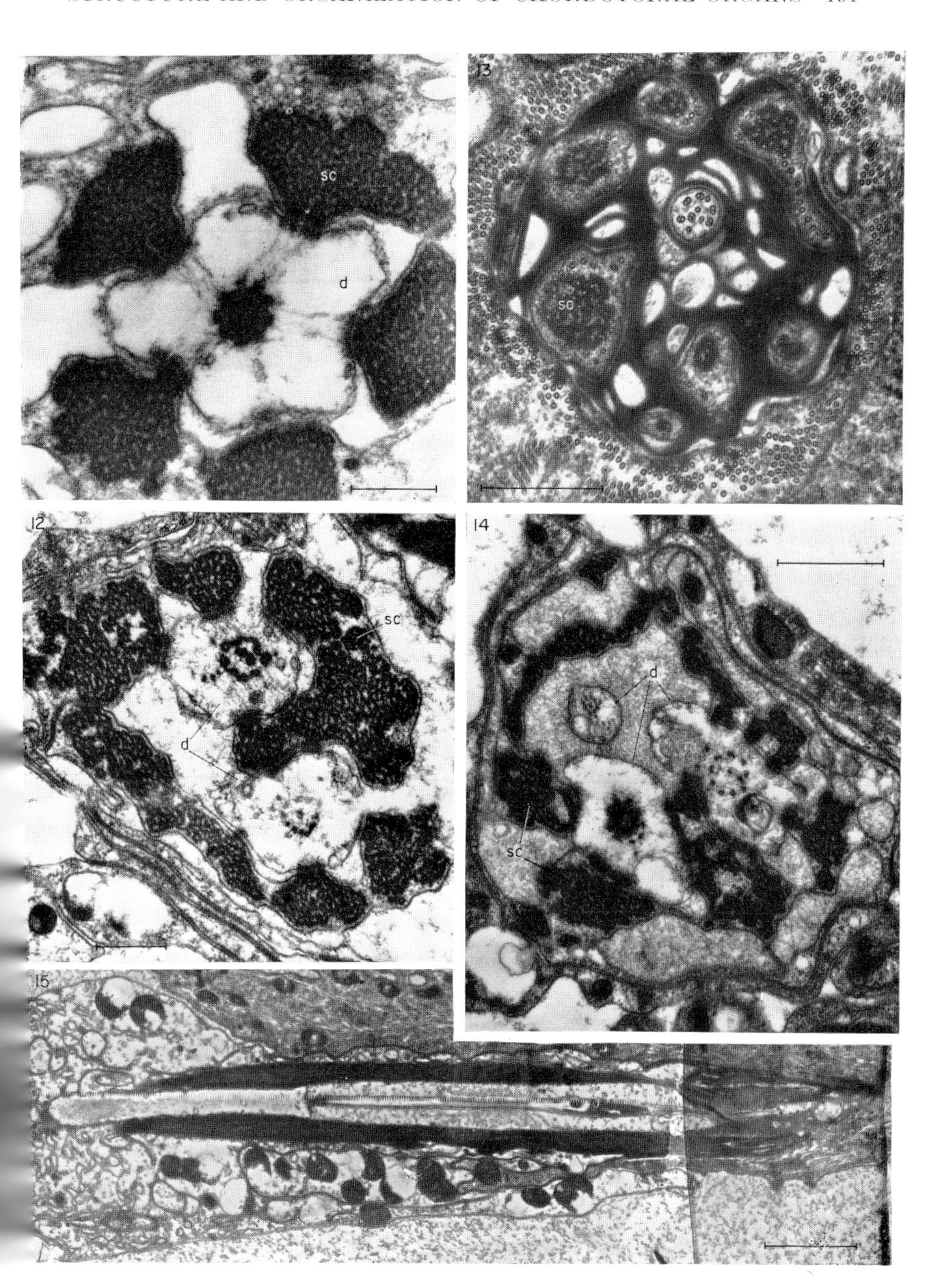

FIGS 11–15.

organ sensillum and connective chordotonal sensillum, which share quite a number of common features. Comparisons between different species may not be valid; for example, while the Johnston's organ sensillum of *Drosophila* is in many respects unlike that of *Zootermopsis*, it seems to be very similar to that found in dipteran halteres, where Pflugstaedt's figures clearly show the cap as a ring within the scolopale ring in transverse section.

Thurm (1966 and personal communication) has equated the cap of elastic material around the terminations of campaniform and chaetic sensilla processes with the cap of the scolopidium and suggested that the latter is stimulated by tension on the cap producing a compression of the ciliary terminal within it. While this remains a possibility it seems unlikely as the cap may be inelastic: it stains like exocuticle and usually has scolopale fibres running into it. Further, the cap of the connective sensillum is typically long and pointed, which is not the ideal shape to produce the pincer action on the terminals that Thurm has in mind.

The functions of the connections that have been demonstrated between the centriole and the scolopales in the subgenual organ sensillum (Fig. 11) may now be considered. Similar connections occur in all the scolopidia examined by the writer, and they are also shown in micrographs of locust tympanal organ scolopidia, which Dr E. G. Gray kindly allowed the writer to examine. The bases of the scolopales of tympanal and subgenual organ sensilla are to some extent blade-like, and it is clear that an axial displacement of the scolopales would result in compression of the dendrite tip if the connections were inelastic. The cilium might play no role in transduction of the stimulus and might

---

FIG. 16. T.S. through a scolopidium of the antennal connective organ of the termite at the level of the ciliary dilatations. The two enlarged cilia are within the scolopale ring (*sc*) and have a number of axial structures within the ring of nine double outer filaments. × 26 000.

FIG. 17. The ciliary dilatation of the cockroach subgenual organ scolopidium in T.S. Granular material lies between the outer filaments. Scale mark = 0·5 μ.

FIG. 18. T.S. through the scolopale–cap articulation in the cockroach subgenual organ scolopidium. The cap (*c*) contains channels into which run processes from the scolopales (*sc*). Microtubules (*m*) arise from the cell membrane of the attachment cell where it meets the scolopale sheath cell. Scale mark = 0·5 μ.

FIG. 19. T.S. (slightly oblique) through the termite Johnston's organ scolopidium in the proximal region of the cap. The cap (*c*) arises within the scolopale ring (*sc*), and the three cilia are seen, at the region of the dilatations, within this. Scale mark = 0·5 μ.

FIG. 20. L.S. through part of the pedicelar–flagellar joint in the termite antenna, showing part of a terminal process of a Johnston's organ sensillum running through the cuticle. The part of the cap shown (*c*) contains a ciliary terminal process. The attachment cell, which contains many microtubules (*m*) also runs some way into the cuticle. Scale mark = 1 μ.

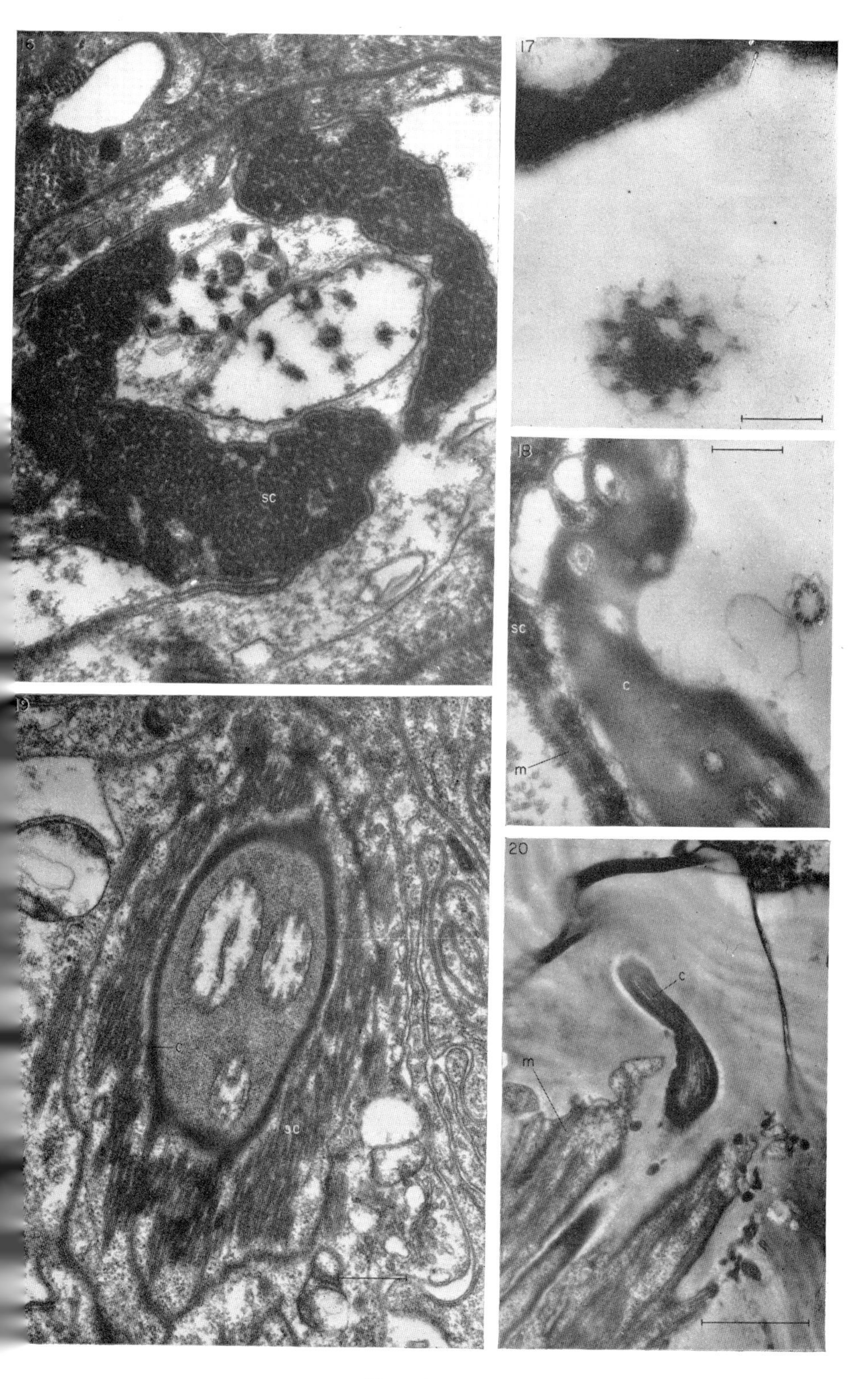

FIGS 16–20.

be purely an organizer in cell development. Such a function is very possible, since it is an axial structure, and in developing scolopidia of the subgenual organ the extracellular space around it is much narrower than normal and the cap appears to form from a condensation of finger-like processes that abut on to the cilium.

If the above explanation is correct, the subgenual organ and tympanal organ scolopidia would make ideal stretch receptors. But in connective scolopidia, although the centriole–scolopale connections are in many cases still clearly present, the blade-like nature of the scolopale bases is not nearly so pronounced. More importantly, the physiological evidence outlined indicates that the receptors in the crab respond to flexion rather than to longitudinal stretch, which would be difficult to reconcile with this argument.

All the scolopidia investigated ultramicroscopically show some form of ciliary dilatation, and if this corresponds with the *spülenförmige Körperchen* of the earlier microscopists these dilatations may be common to all scolopidia. The dilatation takes a different form in every species in which the fine structure has been studied, which seems to suggest that the precise nature of the swelling or of its inclusions does not matter as long as they are there. The ciliary membrane where the dilatation occurs should be less resistant to stretch than elsewhere if there is some internal skeleton that prevents changes in diameter. It is suggested, then, that part of the transduction process is stretch of the ciliary membrane at the dilatation, by flexion between the cap and scolopales. The scolopales can be envisaged as distance pieces maintaining a certain tension on the cilium by virtue of their connections with the centriolar apparatus proximally and with the cap, into which they are locked, distally. The cilium is anchored by the cap at one end and by the ciliary root, which is probably a skeletal structure, at the other. In the sensillum of *Drosophila*, without neurophysiological evidence, one is forced to the view that ciliary stretch is the first part of the transduction process. This theory may not apply to the Johnston's organ scolopidium of the termite or to the sensillum described by Taylor (1967) in the hermit crab. If, as appears likely, these are similar in structure, the same movement that produces flexion of the receptor will produce compression of the ending in the cuticle. Lateral compression is the adequate stimulus in the insect chaetic and campaniform sensilla (Thurm, 1966).

One can only speculate on how conduction might occur along the ciliary process and on what significance the presence of two or more ciliary processes in connective sensilla might have. Possibly additional processes confer a sensitivity to flexion of the sensillum in particular directions, but the investigation of possible inhibitory or excitatory interactions between the processes may answer this problem.

Acknowledgements

I am most grateful to Dr M. F. Claridge for reading parts of the manuscript of this paper, to Drs E. G. Gray and Mary Whitear for kindly letting me see their unpublished photographs, and to Drs B. M. H. Bush and M. A. Sleigh for valuable discussions of some of the views I have put forward here. I am also very grateful to Mr T. Davies and Miss M. Williams for their technical and photographic assistance in connection with the electron-microscopy I have done.

References

Alexandrowicz, J. S. and Whitear, M. (1957). Receptor elements in the coxal region of Decapoda Crustacea. *J. mar. biol. Ass. U.K.* **36**: 603–628.

Autrum, H. (1941). Über Gehör und Erschütterungssinn bei Locustiden. *Z. vergl. Physiol.* **28**: 580–637.

Autrum, H. and Schneider, W. (1948). Vergleichende Untersuchungen über den Erschütterungssinn der Insekten. *Z. vergl. Physiol.* **31**: 77–88.

Barber, S. B. (1960). Structure and properties of *Limulus* articular receptors. *J. exp. Zool.* **143**: 283–322.

Barth, G. (1934). Untersuchungen über Myochordotonalorgane bei dekapoden Crustaceen. *Z. wiss. Zool.* **145**: 576–624.

Bässler, U. (1958). Versuche zur Orientierung der Stechmücken: die Schwarmbildung und die Bedeutung des Johnstonschen Organs. *Z. vergl. Physiol.* **41**: 300–330.

Burke, W. (1954). An organ for proprioception and vibration sense in *Carcinus maenas*. *J. exp. Biol.* **31**: 127–128.

Burkhardt, D. and Gewecke, M. (1966). Mechanoreception in Arthropoda: the chain from stimulus to behaviour pattern. *Cold Spring Harb. Symp. quant. Biol.* **30**: 601–614.

Burkhardt, D. and Schneider, G. (1957). Die Antennen von *Calliphora* als Anzeiger der Fluggeschwindigkeit. *Z. Naturf.* **12**: 139–143.

Bush, B. M. H. (1965a). Proprioception by the coxo-basal chordotonal organ, CB, in the legs of the crab *Carcinus maenas*. *J. exp. Biol.* **42**: 285–297.

Bush, B. M. H. (1965b). Proprioception by chordotonal organs in the mero-carpopodite and carpo-propodite joints of *Carcinus maenas*. *Comp. Biochem. Physiol.* **14**: 185–199.

Busnel, R. G. (1963). On certain aspects of animal acoustic signals. In *Acoustic behaviour of animals*: 69–111. Busnel, R. G. (ed.). New York: Elsevier.

Busnel, M. C. and Burkhardt, D. (1962). An electrophysiological study of the photokinetic reaction in *Locusta migratoria migratorioides* (L.). *Symp. zool. Soc. Lond.* No. 7: 13–44.

Debaisieux, P. (1935). Organes scolopidiaux des pattes d'insectes. I. Lépidoptères et Trichoptères. *La Cellule* **47**: 271–314.

Debaisieux, P. (1938). Organes scolopidiaux des pattes d'insectes. II. *La Cellule* **47**: 77–202.

Debauche, H. (1935). Recherches sur les organes sensoriels antennaires de *Hydropsyche longipennis*. *La Cellule* **44**: 43–83.

Eggers, F. (1924). Zur Kenntnis der antennalen stiftführenden Sinnesorgane der Insekten. *Z. Morph. Ökol. Tiere* **2**: 259–349.

Eggers, F. (1928). Die stiftführenden Sinnesorgane. *Zool. Bausteine* **2**: 1–353.

Erhardt, E. (1916). Zur Kenntnis der Innervierung und der Sinnesorgane der Flügel von Insekten. *Zool. Jb.* (Anat.) **39**: 293–334.

Finlayson, L. H. and Lowenstein, O. (1958). The structure and function of abdominal stretch receptors in insects. *Proc. R. Soc.* (B.) **148**: 443–449.

Friedrich, H. (1929). Vergleichende Untersuchungen über die tibialen Scolopalorgane einiger Orthopteren. *Z. wiss. Zool.* **134**: 84–148.

Gettrup, E. (1962). Thoracic proprioceptors in the flight system of locusts. *Nature, Lond.* **193**: 498–499.

Graber, V. (1882). Die chordotonalen Sinnesorgane und das Gehör der Insekten. *Arch. mikrosk. Anat. EntwMech.* **20**: 506–640.

Gray, E. G. (1960). The fine structure of the insect ear. *Phil. Trans. R. Soc.* (B.) **243**: 75–94.

Hagemann, J. (1910). Beiträge zur Kenntnis von Corixa. *Zool. Jb.* (Anat.) **30**: 394–414.

Hartman, H. B. and Boettiger, E. G. (1965). Localization of receptor types in the propodite-dactylopodite organ of the crab *Cancer irrotatus* Say. *Am. Zool.* **5**: 651.

Haskell, P. T. (1956). Hearing in certain Orthoptera. 2. The nature of the responses of certain receptors to natural and imitation stridulation. *J. exp. Biol.* **33**: 767–776.

Heran, H. (1959). Wahrnehmung und Regelung der Flugeigengeschwindigkeit bei *Apis mellifera* L. *Z. vergl. Physiol.* **42**: 103–163.

Hertweck, H. (1931). Anatomie und Variabilität des Nervensystems und der Sinnesorgane von *Drosophila melanogaster* (Meigen). *Z. wiss. Zool.* **139**: 559–663.

Hollick, F. S. J. (1940). The flight of the dipterous fly *Muscina stabilans* Fallen. *Phil. Trans. R. Soc.* (B.) **230**: 357–390.

Howse, P. E. (1963). Zur Evolution der Erzeugung von Erschütterungen als Benachrichtigungsmittel bei Termiten. *Revue suisse. Zool.* **70**: 258–267.

Howse, P. E. (1964). An investigation into the mode of action of the subgenual organ in the termite, *Zootermopsis angusticollis* Emerson, and in the cockroach, *Periplaneta americana* L. *J. Insect Physiol.* **10**: 409–424.

Howse, P. E. (1965). The structure of the subgenual organ and certain other mechanoreceptors of the termite *Zootermopsis angusticollis* (Hagen). *Proc. R. ent. Soc. Lond.* (A) **40**: 137–146.

Howse, P. E. (1967). Mechanism of the insect ear. *Nature, Lond.* **213**: 367–369.

Janet, C. (1894). Sur les nerfs de l'antenne et les organes chordotonaux chez les Fourmis. *C. r. hebd. Séanc. Acad. Sci., Paris* **118**: 814–817.

Johnston, C. (1855). Auditory apparatus of the *Culex* mosquito. *Q. Jl microsc. Sci.* **3**: 97–102.

Larsen, O. (1957). Truncale Scolopalorgane in den pterothorakalen und den beiden ersten abdominalen Segmenten der aquatilen Heteropteren. *Acta Univ. lund.* **53**: 1–68.

Laverack, M. S. (1964). The antennular sense organs of *Panulirus argus*. *Comp. Biochem. Physiol.* **13**: 302–321.

Lehr, R. (1914). Die Sinnesorgane der beiden Flügelpaare von *Dytiscus marginalis*. *Z. wiss. Zool.* **110**: 87–150.

Lukoschus, F. (1962). Über Bau und Entwicklung des Chordotonalorgans am Tibia-Tarsus-Gelenk der Honigbiene. *Z. Bienenforsch.* **6**: 48–52.

Mendelson, M. (1963). Some factors in the activation of crab movement receptors. *J. exp. Biol.* **40**: 157–169.

Michelsen, A. (1966). Pitch discrimination in the locust ear: observations on single sense cells. *J. Insect Physiol.* **12**: 1119–1131.

Pflugstaedt, H. (1912). Die Halteren der Dipteren. *Z. wiss. Zool.* **100**: 1–59.

Popham, E. J. (1961). The function of Hagemann's organ in *Corixa punctata* (Illig.) (Hemiptera: Corixidae). *Proc. R. ent. Soc. Lond.* (A) **36**: 119–125.

Pringle, J. W. S. (1954). A physiological analysis of cicada song. *J. exp. Biol.* **31**: 525–560.

Pringle, J. W. S. (1957). *Insect flight*. Cambridge: University Press.

Pringle, J. W. S. (1961). Proprioception in arthropods. In *The cell and the organism*: 256–282. Ramsay, J. A. and Wigglesworth, V. B. (eds.). Cambridge: University Press.

Pumphrey, R. J. and Rawdon-Smith, A. F. (1939). Frequency discrimination in insects; a new theory. *Nature, Lond.* **143**: 806–807.

Richard, G. (1950). L'innervation et les organes sensoriels de la patte du termite à cou jaune. *Annls Sci. Nat.* (Zool.) (11) **12**: 65–83.

Richard, G. (1957). L'ontogénèse des organes chordotonaux antennaires de *Calotermes flavicollis* Fab. *Insectes. soc.* **4**: 107–111.

Risler, H. (1955). Das Gehörorgan der Männchen von *Culex pipiens* L. *Aedes aegypti* L. und *Anopheles stephensi* Liston (Culicidae), eine vergleichend morphologische Untersuchung. *Zool. Jb.* (Anat.) **74**: 498–490.

Roeder, K. D. (1966). Auditory system of noctuid moths. *Science, N.Y.* **154**: 1515–1521.

Roth, L. M. (1948). A study of mosquito behaviour. An experimental study of the sexual behaviour of *Aedes aegypti* (Linnaeus). *Am. midl. Nat.* **40**: 451–487.

Schön, A. (1911). Bau und Entwicklung des tibialen Chordotonalorgans bei der Honigbiene und bei Ameisen. *Zool. Jb.* (Anat.) **31**: 439–472.

Schwabe, J. (1908). Beiträge zur Morphologie und Histologie der tympanalen Sinnesapparate der Orthopteren. *Zoologica, Stuttg.* **50**: 1–154.

Swihart, S. L. (1967). Hearing in butterflies (Nymphalidae: *Heliconius, Ageronia*). *J. Insect Physiol.* **13**: 469–476.

Taylor, R. C. (1967). The anatomy and adequate stimulation of a chordotonal organ in the antennae of a hermit crab. *Comp. Biochem. Physiol.* **20**: 709–717.

Thurm, U. (1964). Mechanoreceptors in the cuticle of the honey bee: fine structure and stimulus mechanism. *Science, N.Y.* **145**: 1063–1065.

Thurm, U. (1966). An insect mechanoreceptor. Part I: fine structure and adequate stimulus. *Cold Spring Harb. Symp. quant. Biol.* **30**: 75–82.

Uga, S. and Kuwabara, M. (1965). On the fine structure of the chordotonal sensillum in antenna of *Drosophila melanogaster*. *Jap. J. Electron. microsc.* **14**: 173–181.

Vogel, R. (1912). Über die Chordotonalorgane in der Würzel der Schmetterlingsflügel. *Z. wiss. Zool.* **100**: 210–244.

Vogel, R. (1923). Über ein tympanales Sinnesorgan, das mutmassliche Hörorgane der Singzikaden. *Z. Anat. EntwGesch.* **67**: 190–231.

Wetzel, A. (1934). Chordotonalorgane bei Krebstieren (*Caprella dentata*). *Zool. Anz.* **105**: 125–132.

Whitear, M. (1962). The fine structure of crustacean proprioceptors. I. The chordotonal organs in the legs of the shore crab, *Carcinus maenas. Phil. Trans. R. Soc.* (B.) **245**: 291–325.

Whitear, M. (1965). The fine structure of crustacean proprioceptors. II. The thoracico-coxal organs in *Carcinus*, *Pagurus* and *Astacus*. *Phil. Trans. R. Soc.* (B.) **248**: 437–456.

Wiersma, C. A. G. and Boettiger, E. G. (1959). Unidirectional movement fibres from a proprioceptive organ of the crab, *Carcinus maenas. J. exp. Biol.* **36**: 102–112.

Wigglesworth, V. B. (1965). *The principles of insect physiology.* London: Methuen.

de Wilde, J. (1941). Contribution to the physiology of the Johnston organ and its part in the behaviour of the Gyrinus. *Archs néerl. Physiol.* (3) **25**: 381–400.

Wilson, D. and Gettrup, E. (1963). A stretch reflex controlling wing beat frequency in grasshoppers. *J. exp. Biol.* **40**: 171–185.

Symp. zool. Soc. Lond. (1968) No. 23, 199–216.

STEPS IN THE TRANSDUCER PROCESS OF MECHANORECEPTORS

ULRICH THURM

*Max-Planck-Institut für Biologie, Abteilung Reichardt
Tübingen, Germany*

SYNOPSIS

In motile cilia a distinct mechanosensitivity is found and analysed. Sensitivity of mechanosensory epithelial cells, e.g. lateral-line receptors, is shown to be closely related to the sensitivity found in motile cilia. Sensitivity of bipolar mechanosensory cells of arthropods, e.g. hair-plate receptors, in contrast, appears not to be directly related to the modified ciliary structure enclosed. A specialized microtubular bundle is found at the stimulus site within hair-plate-receptor cells. Based on a comparison with bipolar chemo- and photoreceptor cells the possibility is discussed that the modified ciliary structure occurring in all these bipolar receptors is reduced in function to participate only in later steps of transducer processes whose functional significance is independent of the nature of the adequate stimulus.

We will consider the transducer process in sensory receptors to be delimited on the input side by the adequate physical stimulus at the level of the sense cell and by some kind of a physiological response needing metabolic energy as its output. The response is characterized by its ability to trigger other kinds of nerve cell responses like axonic impulses or synaptic processes. The most usual output of a sensory transducer in this sense is the local change of the membrane potential known as the receptor potential. The generation of axonic impulses by the receptor potential may be considered as not included in the transducer process.

TYPES OF MECHANORECEPTORS

On the nature of the transducer process within mechanoreceptors only very tentative and crude hypotheses can be made at present. There are two main lines of speculation which are in part stimulated by the different ultrastructural appearances of various types of mechanoreceptors. The starting point is the correlation between a depolarizing receptor potential and the increase in membrane permeability found for several receptors. One of the tentative explanations therefore considers a depolarizing receptor potential as resulting from an extension of the plasma membrane, due to mechanical deformation of the cell; this might cause the increase in permeability of the membrane (Goldman, 1965; Loewenstein, 1965). The ultrastructure, as it is known for mechanoreceptors like Pacinian corpuscles or arthropod stretch receptors, seems to favour this explanation. No structure which might be specialized for the process of transduction is found within the terminal regions of such sense cells; however, the electrically active plasma membrane can reasonably be the medium for mechanosensitivity. It must be mentioned, however, that a stimulating effect has so far only been demonstrated for local compression of Pacinian receptor endings or of axons, and not, as yet, for a stretch of the membrane; longitudinal stretch of axons, resulting in an increase of their membrane area, failed to elicit a depolarization (Goldman, 1964; Gray and Ritchie, 1954; Schneider, 1952).

In a contrasting group of mechanoreceptors the region of the stimulus input is characterized by a considerable number of peculiar structures. To this group belong the sense cells of the acousticolateralis system of vertebrates and bipolar receptors of arthropods like those of scolopidial sensilla. The precise function of these structures is unknown at present. But there are a number of reasons to suspect that they are essential to the process of sensory transduction. The preliminary conception of the transducer process introduced by these structures is, in contrast to that mentioned above, one of a chain of intracellular processes elicited by the distortion of a specialized intracellular structure by the input stimulus; these processes would in turn regulate the permeability of the membrane at a distance from the stimulus site as the final result. This view will be discussed later in more detail.

The transducer processes which correspond to the different constructions of these two groups of mechanoreceptors may really be different. But it is doubtful to me, whether the apparent simplicity of the first group is a real one. We know, for comparison, that the mechanism of amoeboid movement is not really simpler than that of striated muscles, although there is no structure found within the cytoplasm of amoebae comparable in complexity to that in these muscles. It may be fruitful therefore, to tackle receptors of complicated but highly ordered structure to gain an insight into sensory transduction, just as it is fruitful to use striated muscles to study the principles of cellular motility. The knowledge of various structures in their special order can be advantageous for dissecting the chain of processes. The subject of this paper will be to show some results and some perspectives of experimentation with receptors of complex structure.

The typical structural element of the mechanoreceptors of the second group is a cilium or a cilium-like structure. The far-ranging occurrence of the basic plan of this element from motor organelles to highly specialized receptor cells like the photoreceptive rods in vertebrate retinae may indicate a fundamental role of this structure in intracellular processes of excitation. Two types of configuration of ciliary structures within receptor cells especially attracted my interest.

(1) Cilia which occur at the stimulus site in some kinds of mechanoreceptors and which are structurally very similar to motile cilia.

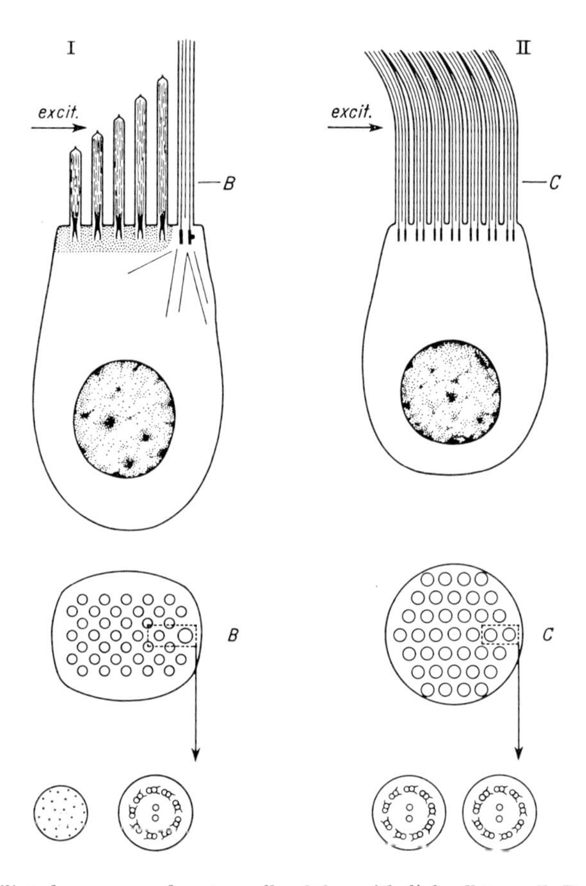

FIG. 1. Ciliated sensory and motor cells of the epithelial-cell type. I: Receptor cell of vertebrate lateral-line and equilibrium organs bearing stereocilia and one kinocilium (like cilia in II not drawn in its total length). II: Cell bearing numerous motile cilia which are lightly coupled (e.g. large abfrontal cilia of *Mytilus*). Arrows (excit.) show the direction of excitatory displacements of the cilia. Enlarged cross sections of cilia are shown below sections *B* and *C*. In the motile cilia studied the direction of structural polarity is unknown with respect to the direction of excitatory displacement; the structural polarity is therefore neglected.

Among the mechanoreceptors whose ultrastructure is known, such structures appear in lateral-line organs and equilibrium receptors of vertebrates, vibration receptors of ctenophores as studied by Horridge (1965a), and equilibrium receptors of cephalopods, Barber (1968). This type is characterized as example I in Fig. 1 based on descriptions given by Engström, Ades and Hawkins (1962), Flock (1965a and b), Lowenstein, Osborne and Wersäll (1964), and Wersäll, Flock and Lundquist (1965) of receptor cells of the acoustico-lateralis system. The single cilium of such a cell has the same number of microtubules in its cross section—the well known $9 \times 2 + 2$ distribution—as motile cilia. A simplified scheme of a cell bearing motile cilia is given as example II of Fig. 1. The peripheral microtubules of both types of cilia bear arms; the basal bodies in both cases show an interesting foot structure. In the organ of Corti and in the statocyst of *Helix* (Laverack, 1968) the cilium is reduced to the basal body. We will call this type of mechanoreceptors for shortness the epithelial-cell type.

(2) Modified ciliary structures which occur in receptors sensitive not only to mechanical but alternatively to chemical or photic adequate stimuli; these structures in many cases lie distant from the stimulus site. This morphological type of receptor cell is called the bipolar-cell type.

For cilia of the first type a contribution to mechanosensitivity has often been suspected, taking into consideration the possibility that reverse relations may exist between active and passive movements of a cilium and the state of excitation of the cell. With this in mind Lowenstein *et al.* (1964) discussed cilia of type (1) as "motile cilia in reverse". For the modified ciliary structure (2) of bipolar receptor cells, in contrast, which occur in both mechano- and chemo- and photoreceptors, a direct role in the primary process of mechanosensitivity seems much less likely. The structural relations of and the functional problems raised by this ciliary structure will be discussed in a final section; the next section will be devoted to cilia of type (1).

MECHANOSENSITIVITY OF MOTILE CILIA

The question arose whether there is a mechanosensitivity inherent in unmodified cilia in general, related perhaps to the mechanism of rhythmic motion. Such a sensitivity might be one reason for the occurrence of ciliary structures within mechanoreceptors. We therefore tested the mechanosensitivity of cilia which perform a *motor* function. Some suggestion of mechanosensitivity of motile cilia is already known (e.g. Gray, 1928; Lucas, 1933; Sleigh, 1962; Horridge, 1965b).

Object and methods

Objects of the experiments were the large abfrontal cilia on the gill of *Mytilus edulis* and similar cilia occurring at the septum between the inhalant and exhalant openings (Fig. 2). Abfrontal cilia of *Mytilus* have been the objects of a number of investigations in motor physiology of cilia (e.g. Gray, 1931; Yoneda, 1960, 1962). Both types of cilia can be described more exactly as bundles of about 40 ciliary units, lightly coupled to beat in unison.

Among the possible expressions of mechanosensitivity, the active movements of the cilia were checked. Observation of ciliary movements was preferred to measurement of membrane potential for the following reason: if mechanosensitivity is primarily a property of a control mechanism within the rhythmic-motion system of the cilia themselves, then it cannot be presupposed that this control works via changes in membrane potential. Such changes, however, might be coupled to the primary sensory mechanism as a secondary process in specialized cells.

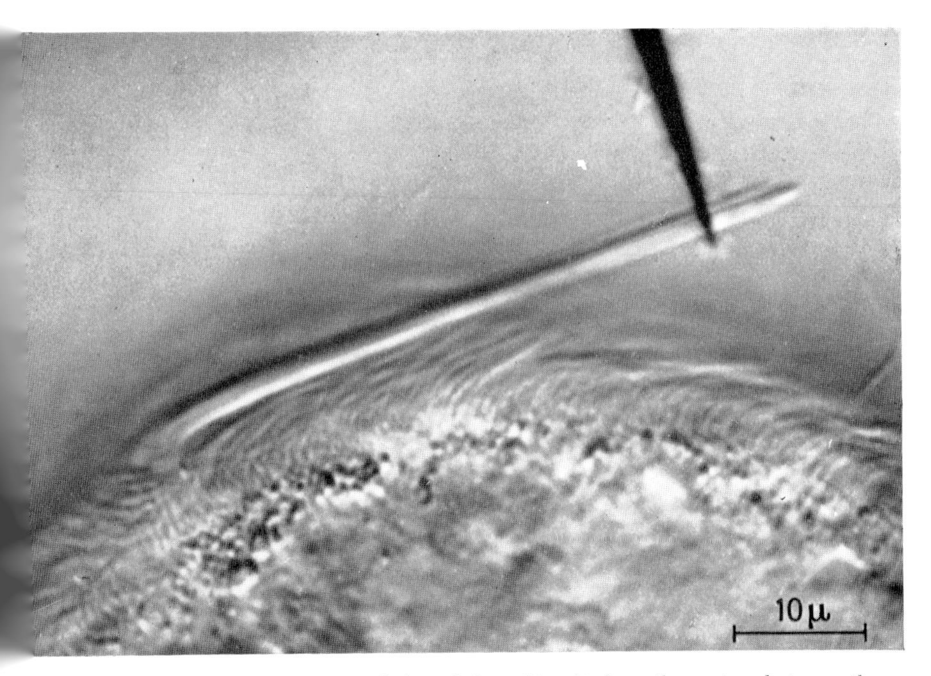

FIG. 2. Large motile cilium consisting of about 50 units from the septum between the inhalant and exhalant openings of *Mytilus edulis*. There are small uncombined cilia on most of the epithelial cells. The tip of a tungsten needle is shown in the upper right. Photomicrograph with interference contrast after Nomarski (Zeiss).

The frequency of spontaneous beats of the cilia was lowered to zero by low potassium concentration, increased carbon dioxide pressure and lowered pH-value of the sea water. These factors control the frequency of spontaneous strokes of cilia over a wide range in a graded, constant, and reversible manner (cf. Gray, 1922). Cilia of the types used come to rest under the conditions mentioned in a position parallel to the surface of the epithelium, as shown in Fig. 2 (post effective phase of the stroke).

Mechanical stimuli were applied by means of a glass or tungsten needle, whose tip was bent through a right angle. Movements of needles were performed by a small loudspeaker and recorded by an inductive pick-up (Wenking) (accuracy 0·1 μ).

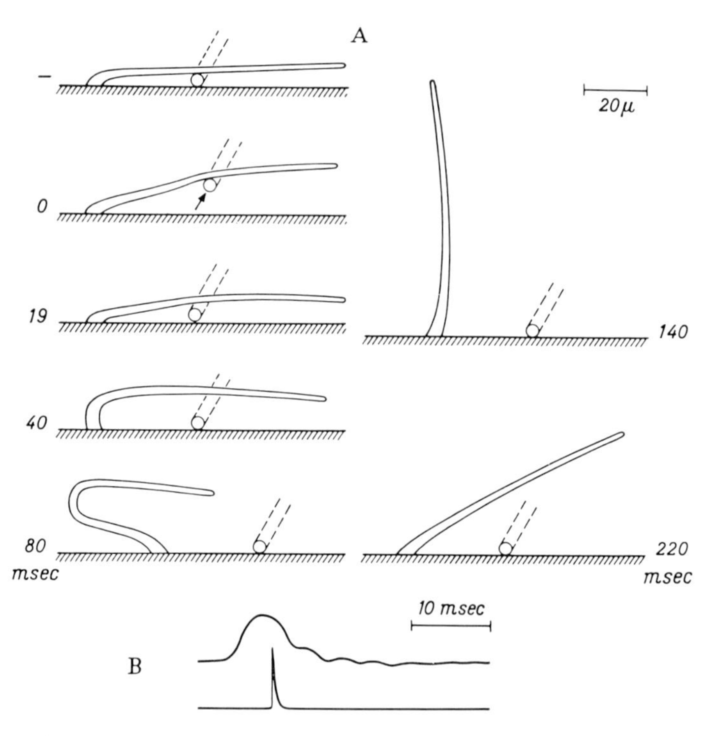

FIG. 3. Sequence of stimulus and response of a large cilium from the septum. A: Drawings from photomicrographs. In the middle of each figure the tip of the needle which is bent through a right angle. The flash photographs were taken at the times indicated after the middle of the stimulus impulse. B: Oscilloscope records of the displacement of the needle (upper trace) and of the light intensity (lower trace), corresponding to the second figure of A.

Mechanically induced ciliary movements

An active stroke of a resting cilium can be evoked by bending the cilium passively through a small angle in the direction in which it would actively move from the resting position. For angles of passive bending between 2° and 15° the probability for the occurrence of a full stroke following displacement increases from 50% to 100%. Fig. 3 shows the sequence of stimulus and response; the figures are drawn from photographs which were taken with a flashlight triggered by the stimulus impulse after various time intervals. As can be seen from the oscilloscope record (B) the whole stimulus movement took about 10 msec. The lower oscilloscope trace shows the time course of the flash, which in the figure occurs at that time of the stimulus which is defined as 0. The first two figures (A) show the cilium in its resting position, and in the passively bent position. The third figure reveals a temporary return of the cilium into the resting position after the end of the stimulus. After a latency of 20–30 msec, however, the cilium begins to erect as in its regular recovery beat, and a movement like an effective stroke finishes the cycle.

This induced movement of the cilium obviously is a response of the cell to its temporary distortion. We may conclude from the movements of the cilium that the amount of energy delivered to the cilium during its passive bending is small with respect to the amount of energy used for the response. Thus the relation, that is characteristic of stimulus-response phenomena in receptors, also occurs in these cells bearing motile cilia.

Localization of the site of mechanosensitivity

Experiments with differential bending of the cilium showed that the probability of response is consistently correlated with the extent of bending of the ciliary base alone. The effective stimulus distortion of the cilium can be reduced to a shearing motion of the ciliary base in parallel to the cell surface, as is indicated in Fig. 4. The angular position of the ciliary shaft remains constant in this case. The shearing amplitude eliciting responses to more than 70% of the stimuli was found to be 0·4 μ, measured about 2 μ above the cell surface in the direction of maximal sensitivity. The centre of the stimulating ciliary displacement can be calculated to lie between 1 and 2 μ below the cell surface, if equal percentages of responses are assumed to be evoked by equal angles of bending independently of the point of attack at the cilium.

Directional sensitivity

The sensitivity varies as the cosine of the stimulus direction. Maximum sensitivity is given in the direction of the start of the active

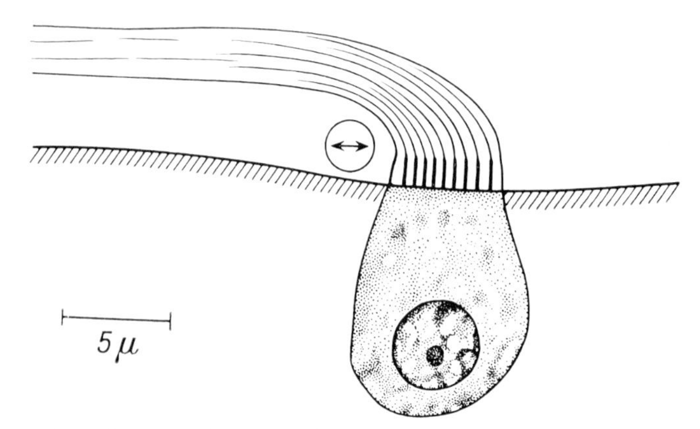

Fig. 4. Cell with large abfrontal cilium drawn from a photomicrograph under interference contrast. The tip of the needle (in cross section) and the direction of movement are shown for shearing distortions of the cilium base.

movement. Stimuli in the opposite half of the cycle do not elicit a stroke but inhibit spontaneous activity.

Adaptation

Long lasting stimuli induce rhythmic beats, whose frequency behaves similarly to the spike frequency of a phasic-tonic receptor: i.e. the frequency of beats has a maximum at the beginning of the stimulus and decays to a lower nearly constant value. (Depending on the position of the needle the effective stimulus is temporarily interrupted during each active beat.)

Conclusions

The results as reported are perhaps not valid for motile cilia in general. The behaviour found seems to be related to the special resting position of this cilium at the end of its effective stroke. Cilia which under identical conditions remain in a resting position at the end of the recovery stroke also responded to mechanical stimuli but in a somewhat different manner, as was found, e.g. for lateral cilia of the *Mytilus* gill. The general question whether mechanosensitivity is a universal property of motile cilia cannot be answered so far without comparative studies. Considerable structural differences exist in the basal parts of motile cilia—especially in cilia of protozoons—which may reflect different functional properties. The general results of this study, however, are:

(1) a distinct mechanosensitivity can be found in certain motile cilia;

(2) this sensitivity is formally related to motile behaviour of the cilia.

The functional significance of the mechanosensitivity of motile cilia possibly lies in the following directions:

(1) regulation within the mechanism of rhythmic motility;
(2) co-ordination of the movements of neighbouring cilia (cf. Sleigh, 1966);
(3) control by exogenic factors of the state of excitation of the cell as a whole and its neighbours (cf. Lucas, 1933; Kaissling, 1967).

So far it is unknown whether a change in membrane potential is involved in the response reported. An observation made by Horridge (1965b) on comb plate cells of a ctenophore indicates that a membrane depolarization can correspond to a mechanically induced ciliary beat. In the comb-plate all cilia of a cell are combined as is the case for the large abfrontal cilia studied here. Lateral cilia of the *Mytilus* gill in contrast are uncombined; cilia of one cell as well as of different cells normally beat metachronally. If a few cilia of a lateral cell are stimulated mechanically in their resting state, only these cilia respond. A tentative interpretation of this result is that the stimulus-response mechanism of motile cilia is basically a local mechanism of the single ciliary unit which does not effectively control the membrane potential of the cell. (Factors which generally depolarize a nerve or muscle cell induce ciliary activity in lateral cells; see Gray, 1928; Aiello, 1960.)

In seeking to delimit the first stage of the transducer process, i.e. the mechanosensitive structure, one can use the distinct directional sensitivity of the stimulus-response mechanism as a kind of tracer. Only a structural polarity of the input stage would explain this directional sensitivity. The experiments reported here point to a site at the ciliary base, which may lie inside the cell, as the locus of the mechanosensitive structure. Electronmicroscopically, the basal body, especially the basal foot (Gibbons, 1961), shows a polar configuration with its axis identical to the axis of maximal sensitivity. In fact, the mechanosensitive structure could be a part of the basal body including the basal foot. Thus the specifically mechanosensitive structure would not be part of the mechanism which is specialized to perform movements, i.e. the ciliary shaft, but of those structures which are thought to be engaged in movement control.

The mechanosensitivity of motile cilia studied show remarkable analogies to the sensitivity of mechanoreceptors such as those of the acoustico-lateralis system of vertebrates. According to investigations of such receptors done by Flock (1965a, b), Görner (1963), Lowenstein *et al.* (1964), Wersäll (1965) and their co-workers, correspondence can be found in the following points:

(1) The kind of effective stimuli: in both types of cells the stimulus of maximal effectiveness is a shearing motion parallel to the cell surface (cf. Holst, 1950).

(2) The amplitude of effective stimuli: effective displacements of the cupula of a lateral-line organ lie between 0·1 μ and 10 μ according to Flock (1965b); in abfrontal cilia the probability of responses increases to 100% with displacements of the ciliary base between 0·2 μ and 1 μ.

(3) The directional sensitivity: for both types of cells, sensitivity varies with the cosine of the stimulus direction; stimuli of contrary directions have contrary effects. In motile cilia maximal sensitivity occurs in the plane of active movements; in sensory cells the corresponding relation is indicated by comparison of the physiological with the ultrastructural axis.

(4) The adaptive behaviour: in both cases long-lasting stimuli induce responses according to the phasic-tonic scheme.

In sensory cells the relation between physiological and morphological polarity is known. In abfrontal cilia this relation so far is unknown; to apply results from other types of motile cilia (Gibbons, 1961) is of uncertain validity because of the differences in motile behaviour mentioned earlier.

The characteristics of mechanosensitivity in both kinds of transducing cells are obviously rather alike. From the studies cited above a corresponding similarity is known for the ultrastructure of motile cilia of mussels and sensory kinocilia. Stereocilia of sense cells in contrast have no analogy in cells of motile cilia that would show a similar configurational polarity; the presence of microvilli is questionable for cells of large abfrontal cilia.

Although the mechanosensitivity of motile cilia in general is not yet clear, a view favoured by present physiological and morphological results may tentatively be formulated: similar structures are responsible for mechanosensitivity in motor and sensory ciliated cells. The mechanosensitive structure is located within the region of the ciliary basal body. The ciliary shaft therefore may be absent in a cell with solely a sensory function, as it is lacking in the organ of Corti. (The converse conclusion has already been arrived at by Engström *et al.*, 1962.) Accessory structures, like stereocilia, appropriate for transmission of the stimulus to the basal body are necessary in this case. They also may be functionally advantageous if only one slender cilium per cell is present, as in labyrinthine and lateral-line receptors. Mechanosensitivity of cilia may be primarily a property of the movement-control mechanism. In ciliated cells specialized for sensory function, an intracellular regulation of cytomembrane permeability is postulated. This mechanism coupled to the mechanosensitive system may control the permeability of a substantial part of the plasma membrane as a function of the external stimulus. A similar mechanism may in some cases be thought to regulate directly the activity of synapses. The postulated mechanism of permeability control may be related functionally to control mechanisms of plasmatic motile systems. The effects of calcium ions known for the control of both motility and membrane permeability may be mentioned in this connection.

RECEPTORS WITH MODIFIED CILIARY STRUCTURES

In this last section I will deal briefly with mechanoreceptors containing a modified ciliary structure. I shall use as an example the hair plate receptor of the honey bee, which we studied a few years ago (Thurm, 1964, 1965).

Results

There are four points which seem to me to be of special interest:

(1) The site of the adequate stimulus. The adequate stimulus is given most probably by a compression of the nerve terminal within a region of only about 1 μ length at a site where a special intracellular structure is located. Bending of the hair as shown in the left-hand diagram of Fig. 5 is the external adequate stimulus. The distortion of the hair base caused by bending is transmitted by spongy cap material to the nerve terminal enclosed. Experiments revealed that only movement transverse to the nerve terminal is effective in stimulating the sense cell. Stretch is quite ineffective. Several additional reasons make the compressional component the most probable stimulating effect. A diminution of the terminal diameter during maximal hair bending was measured in *in vitro* sections to be 0·1 μ. Results of such measurements are shown in the right-hand diagram of Fig. 5. The working range of the receptor can be calculated to lie between changes in diameter of 30 Å and 1000 Å or 0·5% and 15% of the resting diameter.

(2) The intracellular structure at the stimulus site. A specific structure occurs within the nerve terminal at the site where the adequate stimulus affects the terminal. The tip of the terminal enclosed in the cap is shown in the left-hand electronmicrograph of Fig. 6 in longitudinal section and in the middle photograph in cross section This structure consists of 50–100 tubular elements which are combined in parallel by electron dense material. The elements run separately out

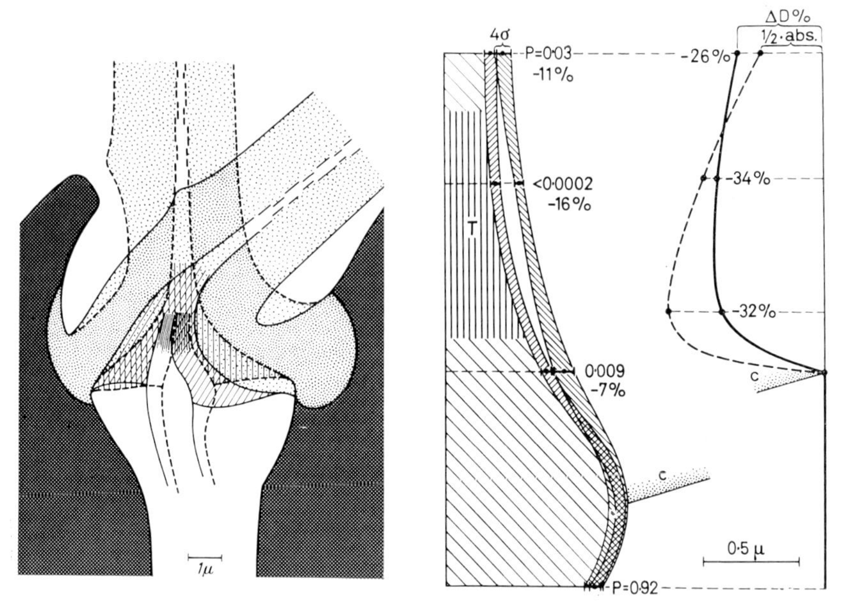

Fig. 5. Hair-plate sensillum of the honey bee. Left: The joint region of the hair with the nerve terminal enclosed showing the distortions induced by bending the hair out of the upright resting position. Right: Changes in diameter of the nerve terminal and of the cap; diameters in the resting position are compared with those of the maximally bent position of the hair. At the left side of the diagram one half of the nerve terminal is shown. The finely hatched regions indicate $\pm 2\sigma$ of the diameter measurements; the entire change is shown. T: the typical position of the tubular body; c: the proximal surface of the cap. At the right side of the diagram are shown 1/2 of the change in diameter of the cap (1/2 abs., dashed line) and the relative change with respect to the resting diameter ($\Delta D\%$, solid line). (Reproduced with permission from Thurm, 1964 (left diagram), and 1965 (right diagram).)

of the bundle in the proximal direction. Here, from their diameters, they can be identified as some kind of plasmatic microtubules (using this term in its narrow sense).

(3) The ciliary structure proximal to the stimulus site. The cross section of the nerve terminal is reduced to a ciliary structure nearly 10 μ distant from the stimulus site. This ciliary cross section—as shown in the right-hand electronmicrograph of Fig. 6—is of modified form, no central filaments and no arms on the peripheral filaments are present. The inner diameter of the tubular filaments appears equal to those in the tubular body. Two basal bodies (centriole-like structures) have been found, whose configuration is shown in the diagram of the distal nerve process given in Fig. 7. These structures show no indication of polarity, in contrast to the corresponding structures in receptors of the epithelial-cell type.

The directional sensitivity of these receptors is different from that of the acoustico-lateralis receptors (Thurm, 1963). This can be easily explained on the basis of a nondirectionally dependent sensitivity of the receptor element, because of the radial asymmetry in the cuticular hair joint.

(4) Sensory structures, unspecific for stimulus modality. A conspicious structural conformity exists within the distal nerve processes of several kinds of mechanoreceptors and of receptors for different modalities of adequate stimuli. The diagrams of Fig. 8 show for comparison two types of mechanoreceptors, namely a scolopidial sensillum

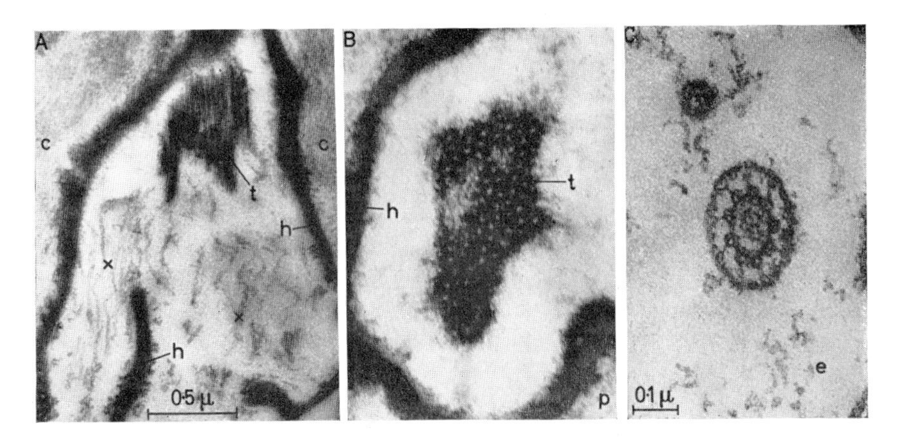

FIG. 6. Electronmicrographs of the nerve terminal of the hair-plate receptor. Longitudinal section (A) and cross section (B) through the tip of the nerve terminal showing the tubular body, *t*. (C) Cross section of the terminal at the level of the ciliary structure. Same scale in B as in C; *e*, extracellular space; *h*, cuticular sheath; *p*, cap; *x*, microtubules. The space between tubular body and cuticular sheath probably is an artifact. (Reproduced with permission from Thurm, 1965.)

according to Gray (1960) and a hair or campaniform sensillum according to the studies just mentioned (Thurm, 1964), next an arthropod chemoreceptor according to Hayes (1966), Laverack and Ardill (1965), and Slifer and Sekhon (1963, 1964a, b), and finally the well known photoreceptor rod of the vertebrate retina (Sjöstrand, 1953a, b). These types of receptors responding to three different modalities of adequate stimuli show the same kind of separation of a terminal segment from an inner segment by a modified ciliary structure like that I have described. The terminal segments show different structural specializations which appear to be adapted to the nature of the adequate stimulus (e.g. membrane lamellae containing photopigment in the photoreceptor). The more proximal structures, in contrast, i.e. the ciliary and accessory structures and an accumulation of mitochondria, are the same in their principal configuration irrespective of the modality of the adequate stimulus. This configuration of structures, however, is probably specific for the receptor function *per se* of these cells. What I want to emphasize is that within that region of the distal nerve process,

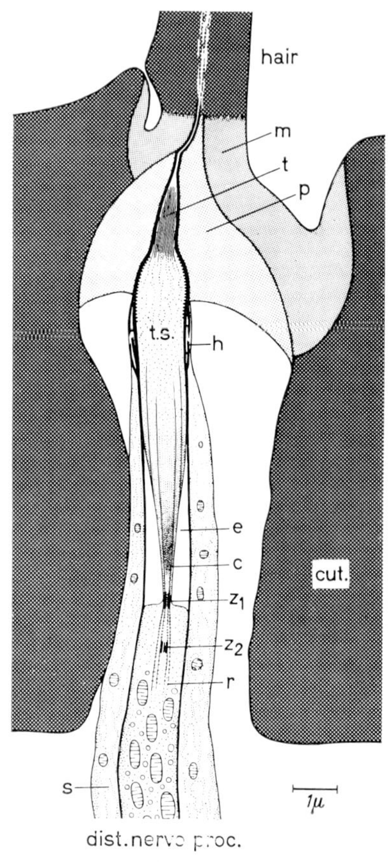

FIG. 7. Diagram of a hair-plate sensillum. *c*, ciliary structure; *cut*, cuticle; *e*, extracellular space; *h*, cuticular sheath; *m*, mitochondria; *n*, joint membrane; *p*, cap; *r*, root fibres; *s*, Schwann cell or trichogen cell; *t*, tubular body; *t.s.*, terminal segment; z_1, z_2, centriole-like structures. (Reproduced with permission from Thurm, 1965.)

which is structurally different from an axon and which can be recognized as specialized for sensory function, modality-specific terminal structures are connected to a modality-unspecific proximal section. The structures so far known to connect modality-specific and modality-unspecific parts are solely microtubules and cytomembrane.

With respect to the function of the modality-unspecific structures only results for photoreceptors of vertebrate retinae can be quoted. According to results by Baumann (1965), Bortoff (1964), Brown, Watanabe and Murakami (1965) and Tomita (1965) the region proximal to the ciliary structure—the inner segment—is a possible site for the generation of the receptor potential. This result casts an interesting light on to the astonishing accumulation of mitochondria within this region in all of these receptor types.

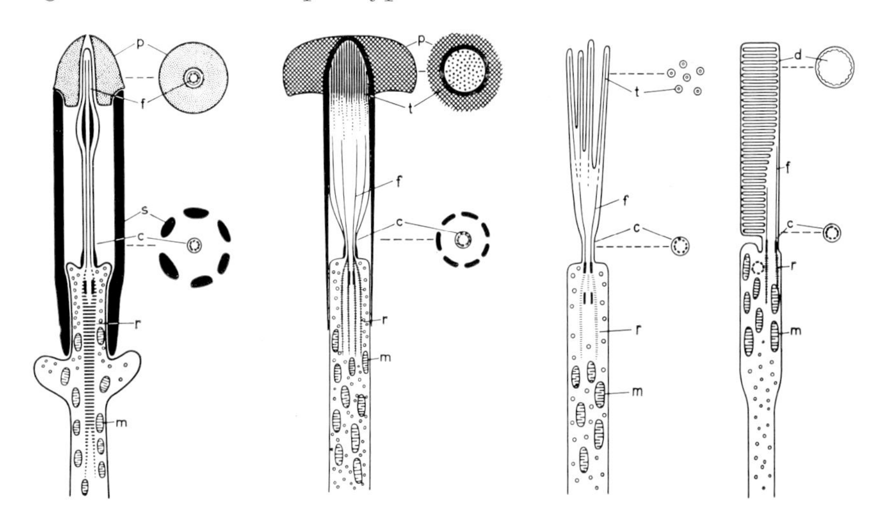

FIG. 8. Diagrams of the distal nerve processes of a scolopidial sensillum, a hair-plate or campaniform sensillum, an arthropod chemoreceptor, and a vertebrate photoreceptor (from left to right). (Authors are named in the text.) *c*, ciliary structure; *d*, membrane disc; *f*, ciliary microtubule; *m*, mitochondrion; *p*, cap; *r*, root structure; *s*, scolopale; *t*, tubule.

Perspectives

These findings are hard to understand on the ground of physiological mechanisms already known. It seems reasonable, however, in attempting to open the unexplored black boxes of sensory transducer processes, to be ourselves receptive to indications of new kinds of mechanisms. Thus the findings appear to give a consistent picture if one may tentatively assume two groups of unknown processes. The modified cilium together with its accessory structures may be working as mechanisms:

(1) of a signal transmission which perhaps is not an ionic process at the cytomembrane but may be a protein-chemical process associated with microtubules (cf. Gliddon, 1965);

(2) of a membrane-permeability control within the inner segment. This permeability control may include amplification and spatial

extension of the signal within the region of the accumulation of mitochondria (cf. Sjöstrand, 1953b).

These functions actually would not be specific for a special modality of stimulus, but would be of functional significance for sensory receptors in general. The assumption therefore is consistent with the occurrence of ciliary structures within sensory cells independent of their modality. The mechanism of permeability control is thought to be identical with that already mentioned for unmodified cilia of the epithelial-cell type. Moreover I have attributed to the basal structures of unmodified cilia the property of mechanosensitivity. Mechanoreceptors of the bipolar-cell type, in contrast, are different from the epithelial-cell type in so far as their site of mechanosensitivity apparently is separated from the ciliary structure. This ciliary structure of the bipolar-cell type appears to be reduced in function, I suppose, to signal transmission and permeability control; in structure it is obviously reduced, lacking the basal foot, central filaments, and peripheral arms. In this reduced form the ciliary structure is found connected alternatively to different specialized input structures for adequate stimuli of chemical, photic or mechanical nature.

It is not yet appropriate to discuss the problems I have raised in more detail. In the present state of transducer physiology a tentative synopsis like that set out may be of some value, however, as a working base on which to plan lines of experimentation.

Acknowledgements

My thanks are due to Dr J. Scholes for linguistic improvement of the manuscript. The permission of the *Cold Spring Harb. Lab.* of *Quant. Biol.* to reproduce Figs 5 (right hand diagram), 6 and 7 is gratefully acknowledged. Copyright for Fig. 5 (left hand diagram) 1964 by the American Association for the Advancement of Science.

References

Aiello, E. L. (1960). Factors affecting ciliary activity on the gill of the musse *Mytilus edulis*. *Physiol. Zool.* **33**: 120–135.

Barber, V. C. (1968). The structure of mollusc statocysts, with particular reference to cephalopods. *Symp. zool. Soc. Lond.* No. 23: 37–62.

Baumann, Ch. (1965). Receptorpotentiale der Wirbeltiernetzhaut. *Pflüger's Arch. ges. Physiol.* **282**: 92–101.

Bortoff, A. (1964). Localization of slow potential responses in the *Necturus* retina. *Vision Res.* **4**: 627–635.

Brown, K. T., Watanabe, K. and Murakami, M. (1965). The early and late receptor potentials of monkey cones and rods. *Cold Spring Harb. Symp. quant. Biol.* **30**: 457–482.

Engström, H., Ades, A. W. and Hawkins, J. E. (1962). Structure and functions of the sensory hairs of the inner ear. *J. Acoust. Soc. Am.* **34**: 1356–1363.

Flock, Å. (1965a). Transducing mechanism in the lateral line canal organ receptors. *Cold Spring Harb. Symp. quant. Biol.* **30**: 133–145.

Flock, Å. (1965b). Electron microscopic and electrophysiological studies on the lateral line canal organ. *Acta Oto-lar. Suppl.* **199**: 1–90.

Gibbons, I. R. (1961). The relationship between the fine structure and direction of beat in gill cilia of a lamellibranch mollusc. *J. Biophys. Biochem. Cytol.* **11**: 179–205.

Gliddon, R. (1965). Ciliary activity and coordination in *Euplotes eurystomus*. *Int. Conf. Protozool.* **2** Abstr. no. 307.

Goldman, D. E. (1965). The transducer action of mechano-receptor membranes. *Cold Spring Harb. Symp. quant. Biol.* **30**: 59–68.

Goldman, L. (1964). The effects of stretch on cable and spike parameters of single nerve fibers; some implications for the theory of impulse propagation. *J. Physiol., Lond.* **175**: 425–444.

Görner, P. (1963). Untersuchungen zur Morphologie und Elektrophysiologie des Seitenlinienorgans vom Krallenfrosch. *Z. vergl. Physiol.* **47**: 316–338.

Gray, E. G. (1960). The fine structure of the insect ear. *Phil. Trans. R. Soc.* **243**: 75–94.

Gray, J. (1922). The mechanism of ciliary movement. *Proc. R. Soc.* **93**: 104–121.

Gray, J. (1928). *Ciliary movement.* Cambridge: Univ. Press.

Gray, J. (1931). The mechanism of ciliary movement. VI. Photographic and stroboscopic analysis of ciliary movement. *Proc. R. Soc.* **107**: 313–331.

Gray, J. A. and Ritchie, M. (1954). The effect of stretch on single myelinated nerve fibres. *J. Physiol., Lond.* **124**: 84–99.

Hayes, W. F. (1966). Chemoreceptors sensillum structure in *Limulus*. *J. Morph.* **119**: 121–142.

Holst, E. v. (1950). Die Arbeitsweise des Statolithenapparates bei Fischen. *Z. vergl. Physiol.* **32**: 60–120.

Horridge, G. A. (1965a). Non-motile sensory cilia and neuromuscular junctions in a ctenophore independent effector organ. *Proc. R. Soc.* **162**: 333–350.

Horridge, G. A. (1965b). Intracellular action potentials associated with the beating of the cilia in ctenophore comb plate cells. *Nature, Lond.* **205**: 602.

Kaissling, K. E. (1967). Ein mechanisch auslösbarer Transportmechanismus im Flimmerepithel von *Mytilus*. *Naturwissenschaften* **54**: 254.

Laverack, M. S. (1968). On superficial receptors. *Symp. zool. Soc. Lond.* No. **23**: 299–326.

Laverack, M. S. and Ardill, D. J. (1965). The innervation of the aesthetasc hairs of *Panulirus argus*. *Q. Jl microsc. Sci.* **106**: 45–60.

Lœwenstein, W. R. (1965). Facets of a transducer process. *Cold Spring Harb. Symp. quant. Biol.* **30**: 29–43.

Lowenstein, O., Osborne, M. P. and Wersäll, J. (1964). Structure and innervation of the labyrinth in the thornback ray (*Raja clavata*). *Proc. R. Soc.* **160**: 1–12.

Lucas, A. M. (1933). Comparison of ciliary activity under *in vitro* and *in vivo* conditions. *Proc. Soc. exp. Biol. Med.* **30**: 501–506.

Schneider, D. (1952). Die Dehnbarkeit der markhaltigen Nervenfaser des Frosches in Abhängigkeit von Funktion und Struktur. *Z. Naturf.* **7b**: 38–48.

Sjöstrand, F. S. (1953a). The ultrastructure of the outer segments of rods and cones of the eye as revealed by the electron microscope. *J. cell. comp. Physiol.* **42**: 15–44.

Sjöstrand, F. S. (1953b). The ultrastructure of the inner segments of the retinal rods of the guinea pig eye as revealed by electron microscopy. *J. cell. comp. Physiol.* **42**: 45–70.

Sleigh, M. A. (1962). *The biology of cilia and flagella.* Oxford, London, Paris, New York: Pergamon Press.

Sleigh, M. A. (1966). The co-ordination and control of cilia. *Symp. Soc. exp. Biol.* No. 20: 11–31.

Slifer, H. and Sekhon, S. S. (1963). Sense organs on the antennal flagellum of the small milkweed bug, *Lygaeus kalmii* (Hemiptera, Lygaeidae). *J. Morph.* **112**: 165–192.

Slifer, H. and Sekhon, S. S. (1964a). The dendrites of the thin-walled olfactory pegs of the grasshopper (Orthoptera, Acrididae). *J. Morph.* **114**: 393–410.

Slifer, H. and Sekhon, S. S. (1964b). Fine structure on the thin-walled sensory pegs on the antenna of a beetle *Popilius disjunctus* (Coleoptera; Passalidae). *Ann. ent. Soc. Am.* **57**: 541–548.

Thurm, U. (1963). Die Beziehungen zwischen mechanischen Reizgrößen und stationären Erregungszuständen bei Borstenfeld-Sensillen von Bienen. *Z. vergl. Physiol.* **46**: 351–382.

Thurm, U. (1964). Mechanoreceptors in the cuticle of the honey bee: Fine structure and stimulus mechanism. *Science, N.Y.* **145**: 1063–1065.

Thurm, U. (1965). An Insect mechanoreceptor. Part I: Fine structure and adequate stimulus. *Cold Spring Harb. Symp. quant. Biol.* **30**: 75–82.

Tomita, T. (1965). Electrophysiological study of the mechanism subserving color coding in the fish retina. *Cold Spring Harb. Symp. quant. Biol.* **30**: 559–566.

Wersäll, J., Flock, E. and Lundquist, P.-G. (1965). Structural basis for directional sensitivity in cochlear and vestibular sensory receptors. *Cold Spring Harb. Symp. quant. Biol.* **30**: 115–132.

Yoneda, M. (1960). Force exerted by a single cilium of *Mytilus edulis*, I. *J. exp. Biol.* **37**: 461–468.

Yoneda, M. (1962). Force exerted by a single cilium of *Mytilus edulis*, II. Free motion. *J. exp. Biol.* **39**: 307–317.

Symp. zool. Soc. Lond. (1968) No. 23, 217–249.

PROPRIOCEPTORS IN THE INVERTEBRATES

L. H. FINLAYSON

Department of Zoology and Comparative Physiology, University of Birmingham, Birmingham, England

SYNOPSIS

Most information about proprioceptors in invertebrates relates to the Arthropoda.

The sensory neurons forming the receptor elements in proprioceptors are uniterminal (formerly Type I) or multiterminal (formerly Type II). Except for the neurons of the chelicerate lyriform organ, arthropod uniterminal neurons are ciliary in ultrastructure, It is not know whether the bipolar sensory neurons of Annelida are uniterminal or multiterminal. Multiterminal neurons have many fine naked dendritic terminations, without any ciliary rudiments or specialised accessory structures.

Proprioceptors with uniterminal neurons include setae, campaniform sensilla, slit sensilla (lyriform organs) and chordotonal organs. Proprioceptors with multiterminal neurons include innervated connective tissue strands, muscle receptor organs of molluscs and arthropods, sub-epidermal neurons, neurons in the appendages of arthropods and various neurons associated with a variety of ordinary tissues of the body in cephalopods and in insects. Neurons in peripheral nerves of insects may be proprioceptors.

The insect abdomen has a range of possible proprioceptors that may include all the above types, except the lyriform organ which is found only in chelicerates, but which is probably the counterpart of the insect campaniform organ. Stretch receptors (with multiterminal neurons) range from innervated connective tissue strands to the complex muscle receptor organ of Lepidoptera. Preliminary studies of a chordotonal organ in the abdomen of the blowfly larva suggest that it is possibly a movement receptor but it may be a vibration receptor. The sub-epidermal neurons of blowfly larvae are slowly-adapting tonic receptors. The innervated connective tissue strands and muscle receptor organs are phasic-tonic receptors.

The insect leg also has a variety of proprioceptors. Muscle receptor organs have been found only in the bee but in cockroaches, termites and grasshoppers there are hair-plates, campaniform organs, multiterminal neurons and chordotonal organs. There is evidence that certain of the chordotonal organs of the leg are tonic receptors and play a part in reflex control of leg movement.

The crab leg has been thoroughly studied by several workers but only chordotonal organs have been reported. They monitor all possible movements of the leg joints and have units of several physiological characteristics (movement, position, unidirectional). The myochordotonal organ is unique among chordotonal organs in having an accessory muscle that can effect reflex adjustment of the receptor.

The chelicerate leg may have only multiterminal neurons, including those innervating a connective tissue organ in the trochanter of *Limulus*. Electron microscope studies are required to establish whether any of the bipolar neurons in chelicerate legs are uniterminal. Electrophysiological studies have shown that the proprioceptors of chelicerate legs, like those of the crab leg, have a wide range of responses to changes in position of the joints.

Several recent studies on moulting and digestion in insects indicate that multiterminal neurons on the wall of the gut which measure food intake are concerned in the control of these processes. In this way proprioceptors are involved in the activity of the neuroendocrine system. In *Rhodnius* unidentified stretch receptors in the body wall play a similar role in the control of moulting. Pupation in Lepidoptera is dependent on the body wall assuming the "normal" shape before the last larval moult. Distortion delays pupation and it is postulated that proprioceptors in the body wall are involved in the regulation of pupation. There is some evidence of a similar system in puparium formation in the tsetse fly.

INTRODUCTION

Most accounts of invertebrate proprioceptors deal with the receptors in the various taxonomic groups and most of the information is for the single phylum, the Arthropoda. Pringle (1961, 1963) and Bullock and Horridge (1965) classify the receptors as well as dealing with their distribution and physiology throughout the invertebrates. Dethier (1963) gives a detailed account of proprioceptors in Insecta; Hoffman gives an account of mechanoreceptors in all groups (1963) and proprioceptors in Arthropoda (1964); Alexandrowicz (1967) reviews muscle receptor organs in Crustacea. The aim of this review will be to survey recent work which adds to our knowledge of the range of proprioceptors in the invertebrates. Features of significance in the structure and topography of the receptors will be described and their main physiological properties discussed briefly.

The insect abdomen and the legs of arthropods of several groups will then be considered *in toto* as systems of proprioceptors. Finally, the role of proprioceptors in the control of digestion and developmental processes will be discussed. Statolith organs will not be included, for a discussion of their role in proprioception see Pringle (1963).

TYPES OF PROPRIOCEPTIVE SENSORY NEURONS

Like other invertebrate mechanoreceptors the proprioceptors are innervated by sensory neurons of two types. Type I neurons have a single sensory process and in Arthropoda are closely associated with specialized cuticular components. Type II neurons are typically multipolar and are not so intimately associated with the cuticle. These terms were first used by Zawarzin (1912) for those of Insecta but they have wider application. It is now valid to use more descriptive names for these neurons, based on recent work on their fine structure. Appropriate names are uniterminal sensory neuron (Type I) and multiterminal sensory neuron (Type II). Electron microscope studies of chordotonal organs in *Locusta* (Insecta) by Gray (1960) and in *Carcinus* (Crustacea) by Whitear (1962), of a campaniform organ in *Apis* (Insecta) by Thurm (1964) and of a hair-plate seta in *Apis* (Thurm, 1964, 1965) show that the distal process of the sensory neuron is similar in structure in all these sensilla. It is unbranched and has within it organelles characteristic of a cilium. The neurons of the lyriform organ of Arachnida are not obviously ciliary in structure but they are uniterminal (Salpeter and Walcott, 1960), and may be derived from ciliary neurons.

In Arthropoda uniterminal neurons are always found in association with auxiliary structures (scolopales, tubular bodies etc.). In the vertebrates such auxiliary structures are not found in association with sensory neurons that have ciliary derivatives and it is likely that invertebrate groups other than the Arthropoda possess uniterminal neurons unaccompanied by auxiliary structures. However, in the annelids *Harmothoë* and *Nereis* there are bipolar sensory neurons with a special structure at the end of the distal process, described as a "bifurcation" or "dot" by Horridge (1963), and as a "swelling" or "sensory process" by Dorsett (1964). It would be interesting to see whether these neurons are ciliary in structure and how their terminations compare with the auxiliary structures of arthropodan uniterminal neurons.

Multiterminal neurons are typically many-branched (multipolar) without any obvious association with the cuticle (Zawarzin, 1912). The electron microscope has revealed that the sensory processes of the multiterminal neurons have many fine terminations along their lengths (Osborne, 1963a, 1964; Osborne and Finlayson, 1965; Whitear, 1965) and no trace of any ciliary structure. No Type II neurons with a single sensory process have been studied with the electron microscope but they are presumably also multiterminal.

TYPES OF SENSE ORGANS WITH PROPRIOCEPTIVE FUNCTION

By far the greatest amount of information available on proprioceptors in the invertebrates is about those in Arthropoda. This is not surprising; the nature of the skeleton has contributed to the evolution of highly complex and well co-ordinated movements of body segments and appendages and therefore to the need for an efficient system of proprioceptors. Both uniterminal and multiterminal neurons are involved in proprioception, and in stretch receptors either type of cell may be the sensory element. Uniterminal neurons are found in setae, campaniform organs, chordotonal organs, lyriform organs and possibly in other receptors of Chelicerata. Multiterminal neurons terminate on a variety of ordinary tissues or may form the sensory element in an organized structure such as a connective tissue strand or a muscle receptor organ.

Receptors with uniterminal neurons

Setae

Tactile setae that are stimulated by the relative movements of parts of the body wall may provide proprioceptive information. The groups of fine setae forming the hair plates of insects at the thoracico-coxal and coxo-trochanteral joints of the legs, the neck and the base of the abdomen have been shown to be proprioceptors (see reviews by Pringle, 1961 and Hoffman, 1963, 1964). Similar hair plates have been described recently in the desert locust (*Schistocerca*) at the joints of the maxillary and labial palps and at the junction of mentum and prementum (Thomas, 1966).

Campaniform sensilla and slit sensilla

Pringle (1961) suggested that the campaniform sensilla of insects and the slit sensilla (lyriform organs) of chelicerates may be homologous in the sense of possessing uniterminal neurons and his hypothesis was confirmed by Salpeter and Walcott (1960) who showed that the neuron of the slit sensillum is uniterminal. The essential role played by campaniform sensilla in the control of flight in locusts has been described by Gettrup and Wilson (1964) and Gettrup (1965, 1966). When the campaniform sensilla, which occur in two groups on the subcosta of the forewing and one group on the subcosta of the hindwing, are destroyed or blocked electrically the reflex control of wing twisting and body orientation during flight is affected. The campaniform sensilla of the hindwing are essential for the regulation of forewing twisting during constant-lift and those of the forewing for the maintenance of stability of the body about its three axes.

Chordotonal organs

That the nomenclature of this group of receptors has presented biologists with etymological problems is evident from the variety of names, mostly derived from the Greek word *scolops*, that have been applied to them. The history and etymology of the names are reviewed by Whitear (1962). The following terms are suggested as being sufficient for reference to these receptors.

Scolopale (plural: scolopalia or scolopales)—a structure secreted within that region of the scolopale cell that surrounds the tip of the dendrite of the sensory neuron. This is the refractive "sense rod" of light microscopy.

Scolopidium (plural: scolopidia)—a group of cells, of which one or two are bipolar, ciliary, sensory (Type I) neurons and one is a scolopale cell.

Chordotonal sensillum (plural: sensilla)—one or more scolopidia forming a discrete sense organ or a unit of a larger organ.

Chordotonal organ—a sense organ composed of one or more chordotonal sensilla.

This nomenclature is based on anatomy and makes no physiological assumptions. Any attempt to restrict the use of the term "chordotonal" to vibration receptors would be impractical.

Chordotonal organs that function as proprioceptors consist of a connective tissue strand with one or more scolopidia attached to it. Although chordotonal organs were first described in insects (Graber, 1882) and their histology and topography studied extensively in that group, recent work on the proprioceptive role of chordotonal organs has been carried out mainly with Crustacea (e.g. Bush, 1962a, b, 1963, 1965a,b,c; Cohen, 1963, 1965; Wyse and Maynard, 1965).

Although a few chordotonal organs had been known in Crustacea (Wetzel, 1934; Barth, 1934) it was not until the electron microscope was used to study innervated elastic strands in the legs of Decapoda that the chordotonal nature of many of the sense organs was revealed (Whitear, 1962). In insects there are chordotonal organs in antennae, legs and all body segments; in crustaceans they have been found so far only in antennae and legs but their distribution may be wider.

Receptors with multiterminal neurons

Multiterminal neurons may be associated with a variety of ordinary tissues of the body or they may form part of a distinct organ which has an accessory structure or structures as well as one or more multiterminal neurons. Commonly the endings of the neuron are embedded in connective tissue, for example at a limb joint (D. M. G. Guthrie, personal communication) or in the connective tissue component of a muscle receptor organ (Whitear, 1965; Osborne and Finlayson, 1965). Although some of the multiterminal neurons in the classification that follows have not been investigated by physiological methods they are included because they would appear from their position in the body to be proprioceptors.

Multiterminal receptors with a special accessory component

The accessory component consists of a strand of connective tissue or a special receptor muscle or a combination of connective tissue and muscle (Fig. 1). Sense organs of this type are found in Crustacea (Alexandrowicz, 1967), Chilopoda (Rilling, 1960), Insecta (Slifer and Finlayson, 1956; Finlayson and Lowenstein, 1958; Osborne and Finlayson, 1962), Chelicerata (Barber and Hayes, 1964) and Mollusca (Alexandrowicz, 1960).

With the exception of the thoracico-coxal organs of decapod crustaceans in which the cell body is presumed to lie in the central nervous system (Fig. 1, f, h), these organs are innervated by neurons in which

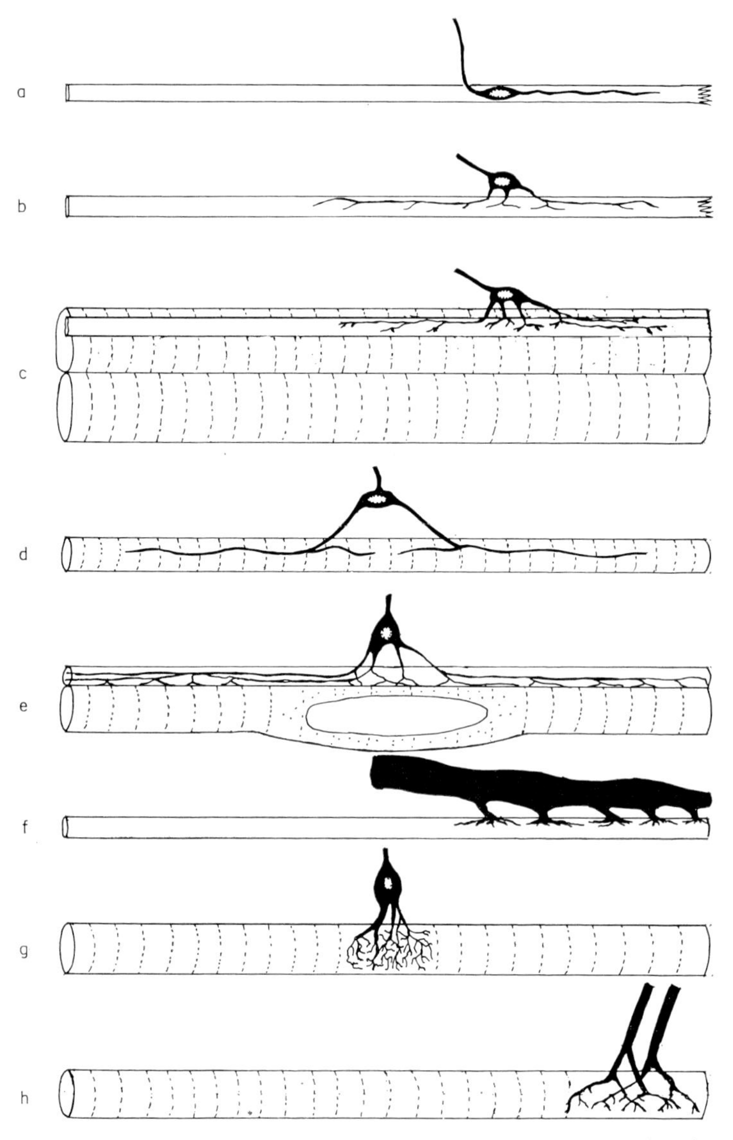

FIG. 1. Diagrams to show the relationships of sensory neuron, connective tissue and muscle in stretch receptors in Insecta (a–e) and Crustacea (f–h).

a, connective tissue strand, bipolar neuron (Plecoptera): b, connective tissue strand, multipolar neuron (Odonata, Dictyoptera, Phasmida, Dermaptera, Hymenoptera, Diptera): c, connective tissue on longitudinal muscle band (Orthoptera): d, receptor muscle (Neuroptera): e, connective tissue and receptor muscle, giant muscle nucleus (Trichoptera, Lepidoptera): f, connective tissue, neuron cell body in CNS (Crustacea): g, receptor muscle, connective tissue at central neuronal attachment (Crustacea): h, receptor muscle, connective tissue terminal in region of dendrites, neuron cell body in CNS (Crustacea).

the cell body is in the periphery (Alexandrowicz, 1967). All the organs of this type that have been investigated experimentally are stretch receptors.

Since the structure and topography of stretch receptors with multi-terminal neurons in Crustacea has been reviewed by Pringle (1961), Hoffman (1964), Bullock and Horridge (1965) and by Alexandrowicz (1967), they will be treated briefly. There are two distinct kinds of muscle receptor organ; one situated at the articulation of the coxa with the thorax and the other in a dorsal, longitudinally arranged series in the abdominal and posterior thoracic segments. The dorsal series consists typically of a pair of organs in each segment. An organ consists of two receptor muscles each bearing a large multipolar neuron at its central region. In addition to excitatory motor fibres each receptor receives one or more inhibitory fibres.

In addition to the muscle receptor organ there is a system of elastic strands. In the crabs, *Carcinus* and *Cancer*, a pair of these strands flank the muscle receptor organ and, like it, are innervated by sensory processes originating in the central nervous system. In the lobster, *Homarus*, and the crayfish, *Astacus* there is a complex of four innervated elastic strands, not so closely associated with the muscle receptor organ.

Muscle receptor organs in centipedes (Chilopoda) have been described by Rilling (1960). They resemble certain crustacean muscle receptors (Alexandrowicz, 1967) in having several neurons on one receptor muscle. In *Lithobius* muscle receptor organs are found throughout the body; each has about 10 neurons with axons running to two ganglia.

In most of the insects that have been studied (Finlayson and Lowenstein, 1958; Slifer and Finlayson, 1956; Osborne and Finlayson, 1962; Osborne, 1963a) the stretch receptors of the body segments consist of a strand of connective tissue with the dendrites of a single sensory neuron "ramifying" within it. The cell body is normally situated in or on the connective tissue strand but it may lie on the neighbouring epidermis or in the nerve itself (Osborne and Finlayson, 1962). The neurons of both vertical and longitudinal receptors in *Dictyopterygella* (Plecoptera) differ from those described for all other insect stretch receptors in being bipolar (Fig. 1a), but this difference is probably not of great significance, being merely a reduction of the several processes of the typical multipolar cell to a single one.

In all the insects in which stretch receptors have been found there is invariably a dorsal, longitudinal pair in each full-size, abdominal segment. Although the thorax has not been so intensively studied, dorsal longitudinal stretch receptors have been found in the thorax of adult stick insects (Slifer and Finlayson, 1956) stonefly larvae (Osborne and Finlayson, 1962) moth larvae (Lowenstein and Finlayson, 1960) and blowfly larvae (Osborne, 1963b). The stick insect, being apterous, is not typical of adult insects and so the evolutionary fate of the series of dorsal longitudinal receptors in the thorax of winged insects is not known. Gettrup (1962, 1963) has described stretch receptors in the thorax of the desert locust (*Schistocerca*) and Wilson and Gettrup (1963) suggest that they may be homologous with the dorsal abdominal series. Moulins (1966) has described a complex structure in the hypopharynx of the cockroach (*Blaberus*) that may be a stretch receptor. It consists of a horizontal sheet of connective tissue, an unpaired multipolar neuron and two lateral groups of three bipolar multiterminal neurons. The single process from each of these three cells is short and gives rise to dendrites that ramify on the connective tissue sheet near its point of anchorage to the hypopharyngeal epidermis. At least one of the dendrites of the single multipolar neuron runs on the connective tissue sheet and the others run dorsally to the epidermis of the hypopharynx. Moulins did not find a corresponding neuron on the left side. This asymmetrical arrangement is unique among peripheral sensory systems The axon of the multipolar neuron (A-cell) terminates in the brain, that of the lateral groups (B-cells) in the sub-oesophageal ganglion. Moulins suggests a possible mode of action of this unusual receptor complex but there is not yet any physiological information. His hypothesis is that the curvature of the hypopharyngeal cuticle varies during the ingestion of food and the sheet of connective tissue is stretched or relaxed by these changes in curvature. He suggests that the A-cell and the B-cells are all stimulated when the connective tissue is stretched but that relaxation of the connective tissue stimulates only the A-cell via its dendrites that run beyond the sheet of connective tissue to the body wall. The B-cells will cease to be stimulated when the connective tissue is slackened.

In the coxa of each leg of the honey bee there is a muscle receptor organ of an unusual type (Markl, 1965). The muscle, which consists of four to six fibres, is inserted on the anterior, proximal wall of the coxa and on the joint membrane at the junction of coxa and trochanter. At the joint the muscle is attached to the joint membrane by 15 to 25 spindle-shaped cells. Surrounding them is a compact group of 40 to 50 "sheath" cells. A single axon runs through the group of sheath cells, to a neuron which terminates among the spindle-shaped attachment cells. Markl does not describe the neuron but it is presumably multiterminal.

In *Limulus* (Chelicerata) there is an innervated connective tissue strand spanning the trochanter of the legs (Barber and Hayes, 1964) and attached to the cuticle at each end. Because a small accessory muscle is attached to the strand and it resembles a vertebrate tendon Barber and Hayes called the organ a "tendon receptor organ". About 30 large neurons, some multipolar, are embedded in the anterior end of the organ. In the same region are embedded about 20 isolated bodies measuring between 30 μ and 200 μ in cross-section. They are ovoid to cylindrical in shape and form a band across the tendon near its anterior insertion. They appear to be cuticular in origin and may be detached apodemes.

Multipolar neurons associated with receptor muscles have been found in the mantle of *Octopus* (Sereni and Young, 1932) and *Eledone* (Alexandrowicz, 1960). The mantle neurons of *Eledone* and *Octopus* are scattered over a plexus of thin muscles near the stellate ganglion. There are at least 50 neurons on each side of the body. Alexandrowicz (1960) considers that the muscle plexus plus neurons constitutes a muscle receptor organ and called it the "substellar organ".

Multiterminal receptors without a special accessory component

Sub-epidermal multiterminal neurons. It has been known for a long time that there are multiterminal neurons on the inner surface of the epidermis of the body wall in arthropods (Viallanes, 1882) but the first comprehensive description of their topography was given by Osborne (1963b) for the larva of the black blowfly (*Phormia*). Other workers have described parts of the system; references to their work are given by Alexandrowicz (1957) and Osborne (1963b, 1964).

The function of the sub-epidermal plexus of neurons is not known but it is presumed to be proprioceptive as well as exteroceptive. Where-ever it has been possible to record from multiterminal neurons they have proved to be mechanoreceptors and to be mainly proprioceptive in function: Insecta (Finlayson and Lowenstein, 1955), Crustacea (Wiersma, Furshpan and Florey, 1953), Annelida (Horridge, 1963).

The sub-epidermal multiterminal neurons of the blowfly larva are remarkably constant in their topography (Fig. 2). The number of neurons in each segment is always the same; 24 in the prothorax, 28 in both the mesothorax and the metathorax and 30 in each abdominal segment. The cell bodies fall roughly into 3 groups on each side of the body. A similar arrangement is found in the larva of the tsetse fly (L. H. Finlayson, unpublished).

Multiterminal neurons in the appendages of arthropods. Richard (1950, 1951) found multiterminal neurons in the femur and tibia of the leg, and in the labrum, labium and hypopharynx of the termite, *Calotermes*. Some are in positions that suggest a proprioceptive function but others seem more likely to be chemoreceptors. Denis (1958) found another in the coxa of the same termite. Similar neurons have been described in the coxa and femur of a caddis larva (Trichoptera) by

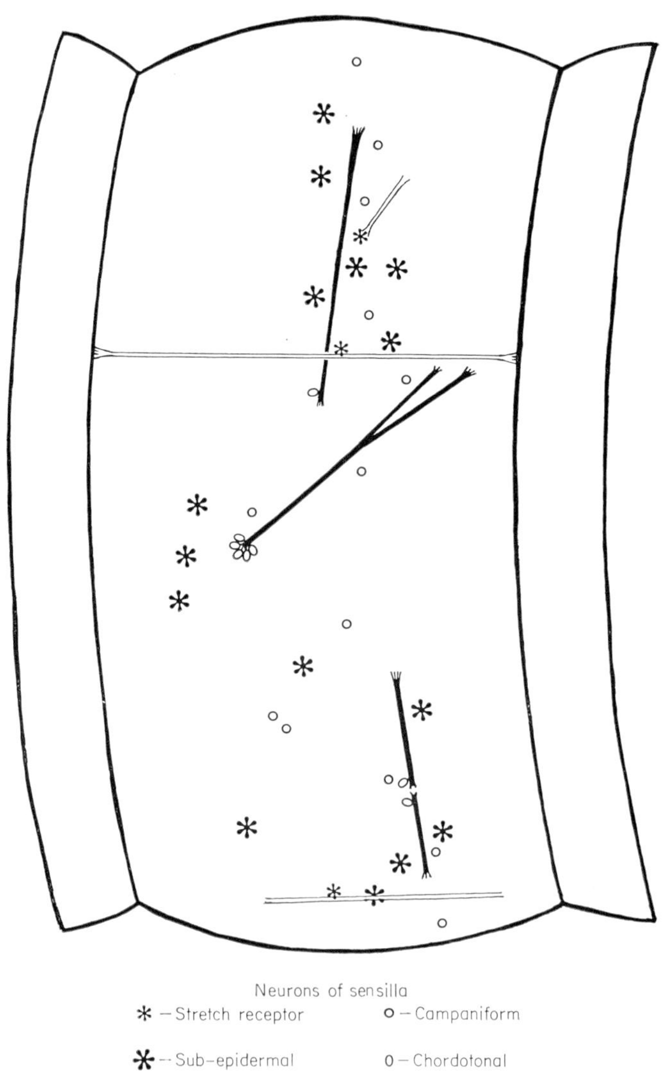

FIG. 2. Diagram to show location of possible proprioceptors in one side of one abdominal segment of blowfly larva (based on Osborne, 1963b). Chordotonal organs shaded, multiterminal stretch receptors unshaded.

Barbier, 1961. Their processes run towards the coxo-trochanteral and femoro-tibial joints. Multiterminal neurons have also been found in the leg of the cockroach (*Periplaneta*) by D. M. G. Guthrie (personal communication). A large neuron with at least 20 branching processes lies at the end of the femur near the anterior ventral condyle. Its processes appear to penetrate the soft cuticle of the joint. Two smaller cells with two or three unbranched processes lie above the postero-dorsal condyle. At the trochanter-femoral joint there is one multi-polar neuron with three or four processes which divide into many fine branches that end in small clubbed terminations.

Multipolar neurons are present in the leg joints of *Limulus* (Pringle, 1956; Barber, 1960). In the femoro-tibial joint Barber and Segel (1960) found two groups of sensory neurons, about 30 embedded in the connective tissue of the joint and sending processes among the epithelial cells, and about 10 within the nerve trunk. In the centipede (*Lithobius*) similar multipolar neurons are found in the legs (Rilling, 1960).

Multiterminal neurons with endings on "ordinary" tissues. In this category the best-known examples are the N-cells of Crustacea Alexandrowicz, 1952, 1956; Wiersma and Pilgrim, 1961; Pilgrim, 1964), which terminate on various thoracic muscles. Alexandrowicz (1956) and Wiersma and Pilgrim (1961) consider that on morphological and physiological grounds the N-cells represent degenerated muscle receptor organs in which the specialised receptor muscle (RM) has disappeared and the cells have become attached to ordinary muscles.

Multiterminal neurons occur in large numbers in the muscles, arms. lips and suckers of Cephalopoda (Graziadei, 1965a,b) but no physiological studies of them have been made.

In the insects there are peripheral multipolar neurons with dendrites terminating on a variety of tissues. Although these cells resemble the N-cells of crustacea in that they are not associated with a special receptor muscle or any orientated connective tissue component there is no reason to suppose that they are derived from more highly organized receptors as Wiersma and Pilgrim (1961) suggest for the N-cells in crustacea. Neurons of this category have been described by Osborne (1963b) and Whitten (1963) in the lateral region of each abdominal segment of the fly larva. The basic number of these cells is three and their topographical distribution in the tsetse larva (L. H. Finlayson, unpublished) is shown in Fig. 3. The axon of one of these cells (mn^1) enters a branch of the median (unpaired) nervous system (Osborne, 1963b). One neuron (mn^2) has no less than five processes, one of which innervates the spiracular trachea (Whitten, 1963; Osborne, 1963b). The second innervates a trachea, the third runs into a motor nerve and

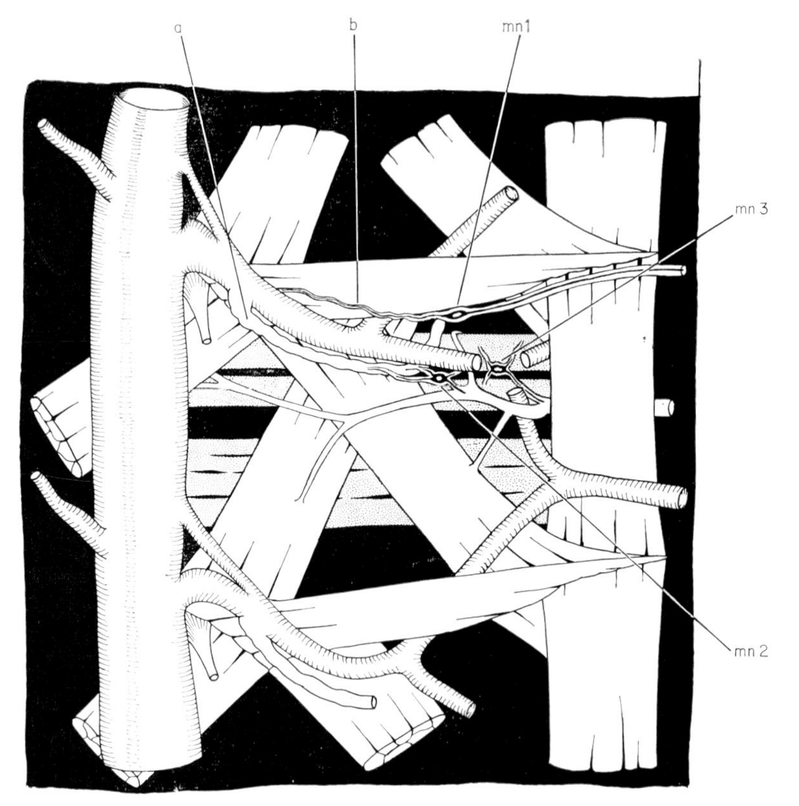

FIG. 3. Diagram of part of an abdominal segment of the larva of the tsetse fly (*Glossina morsitans*) to show the location of three multiterminal neurons (mn 1, mn 2, mn 3).

a, spiracular trachea; b, distal process of mn 1.

the fourth runs towards cell mn^1. The third neuron (mn^3) lies within the nerve that innervates the lateral transverse muscles. One dendrite of mn^3 runs over the epithelium of a trachea and the other processes apparently innervate the nerve and its branches. Nothing is known of the function of these lateral neurons of fly larvae. The spiracular muscle neuron of the adult tsetse fly may be the homologue of mn^2 (Finlayson, 1966).

In the locust (*Schistocerca*) Guthrie (1962) found a pair of large multipolar neurons with dendritic endings on 15–20 muscle fibres of the ventral diaphragm. He obtained records of small groups of impulses at intervals of four or five seconds from the ventral nerve cut at the ganglion. These impulses disappeared when the preparation was cut in the region of the axon of the sensory cell and may have originated in the cell. L. H. Finlayson and M. P. Osborne (unpublished) found in the stick insect (*Carausius*) a pair of large multipolar neurons in a similar position to those in *Schistocerca* but without any connection with muscle fibres. There is no ventral diaphragm in *Carausius*. In *Carausius* the cell body lies near the inner margin of the ventrolateral sheet of fat body and its dendrites ramify for a long distance over the fat body.

In the auditory organ of noctuid moths there is a single multipolar neuron lying in a cuticular structure known as the Bügel (Treat and Roeder, 1959). The lepidopteran auditory organ differs from the auditory organs of most insects in possessing only two chordotonal organs. They are attached to the Bügel via the nerve that carries their axons and the axon of the Bügel neuron. The function of this neuron is not known but Treat and Roeder (1959) suggest that it may be proprioceptive. The discharge frequency changes with alterations in the shape of the tympanic membrane. In life these changes could be brought about by the action of the two tympanic muscles or by distortions produced in the thoracic skeleton during flight.

Multiterminal neurons on the alimentary canal. Zawarzin (1916) and Orlov (1924) described multiterminal neurons on the wall of the gut in insects. Little work has been done, since then, on this system of neurons but there is evidence that they function as stretch receptors. In the locust there are two rows of three multiterminal neurons on the dorsolateral wall of the pharynx (Clarke and Langley, 1963). Their axons enter the paired posterior pharyngeal nerves which originate in the frontal ganglion.

Langley (1965) has described a similar system consisting of one medial and two lateral sensory neurons in the proventricular region of the gut in the adult tsetse fly (*Glossina*).

Multiterminal neurons in nerves. Alexandrowicz (1953) described bipolar and tripolar neurons in the trunks of motor nerves of Stomatopoda (Crustacea). Similar cells occur in the nerves of insects such as the tsetse fly (*Glossina*) and the stick insect (*Carausius*). It is not known whether the processes terminate within the nerves or pass out of the nerves to other tissues except in the case of the spiracular muscle neuron of the adult tsetse fly (Finlayson, 1966). It lies in the nerve and in some segments at a considerable distance from the muscle. One of its processes runs to the spiracular muscle, but it has at least three others whose destination is unknown. While it is possible that these neurons may be trophic rather than sensory in function, or may even be internuncials, there is evidence from work on similar neurons in the labellar nerve of the blowflies *Phormia* and *Calliphora* that they are sensory. In the labellar nerve of blowflies there are about 30 multipolar neurons

(Peters, 1962). In oscillographic studies Stürkow, Adams and Wilcox (1967) recorded responses from these neurons to mechanical stimulation in the form of movements of the proboscis, bending hairs or changing the rate of flow of solutions being used to stimulate chemoreceptors.

Near the femoro-tibial joint of *Limulus* (Chelicerata) there are about 10 neurons within the nerve trunk. It is not known where their dendrites terminate (Barber and Segel, 1960).

PROPRIOCEPTIVE SYSTEMS IN ARTHROPODA

Most anatomical and physiological work on proprioceptors has been concerned with one sense organ or the control of movement at one articulation of an arthropod limb. Although it is too early to attempt a synthesis of proprioceptive function which includes all the sense organs involved in the control of the movements of a limb or a body segment, it may be worthwhile looking at some of the information that is available as a guide to further investigation. Again most of our information relates to Arthropoda and it is from that phylum that examples will be drawn.

The insect abdomen

It is likely that all the sensilla involved in proprioception in abdominal segments of insects have been identified. They include campaniform organs, chordotonal organs, stretch receptors (innervated connective tissue strands and muscle receptor organs) and probably sub-epidermal and other multiterminal neurons. Of these only the stretch receptors have been the subject of neurophysiological studies (Finlayson and Lowenstein, 1958; Lowenstein and Finlayson, 1960; Weevers, 1965, 1966a,b,c,d).

The dorsal longitudinal series of stretch receptors is so widespread in insects that it must have been derived from their ancestors or have originated at an early stage in the evolution of the group. It would be of interest to see whether it is present in the Apterygota. A similar series of longitudinal receptors is found in Crustacea (Alexandrowicz, 1967) but it is doubtful if strictly morphological criteria can be applied in any attempt to homologize such structures within the Arthropoda because peripheral multiterminal neurons are common and may have become incorporated into orientated stretch receptors independently on many occasions in the course of evolution.

The other series of stretch receptors that is widespread but not universal in insects is the "vertical" series. The vertical receptor is slung between a point of attachment on the epidermis of the dorsal body wall and a dorsal nerve or dorsal longitudinal muscles and it lies at a considerable angle to the dorsal longitudinal receptor. Although both dorsal longitudinal receptor and vertical receptor commonly consist of connective tissue plus neuron (Fig. 1) it is significant that only the dorsal longitudinal receptor ever has a muscular component (as in Orthoptera, Neuroptera, Trichoptera and Lepidoptera). It is also of interest, and no doubt of functional significance, that in those insects in which the dorsal longitudinal receptor has a muscular component the vertical receptor is absent except in the larva of the alder-fly *Sialis* (Neuroptera) and that in *Sialis* the vertical receptor is reduced to the neuron alone, without even a connective tissue component. It seems that the vertical receptor tends to disappear when the longitudinal receptor becomes associated with muscle. Another morphological variant is the position of attachment of the two ends of the dorsal longitudinal receptor. In some insects the receptor is slung between the intersegmental folds and in others the anterior attachment is to the tergal epidermis and the posterior attachment to the intersegmental fold. All the receptors with a muscle component are slung between the intersegmental folds.

Although there is physiological evidence (Wigglesworth, 1934; 1957; Van der Kloot, 1961; Bennet-Clark, 1966) for the existence of sense organs that respond to stretch in *Rhodnius* (Hemiptera) they have not been identified anatomically (Osborne and Finlayson, 1962).

In addition to the dorsal longitudinal and vertical series there is a series of paired ventral longitudinal receptors in the blowfly larva (Osborne, 1963b). Each consists of a tubular connective tissue strand plus neuron. The ventral receptor also spans the entire segment but differs from the dorsal longitudinal receptor in its mode of attachment. It is attached posteriorly to the outer edge of an oblique muscle and anteriorly by a three-point anchorage, two points on the epidermis of its own segment and one point in the adjacent segment on the inner edge of an oblique muscle. At both anterior and posterior attachment to oblique muscles a single muscle-fibre enters the strand of connective tissue.

When an isolated innervated connective tissue strand or a muscle receptor organ is stretched, the neuron responds by a marked increase in impulse frequency. When the new length is achieved and the stretching movement ceases, the frequency of firing decreases sharply and then settles to a new level higher than the original resting level (Finlayson and Lowenstein, 1958; Tashmukhamedov, 1961; Wendler and Burkhardt, 1961; Wendler, 1963; Weevers, 1966a,b). During the period of movement of the receptor the rate of firing is higher at any point than it would be at that length during maintained stretch (Weevers, 1966b). An increase in velocity of stretching elicits a corresponding increase in discharge frequency and a decrease in velocity, a corresponding reduction in discharge frequency or a temporary cessation of impulses if the deceleration is sufficiently rapid (Lowenstein and Finlayson, 1960; Weevers, 1966b).

In the lepidopteran muscle receptor organ, Weevers (1966b) recognizes three components of the response to stretching, "position", "movement" and "acceleration". He showed that increasing the velocity during stretching produces an increase in discharge frequency and he points out that in this respect the lepidopteran muscle receptor organ differs from the vertebrate muscle spindle in which such an "acceleration" effect is not seen. If stretching continues at constant velocity the acceleration peak discharge is reached and then the impulse frequency drops to the level characteristic of the "movement" response. It is related to the velocity and the amplitude of stretch but is greater than the "position" response for a corresponding maintained length of the receptor.

The discharge rate of the lepidopteran muscle receptor organ is extremely regular and has an adaptation rate which is lower than that of the crustacean muscle receptor organ or the vertebrate muscle spindle. The amount of precise information they can transmit to the central nervous system is, therefore, exceptionally high (Weevers, 1966b).

Insect stretch receptors, whether they are muscle receptor organs or have only connective tissue in the accessory component, are phasic-tonic (Lowenstein and Finlayson, 1960; Weevers, 1966a,b). The proprioceptive role of phasic input from a connective tissue strand organ has been shown by Gettrup (1962, 1963) for the organ at the base of the wing in the locust.

Two sorts of reflex response have been observed following stimulation of the lepidopteran muscle receptor organ (Weevers, 1966d): self-inhibition of the receptor muscle and negative feedback to body muscles which stimulates them to resist extension. The self-inhibitory reflex of the tonic discharge is brief in duration and probably serves to protect the organ from being damaged during violent stretching. Recordings of muscle potentials from body wall muscles of the caterpillar show that stretching the muscle receptor organ leads to contraction of muscles in parallel with the muscle receptor organ and relaxation of others so that the receptor will be "unloaded". The muscles that lie in parallel with the receptor are more strongly excited than any others. Weevers (1966d) found that at least 32 motor units respond to stretching of a single muscle receptor organ. He found also that stretching both the receptors in an abdominal segment simultaneously reinforced the stretch reflex to a level of muscular activity greater than the sum of the activities resulting from independent stimulation of the two receptors.

If the muscle receptor organ is held isometric by an imposed load and the transitory autoinhibitory period has elapsed, the returning tonic motor discharges causing the receptor muscle to contract will have the effect of increasing the sensory discharge, and reflexly of increasing the unloading activity of the body muscles. When unloading takes place the muscle receptor organ slackens off but the tonic reflex discharge to the receptor muscle causes it to "take up the slack".

Ablation of muscle receptor organs in the caterpillar has little effect on the peristaltic crawling waves, which pass over segments deprived of their muscle receptor organs. Weevers (1965, 1966d) concluded that the normal peristaltic crawling waves of muscular activity are independent of muscle receptor organ reflexes but may be modified by them in the intact animal. In this respect the control of crawling in the caterpillar resembles the control of wing-beat frequency in the locust (Wilson and Gettrup, 1963); both are set by a central rhythm but may be modified by proprioceptive input. The principal function of the caterpillar muscle receptor organ appears therefore to be to resist extension of the segments.

Recordings from isolated preparations of chordotonal organs of the insect abdomen have not been reported in the literature. L. H. Finlayson and M. P. Osborne (unpublished) have recorded from the dorsal vertical chordotonal organ (Fig. 2) of the blowfly larva (*Phormia*). This organ has one neuron. Preliminary studies show that it is a rapidly-adapting phasic receptor and is stimulated by vibrations. While these results do not rule out the possibility that this organ is a proprioceptor, responding to movements of the body, its extreme sensitivity to substrate vibration makes a proprioceptive function unlikely.

Recordings have also been obtained from the two most dorsal sub-epidermal multiterminal neurons of the blowfly larva. They are very different in their activity from the chordotonal organ. They fire continuously and can be stimulated to increase the frequency of discharge by pulling on the cell body. We have not shown conclusively that they are stimulated by distortions of the body wall but this would appear to be their function. They may be exteroreceptors in that they will respond to deformations of the body wall caused by external agencies but they will be proprioceptive in responding to alterations in the shape of the body brought about by the activity of the muscular system.

The insect leg

The proprioceptors of the insect leg are hair plates, campaniform organs, chordotonal organs and multiterminal neurons. Figure 4 shows in diagrammatic form the diversity of proprioceptors that may be present in the leg of an insect. The diagram is based on those of the

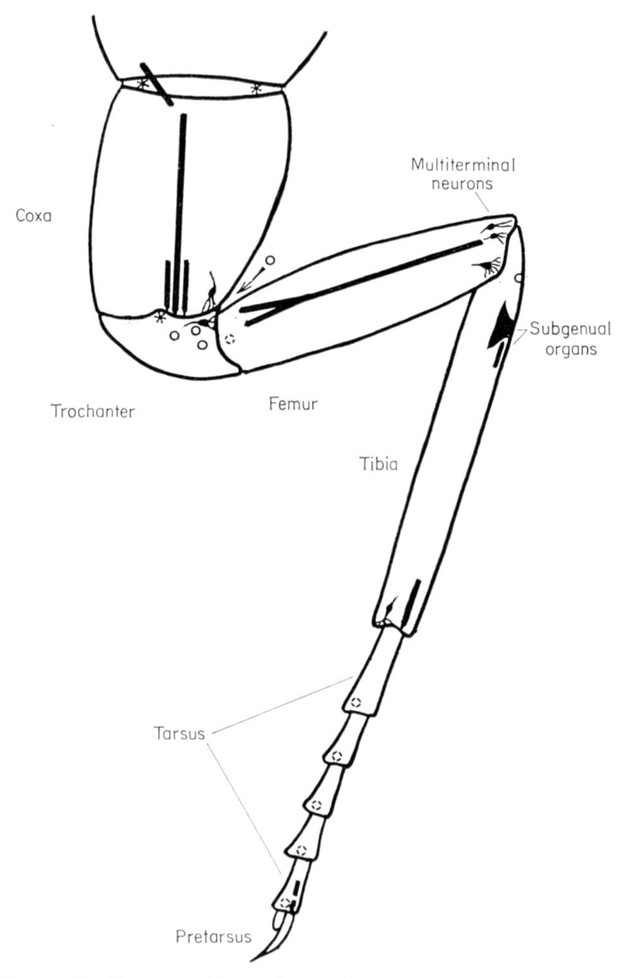

FIG. 4. Composite diagram of insect leg to show all known receptors that may have proprioceptive function (except isolated setae).

*—hair plates (Pringle, 1938b); o—groups of campaniform sensilla (Pringle, 1938a); chordotonal organs (shaded): thoracico-coxal and coxo-trochanteral (Nijenhuis and Dresden, 1955), femoro-tibial, proximal and distal sub-genual and tibio-tarsal (Debaisieux, 1938), tarsal and tarso-pretarsal (Slifer, 1936); multiterminal neurons: coxo-trochanteral (Denis, 1958), trochantero-femoral and femoro-tibial (Guthrie, unpublished), tibio-tarsal (Richard, 1950).

cockroach (Pringle, 1938a,b; D. M. G. Guthrie, personal communication), termite (Richard, 1950; Denis, 1958; Barbier, 1961) and grasshopper (Slifer, 1936).

Hair plates are slowly-adapting, position receptors that register certain positions at the joints of appendages and body articulations (see Hoffman, 1964). On the leg of the cockroach there are three hair plates, the inner coxal, the outer coxal and the trochanteral hair plate (Pringle, 1938b). It is significant that these hair plates are situated at the base of the leg where a small change in position of the joints produces a large change at the distal end of the leg. They are probably especially sensitive to small changes in joint position.

Campaniform organs are more widely distributed but are especially numerous on the trochanter where, in the cockroach, there are three ventral and one dorsal groups, containing a total of about 70 sensilla (Pringle, 1938a). The function of campaniform sensilla is to monitor stresses in the cuticle and not changes in position of the joints, although changes in joint positions may produce stress in the cuticle which will be picked up by the campaniform sensilla. Probably the major function of the campaniform sensilla is to register stresses produced (*a*) by the weight of the insect's body on the limb, (*b*) by the resistance of the cuticle to the actions of muscles and (*c*) by external agencies that tend to alter the normal spatial relationship between different regions of the legs or between the legs, at rest or in motion, and the body. Important among the last mentioned would be unevenness of substrate, restricted passage through crevices, etc. and the differences between leg and body relationships on the ground and during flight or swimming. It would be of interest to discover whether the campaniform sensilla of a group have different sensitivity ranges and could therefore measure varying degrees of deformation of the cuticle.

Chordotonal organs have been found at each major joint of the insect leg (Debaisieux, 1938; Nijenhuis and Dresden, 1952, 1955; Slifer, 1936). Becht (1958) made recordings from three coxo-trochanteral organs and found that they have a resting discharge of about 50 impulses per second. They are stretched by extension of the joint and respond to the movement with an increase in discharge frequency to about 250 impulses per second, dropping to about 100 impulses per second as the extended position is maintained. Hubbard (1959) made a preliminary investigation of the femoro-tibial chordotonal organ of the locust, but reported only that the two parts of the organ appeared to have widely different thresholds and adaptation rates.

Bässler (1965) stimulated the femoro-tibial chordotonal organ of the stick insect (*Carausius*) by passive movement of the leg or by stretching or relaxing the chordotonal organ directly via a hole cut in the distal end of the femur. When the "tendon" (apodeme) of the organ is pulled, the animal extends the femoro-tibial joint; when the tendon is allowed to relax it bends the joint. If the tendons of all the organs are severed the animal, when placed on a vertical surface, sags under the influence of gravity; if the tendons of the three legs on one side are severed the animal leans towards the operated side. A load applied to the leg is resisted when the chordotonal organ is intact to a much greater degree than when the tendon of the organ is severed. The femoro-tibial chordotonal organ of the stick insect resembles those of the crab leg in mediating a "resistance reflex". Bässler (1965) presents a detailed analysis of the interaction of light and gravity in normal and operated animals and shows how important are the chordotonal organs and the hair plates of the legs in proprioception and orientation in the stick insect.

D. M. G. Guthrie (personal communication) made recordings of nerve impulses from the region of the femoro-tibial joint in the cockroach leg. He obtained good evidence that the large multiterminal neuron fires at a low frequency, is stimulated by the extension of the tibia, and is slowly adapting. Other brief, fast-adapting responses to the same stimulus may have originated in the two smaller neurons whose axons enter the same nerve as the axon of the large neuron.

The insect leg has proprioceptors sensitive to changes in length (chordotonal organs) and others sensitive to strain (campaniform organs) (Pringle, 1961). The multiterminal receptors are probably of two types (*a*) position receptors that may be complementary to chordotonal organs or may provide information about angular deflections of the joint which the chordotonal organ cannot discriminate; (*b*) movement receptors which provide the information that the joint is beginning to move, followed by information about the velocity of the movement. The multiterminal neurons of the leg send their processes into the flexible region of the joint and may respond to the first signs of movement before the cuticle is sufficiently stressed to stimulate the campaniform organs.

Much information is still required before a complete picture of the proprioceptive system of the insect leg can be painted. There may be more multiterminal neurons than have been described to date. The main problem is the division of labour between the different types of sensilla and their roles in the reflex control of leg movements and muscle tone.

The crab leg

Most of the recent anatomical and physiological work on proprioceptors in crustacean legs has been carried out on shore crabs (*Carcinus*).

At each joint there is one or more chordotonal organs. All but one of the organs have been investigated by neurophysiological techniques and a fairly complete picture is available of the effective stimuli, sensory neuron discharges and reflex responses from motoneurons.

Each organ is known by the initial letters of the leg segments that articulate at the joint spanned by the organ (Whitear, 1962) but the organ of Barth (1934) which does not span the joint whose movement it monitors is called the myochordotonal organ or MCO (Bush, 1965c). It is coincidental that the joint served by the MCO is the meropodite-carpopodite joint (MC) although the main group of sensory neurons lies within the ischiopodite (Cohen, 1963). The MCO is not to be confused with the two simpler chordotonal organs, MC1 and MC2, that span the meropodite-carpopodite joint. The IM organ has not yet been shown by histological methods to be a chordotonal organ or been investigated physiologically but there is no reason to suppose that it differs from those at the other joints.

Both movement receptors and position receptors have been identified at all the joints between coxopodite and dactylopodite (Burke, 1954; Bush, 1965c). Units responding unidirectionally either to stretch or to relaxation of the sense organ have been found within movement and position receptors except in MC2 and CP2 where Bush (1965a) could find only relaxation units. In other organs there are different relative proportions of stretch and relaxation units (Table I). The proprioceptive information from the meropodite-carpopodite chordotonal

TABLE I

Properties of chordotonal organs of crab leg

Organ	Joint movement stretching organ	Response Phasic	Response Static
MCO (Main organ*)	Flexion	Rs	S
(Proximal organ†)	Extension	S	S
CB	Extension	RS	RS
IM	Extension	Not known	
MC1	Flexion	Rs	RS
MC2	Flexion	R	R
CP1	Production	Rs	Rs
CP2	Reduction	R	R
PD	Flexion	RS	RS

R = Fires on relaxation; S = Fires on stretching.
Relative sizes of initial letters indicate relative sizes of response.
(Bush, 1965a; * Cohen, 1963; † Hwang, 1961.)

organs of the crab leg, including the myochordotonal organ, is greater and more precise for extension than for flexion of the joints. As Bush (1965a) suggests, the flexor muscles of the leg are responsible for bearing the whole weight of the animal during rest and movement and the extensor muscles simply lift the leg during locomotion. It is not surprising, therefore, that the forces tending to extend the leg are monitored more accurately so that the correct position of the leg can be maintained by the action of the flexor muscle.

The myochordotonal organ (MCO), as its name implies, has a muscular component called the "accessory flexor muscle". In this respect it differs from all other chordotonal organs described so far in Arthropoda and bears a functional resemblance to the muscle receptor organs and the vertebrate muscle spindle. Its anatomy and physiology in *Cancer magister*, a species not studied by Barth (1934), has been investigated by Cohen (1963, 1965), Dorai Raj and Cohen (1962) and Hwang (1961). In this species and in other decapod crustaceans there are two parts to the accessory flexor muscle (Cohen, 1963), unlike those described by Barth (1934) which had only one muscle with a single insertion. There are two groups of scolopidia in the myochordotonal organ; the "main organ" and the "proximal organ". Cohen (1963) has shown that the main organ of the MCO contains rapidly-adapting receptor units that respond to changes in position of the IM joint. Within the movement receptors there are some sensitive to stretching and others sensitive to relaxation of the proximal accessory muscle: none respond to stimulation in both directions. The static receptors respond to stretch by an increased discharge; there is no indication of any relaxation receptors in this category. The proximal accessory muscle, the elastic strand and the chordotonal organ form a triangular complex described in detail by Cohen (1963). When the muscle is elongated the elastic strand is stretched and the distal processes of the neurons will be either stretched or relaxed depending on the previous length of the muscle and the position along the elastic strand occupied by the cell body of the neuron. Cohen (1963) suggests that all the neurons are basically sensitive to stretch and explains their separation into two distinct categories on the basis of their location on the elastic strand. The basis for their differential response is not known. In the myochordotonal organ both the movement and the position receptors are sensitive to only a fraction of the total range of possible stimulation attendant on the movement of the leg at the MC joint. Various units are sensitive over different parts of the range and so they provide information about the whole extent of the joint movement by their combined responses. Cohen (1963) suggests that " 'range fractionation' has evolved to permit encoding of a stimulus covering a wide energy range and still allow high sensitivity to small stimulus increments."

Reflex responses mediated by the chordotonal organs in the leg of the crab have been studied by recording the discharges from motor nerves while the chordotonal organs are being stimulated or by recording motor output from joints in which the input from the chordotonal organs has been eliminated by severing the afferent nerve. The lack of response after "ablation" of chordotonal organs suggests that they alone give rise to the reflex motor activity controlling joint movements. The system is a negative feedback one giving rise to "resistance reflexes" that tend to counteract any forces tending to alter the positions of leg joints; passive movement of any joint elicits discharges in the motor axons supplying the muscles stretched by the movement. Identical results can be obtained by stretching the appropriate chordotonal organ alone. Relaxation of a chordotonal organ may also produce a specific reflex response: for example the PD organ, which contains both stretch and relaxation units and is the only organ at the PD joint, gives rise, when stretched, to impulses in the single opener motoneuron and when released, to impulses in the opener inhibitor and the closer motoneurons (Bush, 1962a,b, 1965c).

The myochordotonal organ alone of all the chordotonal organs of the crab leg is capable of reflex adjustment by means of its accessory muscle. It is surprising, therefore, that it apparently plays a smaller part in the reflex control of movements at the MC joint than the elastic strand organs MC1 and MC2. Removal of one myochordotonal organ has little effect on walking in the crab but removal of an increasing number leads to a progressive loss of muscle tone in all legs. Cohen (1965) postulates a general stimulatory role for the myochordotonal organ analogous to that of the vertebrate labyrinth, crustacean statocyst and vertebrate muscle spindle. It is closest in structure to the vertebrate muscle spindle and, as in the muscle spindle, the sensory input can be altered by efferent fibres from the central nervous system acting on the receptor muscle. The length or tension of the limb muscles is influenced by stretch reflexes mediated by the sensory discharge. Cohen (1965) suggests that this general effect on muscle tissues is the principal function of the myochordotonal organ. It must also be recognized (Cohen, 1963) that the MC joint is the most mobile joint of the leg and the one at which the greatest forces are exerted. It is possible that the duplication and complexity of the proprioceptors in that region is correlated with these facts.

There are no campaniform organs in Crustacea and no multiterminal neurons have been described in the legs. In view of the presence of multiterminal neurons in insect legs the possibility of their occurrence in crustacean legs cannot be ruled out at present.

The chelicerate leg

There is some information about the anatomy and distribution of proprioceptors in the legs of *Limulus* (Pringle, 1956; Barber, 1956, 1960; Barber and Hayes, 1964; Barber and Segel, 1960), of spiders (Parry, 1960; Rathmayer, 1967) and of scorpions and whip-scorpions (Pringle, 1955; Laverack, 1966) but it is not possible at present to produce a diagram comparable with that for the insect leg (Fig. 4) and the crab leg (Pringle, 1961; Whitear, 1962). Rathmayer's (1967) description of the proprioceptors at the femoro-patellar joint of the spider leg is the most detailed but there is no comparable description for the other joints. There is, however, a considerable amount of information about the responses of proprioceptors in the leg of chelicerates.

The proprioceptors of the chelicerate leg are slit sensilla (lyriform organs) (Pringle, 1955; Edgar, 1963), multipolar neurons (Pringle, 1956; Barber, 1960; Barber and Segel, 1960) and the "tendon organ" of *Limulus* (Barber and Hayes, 1964). Lyriform organs are comparable in structure and physiology to the campaniform organs of insects: they are stress receptors. They have been found on the legs of phalangids (Edgar, 1963), spiders, scorpions and amblypigids (Pringle, 1955, 1961). Edgar (1963) found structures similar to campaniform organs on the legs of phalangids. Laverack (1966) figures bipolar sensory neurons in his diagrams of the leg of scorpion and describes the processes as ramifying in the region of the joints. If they really ramify the cells are multiterminal, but Rathmayer (in press) describes and figures sensory neurons in the spider leg which could be uniterminal or multiterminal. They appear to have only one distal process and Rathmayer was unable to establish with ordinary histological techniques whether they are uniterminal or multiterminal. Rathmayer states that electron microscopy would be necessary to establish the nature of the sensory terminations of these cells. More detailed investigation of the histology of the sensory neurons is required for all chelicerate proprioceptors except the lyriform organs which have been studied by Salpeter and Walcott (1960) with the electron microscope.

Electrophysiological studies have shown that the proprioceptive neurons of the legs of chelicerates have a wide range of response to mechanical stimulation. They respond to changes in the position of joints, sometimes responding over only part of the arc of movement as in *Limulus* leg (Barber, 1960) and spider leg (Rathmayer, 1967). Some units respond to flexion, some to extension and others are active in the midsector. Within any group there may be a range of response so that sectors of the arc of movement are monitored by different units. Range fractionation and directional sensitivity are shown by both tonic and phasic receptors (Barber, 1960; Rathmayer, 1967).

The "tendon receptor" organ in the trochanter of *Limulus* (Barber and Hayes, 1964) is a connective tissue strand to which is attached an accessory muscle. The precise mode of functioning is not clear but Barber and Hayes suggest that all stimuli are probably effected via the accessory muscle because both ends of the connective tissue strand are attached to rigid cuticle. The sensory units appear to be mainly slowly-adapting position receptors.

Proprioceptors and the endocrine system

It has been known for over 30 years that distension of the body wall in the blood-sucking bug *Rhodnius* is the stimulus to moulting (Wigglesworth, 1934) but the proprioceptors responsible have not yet been identified. Van der Kloot (1961) recorded tonic sensory discharges in the nerve from the body wall and found that the discharge frequency increased when the segment was stretched. The proprioceptors function by controlling the release of brain hormone (Wigglesworth, 1934, 1957). If the abdomen of the bug is kept in a stretched condition by blocking the anus after a meal, moulting is inhibited or delayed for a long time (Wigglesworth, 1934). Recently Bennet-Clark (1966) has extended these observations by blocking the anus or by squeezing the bug between two microscope slides. A further method which avoided blocking the anus or subjecting the bug to pressure was to coat the distended abdomen with cellulose paint. When the insect voided the excess water taken in with its meal the abdominal cuticle remained in its distended position. All these procedures led to a delay in moulting beyond the 15 days that are normal for fourth-instar larvae kept at 25° C. When the abdomen was stretched food transfer from the storage to the digestion region of the midgut was delayed by a maximum of seven days and Bennet-Clark (1966) postulates that moulting is delayed because of lack of nourishment.

In studies of growth and moulting in the migratory locust, Clarke and Langley (1963) found evidence that the three pairs of multiterminal neurons on the foregut provide the proprioceptive link in a chain that relates the distension of the foregut to the neuroendocrine control of protein metabolism and hence of growth and moulting. Breaking this chain at any point along the sequence: sensory neurons of pharynx—posterior pharyngeal nerves—frontal ganglion—posterior pharyngeal nerves—frontal ganglion—frontal connectives—brain (medial neurosecretory cells) has the effect of inhibiting growth and moulting. In the tsetse fly (*Glossina*) Langley (1965) found similar sensory neurons in the proventricular region and on the crop duct and he suggests (Langley, 1966) that the size of the blood meal as measured by these sensory neurons controls the liberation of hormones concerned with the production of enzymes in the digestive region of the midgut. Physiological evidence for the presence of stretch receptors on the gut comes from the work of Gelperin (1966a,b) on the control of crop-emptying in the blowfly, *Phormia*.

Evidence for another link between proprioceptors in the body wall and pupation is presented by Edwards (1966). He found in the wax moth (*Galleria*) that nerve cord section or removal of ganglia in fully-grown larvae prevented or delayed pupation. The farther forward in the nerve cord that the operation was performed the greater was the effect. Similar experiments carried out by Bounhiol (1943) and Finlayson, (1956) on other lepidopterans, are quoted as supporting evidence. Edwards (1966) obtained the same results by merely distorting the shape of the body of the larva without subjecting it to any surgical procedures. Such distortion is alone sufficient to inhibit or delay pupation. He concluded that the surgical operations were effective because they led to distortions in the shape of the larva and that body wall proprioceptors must be involved in the control of the neuroendocrine system involved in pupation. He suggested that the proprioceptors might be sub-epidermal, multiterminal neurons or stretch receptors. The hypothesis of Beck and Alexander (1964) that *Ostrinia* produces a hormone ("protodone") from the hindgut may be unnecessary in the light of the discovery that bodily shape is so important. Edwards (1966) suggests that the phenomena explained by Beck and Alexander (1964) on the basis of a proctodeal hormone can also be explained as consequences of distortion produced by surgical operations or ligatures. He concluded that the process of shortening that precedes pupation is regulated by proprioceptors and that the appropriate body form must be achieved before the neurosecretory cells of the brain release their secretion.

In the larva of the tsetse fly (*Glossina*) Finlayson (1967) has shown that the sensory input from mechanoreceptors is an important factor in determining the onset of puparium formation. Larvae subjected to much mechanical stimulation under natural conditions, as for example during burrowing in sand, form puparia soon after birth. Larvae suspended in the air by their posterior polypneustic lobes also pupate very rapidly despite the lack of mechanical stimuli over the body surface. The proprioceptors of the body wall are probably involved in this latter response. Langley (1967) discovered that a ligature applied round the middle of a tsetse larva had the surprising result of inhibiting puparium formation in the anterior end but not in the posterior end. Although this remarkable phenomenon requires detailed investigation it seems possible that the shape of the two halves of the body in the ligatured larva may be of significance. The anterior end is put under pressure by the ligature and the larva cannot retract its head, the normal prelude to becoming barrel-shaped just before puparium formation (Finlayson, 1967). The posterior half, on the other hand, is near normal in shape for a barrel-shaped, head-retracted stage. It is possible that the proprioceptors signal "barrel shape" for the posterior half but "head not retracted" for the anterior half. This implies that the ligature does not cut off nervous communication between the two parts of the body. If the control of puparium formation is local and not via a blood-borne hormone, an incomplete ligature might have the same effect. There are other possible explanations but they are not relevant to the present discussion.

CONCLUSIONS

The study of proprioception must in the long run be aimed at producing a comprehensive account of the reflex control of locomotory and other processes. The physiology of the proprioceptors is but one link in a complex chain of events. This review will have served a useful purpose if it has drawn attention to the gaps in our knowledge of the morphology and distribution of proprioceptors in the invertebrates. Without a sound anatomical basis it is not possible to build an accurate functional description. At the level of the morphology and distribution of proprioceptors there is some fundamental information required. Are there, for example, any uniterminal sensory neurons in the chelicerate leg corresponding to the chordotonal sensilla of insects and crustacea? Are there any multiterminal sensory neurons in the crustacean leg? Functional problems at the receptor level also require investigation. Perhaps the most obvious is a functional explanation for the variety of proprioceptors, especially in the insect. The terrestrial insects and chelicerates have stress receptors on the cuticle in the form of campaniform organs and lyriform organs (slit sensilla). Is their absence in Crustacea correlated with the less flexible cuticle, or with the properties of the aquatic environment? The chordotonal organs of the insect leg and body require systematic investigation to sort out vibration receptors from stretch receptors or to establish that an organ that responds to both stimuli can transmit proprioceptive information to the central nervous system.

Another fundamental problem involving both morphology and

physiology is the distinction between phasic, phasic-tonic and tonic receptors, both chordotonal and multiterminal. This problem is not confined to proprioceptors but they show very well the whole range of responses from apparently identical receptor neurons. Some of the response characteristics of proprioceptors may be a function of the accessory structures or the precise relationships between dendritic terminations and the tissues on which they occur, rather than a reflection of differences between the neurons. There is evidence in crustacean chordotonal organs that directionality of response may be a function of the topographical relationship between the sensory ending and the accessory structure (Taylor, 1967a,b). Such an explanation is perhaps easier to visualize for uniterminal than for multiterminal neurons such as those in the leg of *Limulus* (Barber, 1960), which also show directionality in their response.

The role of proprioceptors in the control of metabolic and developmental processes is a new field of great importance. There may be a correlation between the proprioceptive control of muscle tonus, the maintenance of the "correct" body shape and the endocrine control of moulting and metamorphosis in insects. Conversely, forced deformation of the body may be signalled by proprioceptors and the neuroendocrine system inhibited from releasing the developmental hormones.

References

Alexandrowicz, J. S. (1952). Receptor elements in the thoracic muscles of *Homarus vulgaris* and *Palinurus vulgaris*. *Q. Jl microsc. Sci.* **93**: 315–346.

Alexandrowicz, J. S. (1953). Notes on the nervous system in the Stomatopoda. III. Small nerve cells in motor nerves. *Pubbl. Staz. zool. Napoli* **24**: 39.

Alexandrowicz, J. S. (1956). Receptor elements in the muscles of *Leander serratus*. *J. mar. biol. Ass. U.K.* **35**: 129–144.

Alexandrowicz, J. S. (1957). Notes on the nervous system in the Stomatopoda. V. The various types of sensory nerve cells. *J. mar. biol. Ass. U.K.* **36**: 603–628.

Alexandrowicz, J. S. (1960). A muscle receptor organ in *Eledone cirrhosa*. *J. mar. biol. Ass. U.K.* **39**: 419–431.

Alexandrowicz, J. S. (1967). Receptor organs in thoracic and abdominal muscles of Crustacea. *Biol. Rev.* **42**: 265–287.

Barber, S. B. (1956). Chemoreception and proprioception in *Limulus*. *J. exp. Zool.* **131**: 51–74.

Barber, S. B. (1960). Structure and properties of *Limulus* articular proprioceptors. *J. exp. Zool.* **143**: 283–322.

Barber, S. B. and Hayes, W. F. (1964). A tendon receptor organ in *Limulus*. *Comp. Biochem. Physiol.* **11**: 193–198.

Barber, S. B. and Segel, M. H. (1960). Structure of *Limulus* articular proprioceptors. *Anat. Rec.* **137**: 336–337.

Barbier, R. (1961). Contribution à l'étude de l'anatomie sensori-nerveuse des insectes trichoptères. *Annls Sci. nat.* (*Zool.*) (12) **3**: 173–183.

Barth, G. (1934). Untersuchungen über Myochordotonalorgane bei dekapoden Crustaceen. *Z. wiss. Zool.* **145**: 576–624.

Bässler, U. (1965). Proprioceptoren am Subcoxal- und Femur-Tibia-Gelenk der Stabheuschrecke *Carausius morosus* und ihre Rolle bei der Wahrnehmung der Schwerkraftrichtung. *Kybenetik* **2**: 168–193.

Becht, G. (1958). Influence of DDT and lindane on chordotonal organs in the cockroach. *Nature, Lond.* **181**: 777–779.

Beck, S. D. and Alexander, N. (1964). Proctodone, an insect developmental hormone. *Biol. Bull. mar. biol. Lab., Woods Hole* **126**: 185–198.

Bennet-Clark, H. C. (1966). Abdominal stretch and inhibition of moulting in *Rhodnius*. *J. Insect Physiol.* **12**: 1019.

Bounhiol, J. J. (1943). Nymphose (partielle) localisée chez des vers à soie divisé en trois parties par deux ligatures. *C. r. hebd. Séanc. Acad. Sci., Paris* **217**: 203–204.

Bullock, T. H. and Horridge, G. A. (1965). *Structure and function in the nervous system of invertebrates.* San Francisco: Freeman.

Burke, W. (1954). An organ for proprioception and vibration sense in *Carcinus maenas*. *J. exp. Biol.* **31**: 127–138.

Bush, B. M. H. (1962a). Peripheral reflex inhibition in the claw of the crab, *Carcinus maenas* (L.). *J. exp. Biol.* **39**: 71–88.

Bush, B. M. H. (1962b). Proprioceptive reflexes in the legs of *Carcinus maenas*. *J. exp. Biol.* **39**: 89–105.

Bush, B. M. H. (1963). A comparative study of certain limb reflexes in decapod crustaceans. *Comp. Biochem. Physiol.* **10**: 273–290.

Bush, B. M. H. (1965a). Proprioception by chordotonal organs in the mero-carpopodite and carpo-propopodite joints of *Carcinus maenas* legs. *Comp. Biochem. Physiol.* **14**: 185–199.

Bush, B. M. H. (1965b). Proprioception by the coxo-basal chordotonal organ, CB, in legs of the crab, *Carcinus maenas*. *J. exp. Biol.* **42**: 285–297.

Bush, B. M. H. (1965c). Leg reflexes from chordotonal organs in the crab, *Carcinus maenas*. *Comp. Biochem. Physiol.* **15**: 567–587.

Clarke, K. U. and Langley, P. A. (1963). Studies on the initiation of growth and moulting in *Locusta migratoria migratorioides* R. & F. III. The role of the frontal ganglion. *J. Insect Physiol.* **9**: 411–421.

Cohen, M. J. (1963). The crustacean myochordotonal organ as a proprioceptive system. *Comp. Biochem. Physiol.* **8**: 223–243.

Cohen, M. J. (1965). The dual role of sensory systems: detection and setting central excitability. *Cold Spring Harb. Symp. quant. Biol.* **30**: 587–599.

Debaisieux, P. (1938). Organs scolopidiaux des pattes d'insectes. *Cellule* **47**: 78–202.

Denis, C. (1958). Contribution à l'étude de l'ontogenèse sensori-nerveuse du termite *Calotermes flavicollis* Fab. *Insectes soc.* **5**: 171–188.

Dethier, V. G. (1963). *The physiology of insect senses.* London: Methuen.

Dorai Raj, B. S. and Cohen, M. J. (1962). Efferent control of a proprioceptive system in the crab leg. *Am. Zool.* **2** Abstracts No. 275.

Dorsett, D. A. (1964). The sensory and motor innervation of *Nereis*. *Proc. R. Soc. Lond.* (B) **159**: 652–667.

Edgar, A. L. (1963). Proprioception in the legs of phalangids. *Biol. Bull. mar. biol. Lab., Woods Hole* **124**: 262–267.

Edwards, J. S. (1966). Neural control of metamorphosis in *Galleria mellonella* (Lepidoptera). *J. Insect Physiol.* **12**: 1423–1433.

Finlayson, L. H. (1956). Normal and induced degeneration of abdominal muscles during metamorphosis in the Lepidoptera. *Q. Jl microsc. Sci.* **97**: 215–233.

Finlayson, L. H. (1966). Sensory innervation of the spiracular muscle in the tsetse fly (*Glossina morsitans*) and the larva of the wax moth (*Galleria mellonella*). *J. Insect Physiol.* **12**: 1451–1454.

Finlayson, L. H. (1967). Behaviour and regulation of puparium formation in the larva of the tsetse fly *Glossina morsitans orientalis* Vanderplank in relation to humidity, light and mechanical stimuli. *Bull. ent. Res.* **57**: 301–313.

Finlayson, L. H. and Lowenstein, O. (1955). A proprioceptor in the body musculature of Lepidoptera. *Nature, Lond.* **176**: 103.

Finlayson, L. H. and Lowenstein, O. (1958). The structure and function of abdominal stretch receptors in insects. *Proc. R. Soc. Lond.* (B) **148**: 433–449.

Gelperin, A. (1966a). Control of crop emptying in the blowfly. *J. Insect Physiol.* **12**: 331–345.

Gelperin, A. (1966b). Investigations of a foregut receptor essential to taste threshold regulation in the blowfly. *J. Insect Physiol.* **12**: 829–841.

Gettrup, E. (1962). Thoracic proprioceptors in the flight system of locusts. *Nature, Lond.* **193**: 498–499.

Gettrup, E. (1963). Phasic stimulation of a thoracic stretch receptor in locusts. *J. exp. Biol.* **40**: 323–333.

Gettrup. E. (1965). Sensory mechanisms in locomotion: the campaniform sensilla of the insect wing and their function during flight. *Cold Spring Harb. Symp. quant. Biol.* **30**: 615–622.

Gettrup, E. (1966). Sensory regulation of wing twisting in locusts. *J. exp. Biol.* **44**: 1–16.

Gettrup, E. and Wilson, D. M. (1964). The lift control reaction of flying locusts. *J. exp. Biol.* **41**: 13–90.

Graber, V. (1882). Die chordotonalen Sinnesorgane und das Gehör der Insekten. *Arch. mikrosk. Anat.* **21**: 65–145.

Gray, E. G. (1960). The fine structure of the insect ear. *Phil. Trans. R. Soc.* (B) **243**: 75–94.

Graziadei, P. (1965a). Muscle receptors in cephalopods. *Proc. R. Soc. Lond.* (B) **161**: 392.

Graziadei, P. (1965b). Sensory receptor cells and related neurons in cephalopods. *Cold Spring Harb. Symp. quant. Biol.* **30**: 45–57.

Guthrie, D. M. G. (1962). Control of the ventral diaphragm in an insect. *Nature, Lond.* **196**: 1010–1012.

Hoffman, C. (1963). Vergleichende Physiologie der mechanischen Sinne. *Fortsch. Zool.* **16**: 269–332.

Hoffman, C. (1964). Bau und Vorkommen von proprioceptiven Sinnesorganen bei den Arthropoden. *Ergebn. Biol.* **27**: 1–38.

Horridge, G. A. (1963). Proprioceptors, bristle receptors, efferent sensory impulses, neurofibrils and number of axons in the parapodial nerve of the polychaete *Harmothoe*. *Proc. R. Soc. Lond.* (B) **157**: 199–222.

Hubbard, S. J. (1959). Femoral mechanoreceptors in the locust. *J. Physiol., Lond.* **147**: 88–108.

Hwang, J. C. L. (1961). The function of a second sensory cell group in the accessory-flexor proprioceptive system of crab limbs. *Am. Zool.* **1**: 453.

Langley, P. A. (1965). The neuroendocrine system and stomatogastric nervous system of the adult tsetse fly *Glossina morsitans*. *Proc. zool. Soc. Lond.* **144**: 415–424.

Langley, P. A. (1966). The control of digestion in the tsetse fly *Glossina morsitans*. Enzyme activity in relation to the size and nature of the meal. *J. Insect Physiol.* **12**: 439–448.

Langley, P. A. (1967). Effect of ligaturing on puparium formation in the larva of the tsetse fly, *Glossina morsitans* Westwood. *Nature, Lond.* **214**: 389–390.

Laverack, M. S. (1966). Observations on a proprioceptive system in the legs of the scorpion *Hadrurus hirsutus*. *Comp. Biochem. Physicl.* **19**: 241–251.

Lowenstein, O. and Finlayson, L. H. (1960). The response of the abdominal stretch receptor of an insect to phasic stimulation. *Comp. Biochem. Physiol.* **1**: 56–61.

Markl, H. (1965). Ein neuer Propriorezeptor am Coxa-Trochanter-Gelenk der Honigbiene. *Naturwissenschaften* **52**: 460.

Moulins, M. (1966). Présence d'un recepteur de tension dans l'hypopharynx de *Blabera craniifer*. *C. r. hebd. Séanc. Acad. Sci., Paris* **262**: 2476–2479.

Nijenhuis, E. D. and Dresden, D. (1952). A micro-morphological study of the sensory supply of the mesothoracic leg of the American cockroach, *Periplaneta americana*. *Proc. K. ned. Akad. Wet.* **55**: 300–310.

Nijenhuis, E. D. and Dresden, D. (1955). On the topographical anatomy of the nervous system of the mesothoracic leg of the American cockroach, *Periplaneta americana*. *Proc. K. ned. Akad. Wet.* **58**: 121–136.

Orlov, J. (1924). Die Innervation des Darmes der Insekten (Larven von Lamellicorniern). *Z. wiss. Zool.* **122**: 425–502.

Osborne, M. P. (1963a). An electron microscope study of an abdominal stretch receptor of the cockroach. *J. Insect Physiol.* **9**: 237–245.

Osborne, M. P. (1963b). The sensory neurones and sensilla in the abdomen and thorax of the blowfly larva. *Q. Jl microsc. Sci.* **104**: 227–241.

Osborne, M. P. (1964). Sensory nerve terminations in the epidermis of the blowfly larva. *Nature, Lond.* **201**: 526.

Osborne, M. P. and Finlayson, L. H. (1962). The structure and topography of stretch receptors in representatives of seven orders of insects. *Q. Jl microsc. Sci.* **103**: 227–242.

Osborne, M. P. and Finlayson, L. H. (1965). An electron microscope study of the stretch receptor of *Antheraea pernyi* (Lepidoptera, Saturniidae). *J. Insect Physiol.* **11**: 703–710.

Parry, D. A. (1960). The small leg-nerve of spiders and a probable mechanoreceptor. *Q. Jl microsc. Sci.* **101**: 1–8.

Peters, W. (1962) Die propriorezeptiven Organe am Prosternum und an den Labellen von *Calliphora erythrocephala* Mg. *Z. Morph. Ökol. Tiere* **51**: 211–226.

Pilgrim, R. L. C. (1964). Stretch receptor organs in *Squilla mantis* Latr. (Crustacea. Stomatopoda). *J. exp. Biol.* **41**: 793–804.

Pringle, J. W. S. (1938a). Proprioception in insects. II. The action of the campaniform sensilla on the legs. *J. exp. Biol.* **15**: 114–131.

Pringle, J. W. S. (1938b). Proprioception in insects. III. The function of the hair sensilla at the joints. *J. exp. Biol.* **15**: 467–473.

Pringle, J. W. S. (1955). The function of the lyriform organs of arachnids. *J. exp. Biol.* **32**: 270–278.

Pringle, J. W. S. (1956). Proprioception in *Limulus*. *J. exp. Biol.* **33**: 658–667.

Pringle, J. W. S. (1961). Proprioception in arthropods. In *The cell and the organism*: 256–282. Ramsay, J. A. and Wigglesworth, V. B. (eds). Cambridge: University Press.

Pringle, J. W. S. (1963). The proprioceptive background to mechanisms of orientation. *Ergebn. Biol.* **26**: 1–11.

Rathmayer, W. (1967). Elektrophysiologische Untersuchungen an Proprioceptoren im Bein einer Vogelspinne (*Eurypelma hentzi* chamb). *Z. vergl. Physiol.* **54**: 438–454.

Richard, G. (1950). L'innervation et les organes sensoriels de la patte du termite à cou jaune. *Annls Sci. nat.* (Zool.) (11) **12**: 65–83.

Richard, G. (1951). L'innervation et les organes sensoriels des pièces buccales du termite à cou jaune. *Annls Sci. nat.* (Zool.) (11) **13**: 397–412.

Rilling, G. (1960). Zur Anatomie des braunen Steinlaufers *Lithobius forficatus* L. (Chilopoda), Skeletmuskelsystem, peripheres Nervensystem und Sinnesorgane des Rumpfes. *Zool. Jb.* (Anat.) **78**: 39–128.

Salpeter, M. M. and Walcott, C. (1960). An electron microscope study of a vibration receptor in the spider. *Expl Neurol.* **2**: 232–250.

Sereni, E. and Young, J. Z. (1932). Nervous degeneration and regeneration in cephalopods. *Pubbl. Staz. zool. Napoli* **12**: 173–208.

Slifer, E. H. (1936). The scoloparia of *Melanoplus differentialis* (Orthoptera, Acrididae). *Ent. News* **47**: 174–180.

Slifer, E. H. and Finlayson, L. H. (1956). Muscle receptor organs in grasshoppers and locusts (Orthoptera, Acrididae). *Q. Jl microsc. Sci.* **97**: 617–620.

Stürkow, B., Adams, J. R. and Wilcox, T. A. (1967). The neurons in the labellar nerve of the blowfly. *Z. vergl. Physiol.* **54**: 268–289.

Tashmukhamedov, B. (1961). Peculiarities of abdominal stretch receptors in insects. *Z. obsc. Biol.* **23**: 76–80.

Taylor, R. C. (1967a). The anatomy and adequate stimulation of a chordotonal organ in the antennae of a hermit crab. *Comp. Biochem. Physiol.* **20**: 709–717.

Taylor, R. C. (1967b). Functional properties of the chordotonal organ in the antennal flagella of a hermit crab. *Comp. Biochem. Physiol.* **20**: 719–729.

Thomas, J. G. (1966). The sense organs on the mouth parts of the desert locust (*Schistocerca gregaria*). *J. Zool., Lond.* **148**: 420–448.

Thurm, U. (1964). Mechanoreceptors in the cuticle of the honey bee: Fine structure and stimulus mechanism. *Science, N.Y.* **145**: 1063–1065.

Thurm, U. (1965). An insect mechanoreceptor. Part I. Fine structure and adequate stimulus. *Cold Spring Harb. Symp. quant. Biol.* **30**: 75–82.

Treat, A. E. and Roeder, K. D. (1959). A nervous element of unknown function in the tympanic organs of moths. *J. Insect Physiol.* **3**: 262–270.

Van der Kloot, W. G. (1961). Insect metamorphosis and its endocrine control. *Am. Zool.* **1**: 3–9.

Viallanes, H. (1882). Recherches sur l'histologie des insectes. *Annls Sci. nat.* (6) **14**: 1–348.

Weevers, R. de G. (1965). Proprioceptive reflexes and the co-ordination of locomotion in the caterpillar of *Antheraea pernyi* (Lepidoptera). In *The physiology of the insect central nervous system*: 113–124. Treherne, J. E. and Beament, J. W. L. (eds). London and New York: Academic Press.

Weevers, R. de G. (1966a). A lepidopteran saline: effects of inorganic cation concentration on sensory, reflex and motor responses in a herbivorous insect. *J. exp. Biol.* **44**: 163–175.

Weevers, R. de G. (1966b). The physiology of a lepidopteran muscle receptor. I. The sensory response to stretch. *J. exp. Biol.* **44**: 177–194.

Weevers, R. de G. (1966c). The physiology of a lepidopteran muscle receptor. II. The function of the receptor muscle. *J. exp. Biol.* **44**: 195–208.

Weevers, R. de G. (1966d). The physiology of a lepidopteran muscle receptor. III. The stretch reflex. *J. exp. Biol.* **45**: 229–249.

Wendler, L. (1963). Über die Wirkungskette zwischen Reiz und Erregung. Versuche an den abdominalen Streckrezeptoren des Flusskrebses. *Z. vergl. Physiol.* **47**: 279–315.

Wendler, L. and Burkhardt, D. (1961). Zeitlich ablingende Vorgänge in der Wirkungskette zwischen Reiz und Erregung (Versuche an abdominalen Streckreceptoren dekapoder Krebse). *Z. Naturforsch.* **16b**: 649.

Wetzel, A. (1934). Chordotonalorgane bei Krebtieren (*Caprella dentata*). *Zool. Anz.* **105**: 125–152.

Whitear, M. (1962). The fine structure of crustacean proprioceptors. I. The chordotonal organs in the legs of the shore crab, *Carcinus maenas*. *Phil. Trans. Roy. Soc.* (B) **245**: 291–325.

Whitear, M. (1965). The fine structure of crustacean proprioceptors. II. The thoracico-coxal organs in *Carcinus*, *Pagurus* and *Astacus*. *Phil. Trans. Roy. Soc.* (B) **248**: 437–456.

Whitten, J. M. (1963). Observations on the cyclorrhaphan larval peripheral nervous system: muscle and tracheal receptor organs and independent Type II neurons associated with the lateral segmental nerves. *Ann. ent. Soc. Am.* **56**: 755–763.

Wiersma, C. A. G., Furshpan, E. and Florey, E. (1953). Physiological and pharmacological observations on muscle receptor organs of the crayfish *Cambarus clarkii* Girard. *J. exp. Biol.* **30**: 136–150.

Wiersma, C. A. G. and Pilgrim, R. L. C. (1961). Thoracic stretch receptors in crayfish and rocklobster. *Comp. Biochem. Physiol.* **2**: 51–64.

Wigglesworth, V. B. (1934). Physiology of ecdysis in *Rhodnius prolixus* (Hemiptera). II. Factors controlling moulting and metamorphosis. *Q. Jl microsc. Sci.* **77**: 191–222.

Wigglesworth, V. B. (1957). The action of growth hormones in insects. *Symp. Soc. exp. Biol.* **11**: 204–227.

Wilson, D. M. and Gettrup, E. (1963). A stretch reflex controlling wingbeat frequency in grasshoppers. *J. exp. Biol.* **40**: 171–185.

Wyse, G. A. and Maynard, D. M. (1965). Joint receptors in the antennule of *Panulirus argus* Latreille. *J. exp. Biol.* **42**: 521–535.

Zawarzin, A. (1912). Histologische Studien über Insekten. II. Des sensible Nervensystem der Aeschnalarven. *Z. wiss. Zool.* **100**: 245–286.

Zawarzin, A. (1916). Quelques données sur la structure du système nerveux intestinal des insectes. *Res. zool. Russe* **1**: 176–180.

Symp. zool. Soc. Lond. (1968) No. 23, 251–261.

THE PECTINES OF SCORPIONS

J. D. CARTHY

*Field Studies Council, London, England**

SYNOPSIS

The fine structure of the sensory pegs on the pectines of the scorpion *Leiurus quinquestriatus* is described. Both their anatomy and their electro-physiology show them to be mechanoreceptors. It is proposed that they function in the selection of substrata with particular sand grain size.

The pectines of scorpions, sense organs peculiar to those animals, have for long been the subject of much discussion concerning their function (e.g. Cloudsley-Thompson, 1955). Cloudsley-Thompson produced evidence that sensitivity to air- and substratum-born vibration is mediated by the pectines, but this could not be confirmed by F. T. Abushama and J. D. Carthy (unpublished). They found scorpions without pectines to show the same responsiveness as intact ones to vibration of the surface upon which they rested, under conditions in which air-movement was reduced as much as possible. (There seems little doubt that air movements can be detected by the trichobothria on the pedipalps (Hoffmann, 1967).) Located just behind the bases of the hindermost legs, they are paired organs, each of which consists of a basal part upon which are articulated a series of teeth (Figs. 1 and 2a). There are numerous stiff spines, both on the base and around the articulations of the teeth; no doubt some of these are sensory. But the main interest has centred on the fields of short blunt pegs (Schröder, 1908) which lie to one side of the mid-line of each tooth, covering a large part of the surface and extending some two-thirds of the way down towards the base (Fig. 2b). When a scorpion is moving it can be seen to turn the whole pectine on its basal articulation, so that the teeth are now vertical and brush the ground surface. Thus, what is morphologically the ventral surface of the pectine when it is at rest, now makes contact with the ground. It is this surface upon which the pegs are distributed. The number of teeth varies according to the species of scorpion.

In *Leiurus quinquestriatus* H. & E. (to which all the detailed description in this paper refers) the field of pegs is approximately 0·5 mm in

* The work was carried out in the Department of Zoology, Queen Mary College University of London.

length. A field contains about 1200 pegs (Fig. 2b); Hoffmann (1964) describes 100 pegs in a sensory field of *Euscorpius carpathicus*. Each lies along the lower side of the tooth, as the pectine is carried at rest beneath the animal. Therefore, when the pectine is lowered to brush the ground the sensory field is brought into contact with the substratum.

Each peg consists of a cylindrical base inserted into a pit in the cuticle, surmounted by a short, blunt-ended projection (Fig. 2c). This projection appears to be oval in cross-section, the long axis of each being aligned across the largest dimension of the sensory field. The tip is slightly concave, but no pores are visible either in section or in surface view. The whole peg is approximately 4 μ in height, the base and the projection being equal in height. However, the projection is about 2 μ in diameter while the base has a diameter of 5 μ. In *Euscorpius carpathicus*, each peg is 12 μ high on a base 6 μ high and 10 μ in diameter (Schröder, 1908).

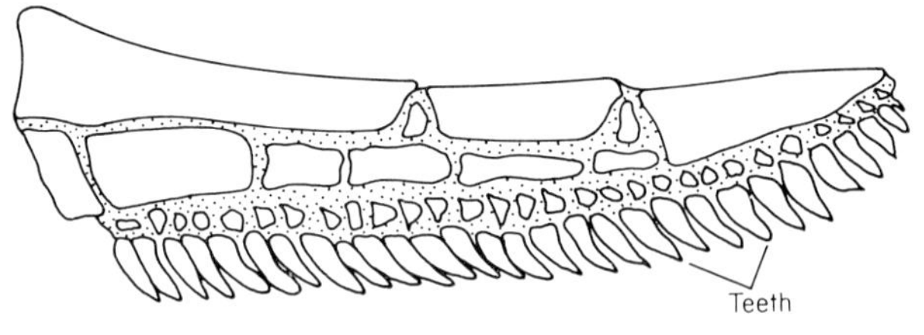

Fig. 1. Pectine of *Leiurus quinquestriatus*.

The cuticle of the peg is thin, being about 0·1 μ thick in most parts, this contrasts with a cuticle thickness of 5–6 μ between the pegs, though the projecting part may be strengthened by thickening of the cuticle. There is no evidence of a hinge joint mechanism around the peg base, and the cuticle is continuous with that of the rest of the tooth.

Within each peg base, and concentric with it is a cylindrical membrane, possibly also cuticular in nature (Fig. 2d). It is pleated, the folds projecting inwards along the radii. Most often there are eight such pleats,

Fig. 2(a) Pectines of scorpion *Leiurus quinquestriatus* (ventral view).

(b) Stereo-electron micrograph of teeth of pectine, showing the sensory field.

(c) Stereo-electron micrograph of single peg. (Stereo-electron micrographs made on Stereoscan instrument by kind permission of Cambridge Instrument Co.)

(d) Electron micrograph of a cross section through a peg base showing the pleated membrane and the attachment of the stereocilia.

(e) Electron micrograph of section through base of a cilium.

(f) Electron micrograph of a vertical section through a peg. The expanded tips of the microtubules are seen, and the extension of the pleated membrane downwards to enclose the whole sensory apparatus.

Figs. 2(d), (e) and (f) fixed in osmium Palade, embedded in Araldite; Siemens Elmiskop 1A.

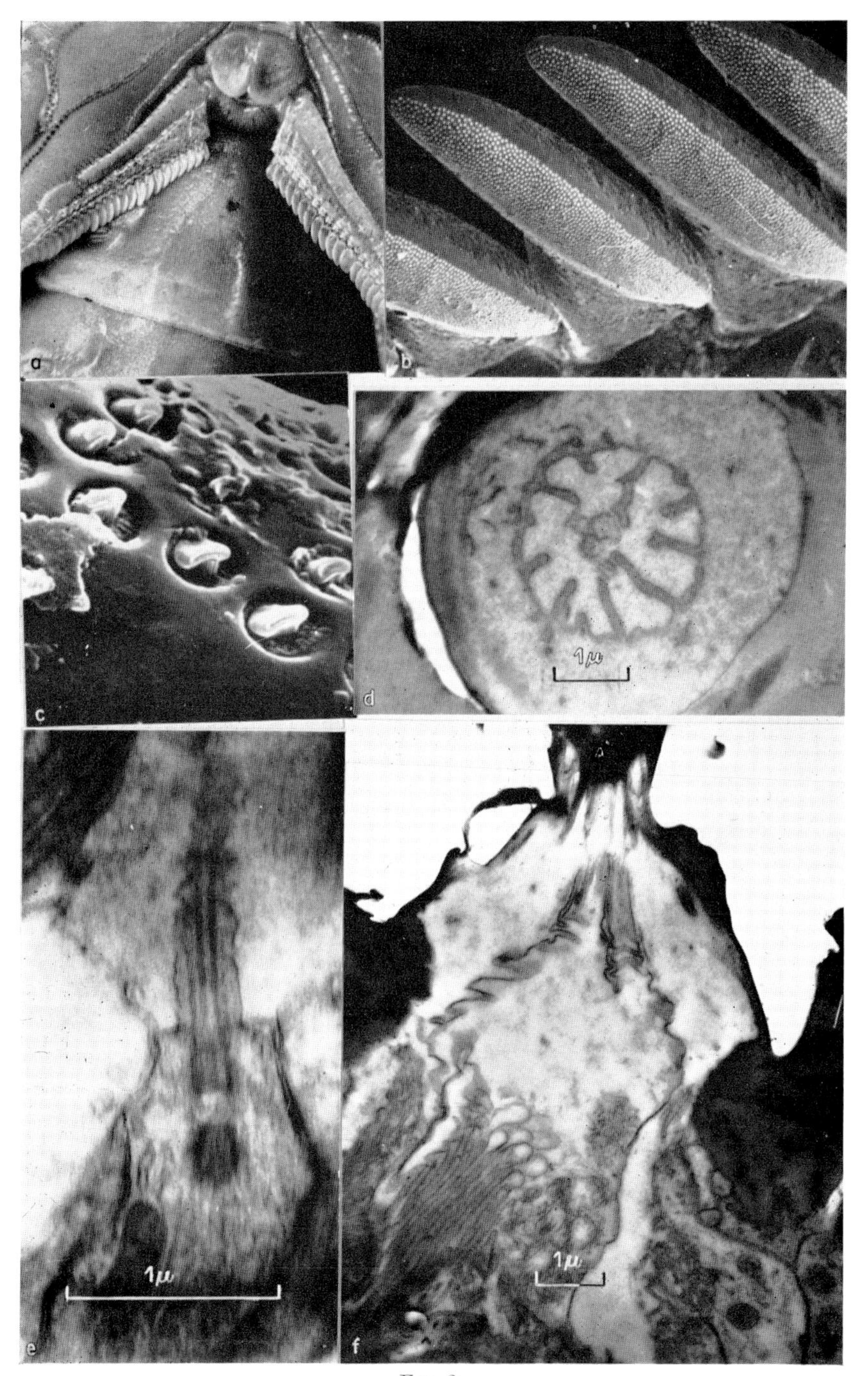

Fig. 2.

though there may be 10 or 11. Both the contents of the cylinder and the material surrounding it within the peg show no evidence of organized structure in electron micrographs. The membrane is continuous with a flange of thickened cuticle forming part of the wall of the projection.

To each pleat is attached a strand (about 11 μ long) which lower down its length can clearly be seen to contain nine peripheral fibrils and thus is a cilium (Fig. 2e). The pleats taper away in the basal part of the peg until the membrane is unfolded and roughly circular at the level of the base of the cuticle.

The ciliary strands pass down within the membrane until each enters a cell. Each of these sensory cells are elongated, measuring 30–37 μ. Their rounded nuclei lie at the lower end of the cells, from which axons take origin, which run together as the sensory nerve of the field. Since there is one cell to a strand, there is a group of eight to ten to be accommodated in the sensory unit of each peg. For reasons of spatial economy, the cell bodies and their nuclei are situated at various levels.

Near their bases but before they enter a cell the cilia can be seen to have a radiating system of finer fibrils connecting the peripheral fibrils (Fig. 5b). Below the cell surface, the fibrils end in a thickening, a typical basal granule, from which lead roots (Fig. 2e). These roots end in elongated osmiophilic thickenings. There is also occasional evidence of much longer finer fibrils penetrating deeper into the cell.

The internal membrane of the peg appears to be continuous with the outer borders of the extensions of the sensory cells into which the ciliary strands enter. Packed close to them within the membrane is a cell whose cytoplasm contains numerous small vesicles. From its upper end and on the medial side, numerous long parallel tubules arise as out-pockets from the cell membrane (Fig. 5a). The tubules form a dense mass beside the ciliary strands. They extend to the base of the peg where they end bluntly and may be slightly swollen. The tubule walls appear to be double, though no other structure is visible within them.

The structures described thus far appear to be concerned with sensory functions, but they are enclosed in sheathing cells, each of which reaches about half the way round the circumference of the bundle. These cells have elongated nuclei which lie at a higher level than the characteristically rounded nuclei of the sensory cells (Fig. 3). They were identified by Schröder (1908) as hypodermal cells, as indeed they appear to be. At their upper extremities they form dense bunches of short microtubular outgrowths which appear to be in direct contact

Fig. 3. Electron micrograph of a vertical section through pegs and the underlying structures. *c*, cuticle; *sh*, sheathing hypodermal cells; *stc*, cilium; *mic*, microtubules. (Fixed in osmium Palade, embedded in Araldite; Siemens Elmiskop 1A.)

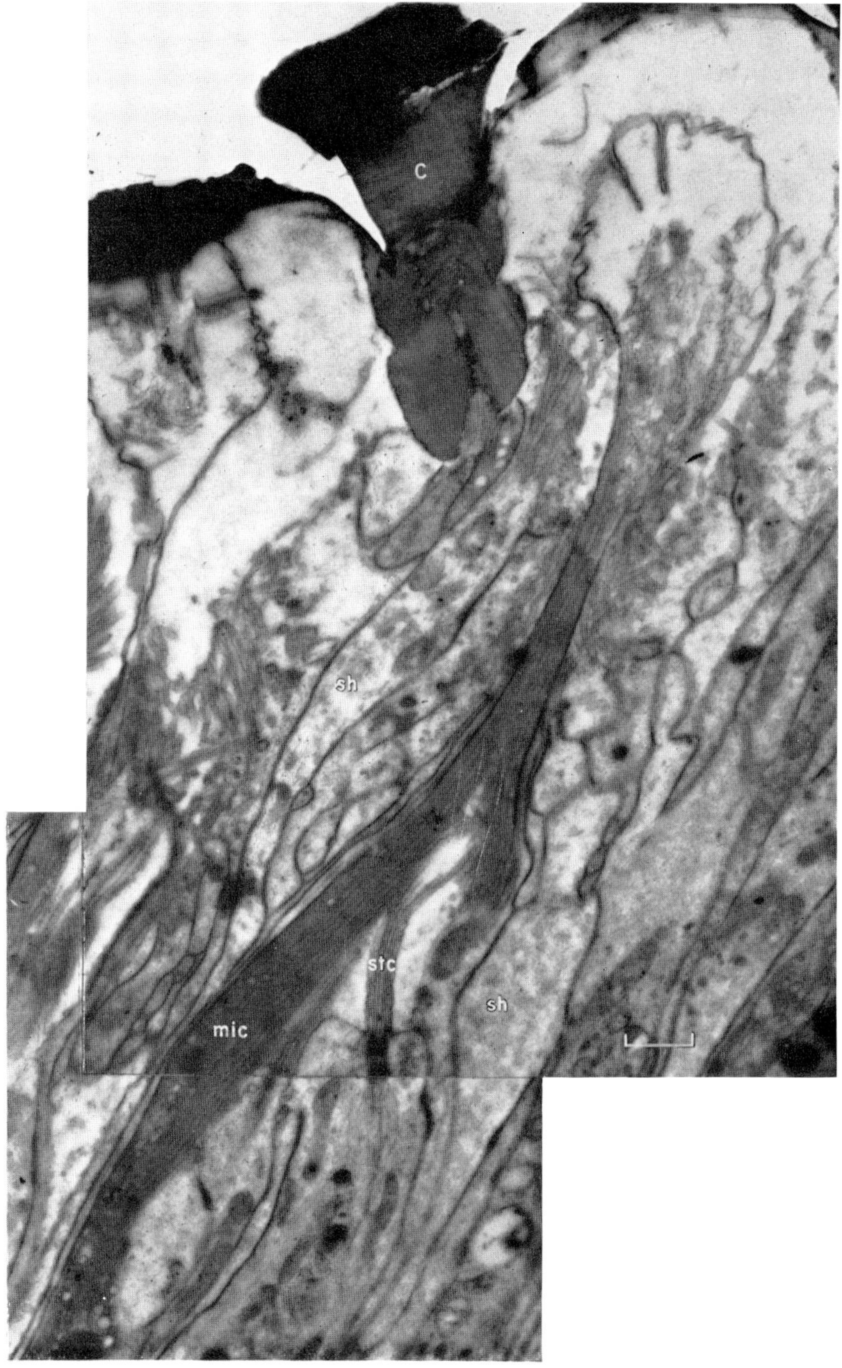

FIG. 3.

with the cuticle (Fig. 4). The dense, deeply osmiophilic rounded inclusions within the cells may well be cuticle-forming materials.

However, at the upper end of some of the cells a confused mat of short microtubules thrust out medially from the cell membranes (Fig. 3) at the level of the upper ends of the larger microtubules which lie beside

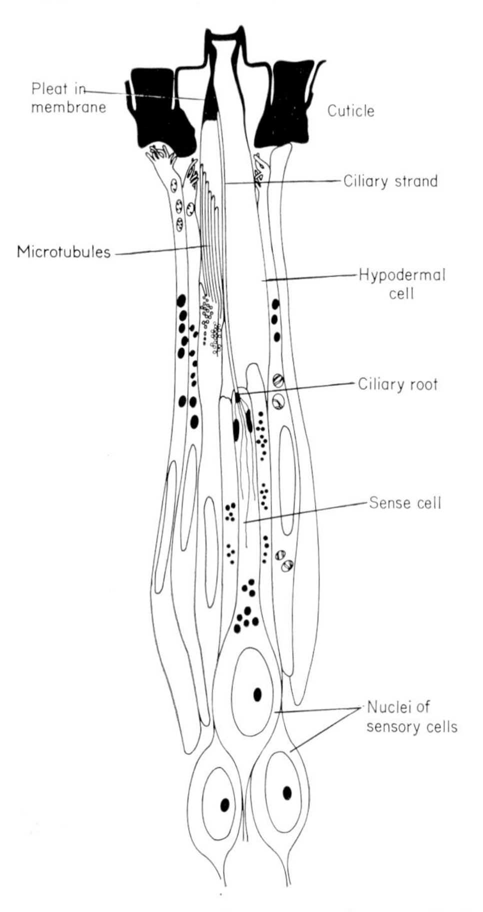

Fig. 4. Diagram of structure of sensory peg from tooth of pectine.

the stereocilia. As these tubules are directed towards and make contact with the pleated membrane, they may be the cells responsible for producing it.

In sections, scattered elongate cavities can be seen which are no doubt sections through intercellular blood spaces.

THE FUNCTION OF PECTINES

Many functions have been ascribed to pectines (e.g. Cloudsley-Thompson, 1955). Though they have been said in the main to be sensory, the collection of water and the creation of air movements over the openings of the lung books have also been mentioned. Electrophysiological studies by Hoffmann (1964) seem to have established beyond doubt that the pegs on the pectines are mechanoreceptors and their fine structure supports this conclusion.

Hoffmann stimulated individual pegs, or small numbers of them, mechanically, causing them to bend. Such stimulation produced electrical changes, while the use of a range of chemicals and of water caused no changes. Therefore a role as a chemosensitive or humidity sensitive organ is eliminated. Though Schröder (1908) considered the thin cuticular endings to the pegs to be proof of a chemoreceptive function in *Euscorpius*, the thickness of the peg tip and the absence of pores in them in *L. quinquestriatus* is morphological evidence against this.

The pegs responded to stimulation in the range of 14 to 200 c/sec. The electrical responses indicated that they were acting phasically for the negative initial phase appeared after four or five cycles, and was followed rapidly by a positive after-potential before returning to the base line. There was no evidence of a tonic response.

There is no anatomical evidence for any hinging system at the junction of peg and cuticle. Therefore bending the projection will move it in relation to the base, held upright in the cuticular socket. This would lead to deformation of the cylindrical membrane and in particular alter the arrangement of the pleats. This in turn could lead to stretching or relaxation of the ciliary strands, which appear taut in all sections. Such stretching would stimulate the cell into which the strands were rooted.

With a radial arrangement such as is found in the pleats, deformations in various directions could be distinguished from each other. As each ciliary strand appears to be connected to a separate cell each with its own axon, the means of detecting such differences exists. Hoffmann draws attention to results which suggest that movement in two directions may be separately sensed, and the disposition of the sensory cells could well lead to this, for deformation to one side would cause the four or so strands on the opposite side to be stretched, while movement in the opposite direction would stimulate the remaining strands.

The role of the fine tubules around the ciliary strands is problematical. If we suppose that stimulation of the ciliary strands involves some change and alteration of the polarisation of their membranes it

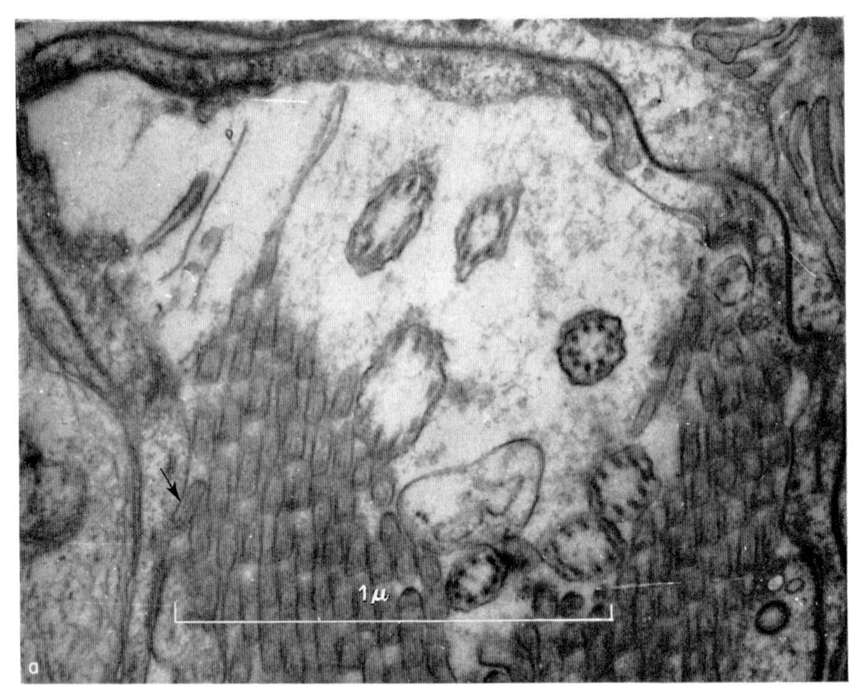

FIG. 5a.

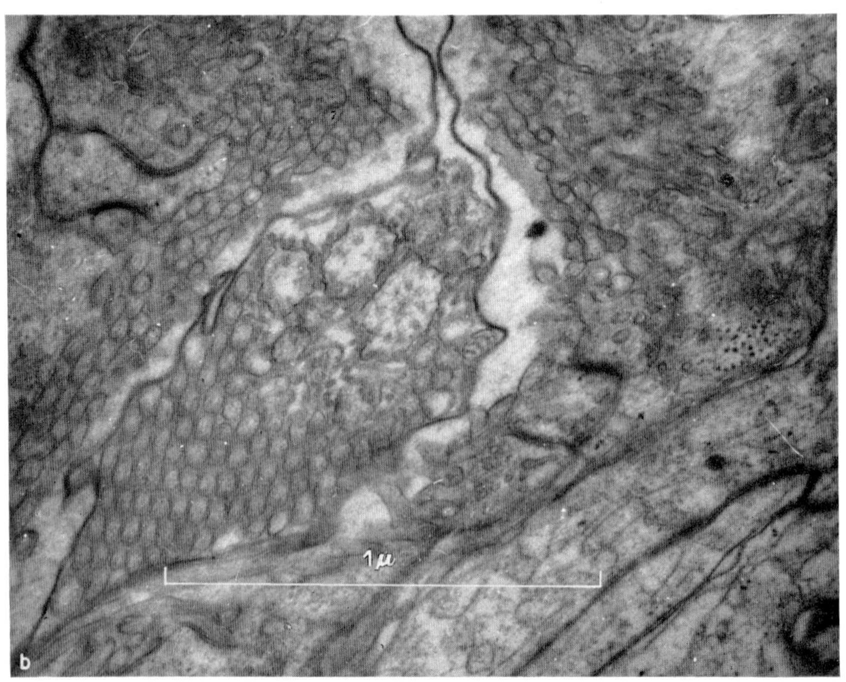

FIG. 5b.

could be that the tubules have some function in restoring rapidly the resting condition of the membranes. This would result in a quickly recovering sense organ such as this appears to be. The enclosure of the tubules within the cylindrical membrane with the ciliary strands might be indicative of a close association of function.

THE USE OF THE PECTINAL PEGS

Descriptions of scorpion courtship (e.g. Alexander, 1959) show the male leading the female to a place where he deposits his spermatophore. Then he guides her over it until it releases the package of sperm into her genital opening. Though at one time it was believed that the male alone exercised selection of the ground upon which this stage of courtship was to take place, Alexander has shown that the female is not without her influence in the choice.

Indeed, Abushama (1964) has shown that, regardless of sex, scorpions (*Leiurus quinquestriatus*) show a distinct preference for certain substrata against others. Given sands of uniform grain size, they will come to rest on that with 0·5 mm diameter grains in preference to those of 7, 4, 2 and 1 mm diameter, though the distinction became more difficult as the difference in size became less. Offered a smooth surface and grains 0·5 mm in diameter the scorpions were randomly distributed. This suggests that the limit had been reached of whatever sensing mechanism was used in this preference.

Thus this species show a preference for certain grounds even when the animals are not courting. Whatever sense organ is involved must exist in both males and females. Abushama also demonstrated that scorpions whose pectines had been covered by varnish, or had been amputated, could no longer make the distinction between grains of two sizes. He concluded that it was the tactile hairs which occur over the pectine which were responsible but it is equally likely on the evidence presented here that the pegs, now shown to be mechanoreceptive, could be implicated.

It is a feature of desert sands that the grains are well-rounded, indeed they are almost spherical. More, they are sorted by wind action into patches in which the grains are closely uniform in size. It is therefore

---

FIG. 5(a) Electron micrograph of cross-section of cilia and microtubules. Arrow indicates point of origin of a tubule.

(b) Electron micrograph of cross-section lower down sensory unit than (a), showing the fibrillar cross-sections within the cilium near its base. (Fixed osmium Palade, embedded Araldite; Zeiss E.M.9.)

not unreasonable to use a model in which the grains are replaced by perfect spheres. If these spheres are packed closely on a flat surface, the highest points of two neighbouring sand grains 0·5 mm in diameter will be 0·5 mm apart. In other words the sensory field on one peg would just bridge the gap, its ends making contact with the spheres (Fig. 6). However, grains of a larger size will have a greater distance between neighbouring grains, distances too great to be bridged by a single sensory field.

FIG. 6. Diagram showing suggested relationship between sand-grains and sensory field.

A smooth surface will ideally stimulate the whole sensory field. One can therefore postulate that the preferred ground is the one whose grains cause the maximum stimulation of the pegs on the pectine teeth. Since the pectines are swept across the ground by the moving animal, a mechanoreceptor which is phasic in its response would be advantageous, and this Hoffmann (1964) has demonstrated the pegs to be.

Sreenivasa-Reddy (1959) has found a strong correlation between the number of pectinal teeth and the habitat of the scorpions. Those with few teeth (e.g. the genus *Belisarius* with four teeth, inhabiting caves in the Pyrenees) can be grouped as the chactoid group while the buthoid group (consisting of the Buthidae only) have numerous teeth. The buthoid group typically inhabits dry habitats (often desert) and are poorly adapted for burrowing; the chactoid scorpions live in moist habitats and usually burrow. There is a definite relationship between the number of pectinal teeth and the humidity of the habitat in which the scorpion lives. Since the dry habitats will be sandy ones, it is possible that the correlation is with the nature of the soil rather than the wetness of the air. There is no information on the comparative sizes of the sensory fields on the pectinal teeth of animals of the two groups, nor of the numbers of pegs in each field.

It only remains to produce a plausible explanation for the choice of substratum, and this is far more difficult. It may be that the biologically adaptive function of the behaviour is to ensure that courtship takes place where there is least likelihood of the spermatophore being lost between grains. But it is very difficult to explain the remarkably dense array of individual receptors which occur on the teeth of the pectines of this scorpion and which suggest that they could be involved in some tactile discrimination of fine detail.

Acknowledgements

I should like to express my thanks to Professor D. Lacey of the Department of Zoology, St Bartholomew's Hospital Medical College, London, for making available his Electron-Microscope Unit to me in the early stages of this investigation and to Mrs Juliet Pettit for her help at that time.

References

Abushama, F. T. (1964). On the behaviour and sensory physiology of the scorpion (*Leiurus quinquestriatus* H. & E.). *Anim. Behav.* **12**: 140–153.

Alexander, A. (1959). Courtship and mating in the buthid scorpions. *Proc. zool. Soc. Lond.* **133**: 145–169.

Carthy, J. D. (1966). Fine structure and function of the sensory pegs on the scorpion pectine. *Experientia* **22**: 89.

Cloudsley-Thompson, J. L. (1955). On the function of the pectines of scorpions. *Ann. Mag. nat. Hist.* (12) **8**: 556–560.

Hoffmann, C. (1964). Zur Function de Kammförmigen Organe von Skorpionen. *Naturwissenschaften* **7**: 172.

Hoffmann, C. (1967). Bau und Funktion der Trichobothrien von *Euscorpius carpathicus* L. *Z. vergl. Physiol.* **54**: 200–352.

Schröder, O. (1908). Die Sinnesorgane der Skorpionskämme. *Z. wiss. Zool.* **33**: 436–444.

Sreenivasa-Reddy, R. P. (1959). A contribution towards the understanding of the functions of the pectines of scorpions. *J. Anim. Morph. Physiol.* **6**: 75–80.

Symp. zool. Soc. Lond. (1968) No. 23, 263–268

ANATOMICAL AND ELECTROPHYSIOLOGICAL STUDIES ON THE GASTROPOD OSPHRADIUM

D. F. BAILEY and P. R. BENJAMIN

Department of Pharmacy, University of Aston in Birmingham, and Department of Zoology, University of Durham, England

SYNOPSIS

Studies on the osphradium of the aquatic pulmonate *Planorbarius corneus* have indicated similarities between the organization of this organ and the osphradium of the marine prosobranch *Buccinum undatum*, in spite of the considerable differences in the position and form of the organ in the two species. Sensory nerve endings appear to penetrate specialized regions of the epithelium of both organs, but whilst the large bipectinate osphradium of *Buccinum* has been shown to be a chemoreceptive organ involved in the location of food, the small tubular osphradium of *Planorbarius* does not serve the same function, although the latter species has been shown to be capable of actively seeking its plant food. All attempts to determine the function of the osphradium of *Planorbarius* have so far failed.

The function of the gastropod osphradium has been the subject of theoretical speculation for many years. Experimental evidence as to its function is, however, scattered, and in some cases rather ambiguous. The histology of these organs as studied by Bernard (1890), Dakin (1912) and Stork (1935) and their position, in the prosobranchs at least in the line of the inhalant current into the mantle cavity (Yonge, 1947), have led to the general conclusion that they are organs with sensory functions. The modality of the sensory function of the osphradium has remained in doubt. Hulbert and Yonge (1937) and Yonge (1947) favour a sediment detecting function, i.e. mechanoreception, basing their argument on the fact that many herbivorous and plankton feeding gastropods possess well developed osphradia. Other workers, such as Copeland (1918), Henschel (1932), Brock (1936), Wolper (1950) and Brown and Noble (1960), after carrying out a variety of choice chamber and excision experiments on a variety of gastropods, favour the chemosensory modality.

The work discussed in this present paper involves experiments upon two widely differing species of gastropod. The marine prosobranch *Buccinum undatum*, the common whelk, has a typically situated osphradium within the mantle cavity and, according to Dakin (1912) represents the acme of molluscan osphradial development. In contrast the osphradium of the fresh water pulmonate *Planorbarius corneus* lies outside the mantle cavity and is small in comparison to that in *Buccinum*. The feeding habits of these species also differ widely. *Buccinum* feeds upon carrion or sedentary molluscs such as *Mytilus*. Both the carnivorous habit and food locating ability of this animal are easily observed in sea water tanks with normal live animals. *Planorbarius*, on the other hand, is a browsing herbivore. It has been shown, however, using Y-maze trials, that this species is also capable of detecting food at a distance. Statistically significant correlation ($p = 0{\cdot}01$) was obtained for the orientation and movement of the animal into one branch of a Y-maze containing a homogenate of algae.

The osphradium in *Buccinum* is large, bipectinate and situated in a gutter in the mantle cavity into which runs the main inhalent siphonal current. This current therefore runs over the length of the osphradium and the ciliated cells situated on the faces of osphradial leaflets provide a subsidiary current which passes over the lateral sensory regions of the leaflets. The blind, tubular osphradium of *Planorbarius* is situated on the anterior mantle outside the mantle cavity, close to the base of the folds which constitute the opening of the pulmonary cavity (the modified mantle cavity). The main water current over the opening of the osphradial canal is provided by the mass of ciliated cells which are present on the anterior mantle surface. In addition to this general current, the ciliated cells within the osphradial canal provide local movements by which small particles of about 1μ diameter can be carried to the base of the canal, i.e. to the area of the supposed sensory endings.

The 90–100 leaflets which make up the osphradium in *Buccinum* are arranged on either side of a central axis 1–2 cm long which comprises an elongated nervous ganglion covered by connective and epithelial tissue. Each leaflet is roughly triangular in shape and has three regions, the sensory region which extends over the greater part of the free lateral surface; the glandular region which is restricted to the leaflet edges, particularly the lateral edge; and the ciliated region which forms a narrow band between the lateral glandular region and the sensory area. The sensory area lacks cilia and has a thickened epithelium penetrated by neurofibrillae of multipolar nerve cells lying in the nerves beneath the epithelium. These are the supposed sensory nerve endings. The nerves of the osphradial leaflets are derived from the axial osphradial ganglion which consists of a central neuropile and a cortex of ganglion cells (Dakin, 1912). In comparison the osphradium of *Planorbarius* is about 0·5 mm long and consists of a blindly ending ciliated canal with a basal nervous ganglion. The epithelial layer of the canal contains ciliated, secretory and basal cells similar to those in *Buccinum* (Fig. 1). The ciliated cells occur along the entire length of the canal and in the living state beat continuously. The secretory cells also occur along the entire canal surface but are more numerous in the central region. The nerve cell bodies of the osphradial ganglion send processes through the epithelial layer, to make contact with the contents of the canal, only in the basal region of the canal. The osphradial ganglion is arranged in typical molluscan fashion with a central neuropile and

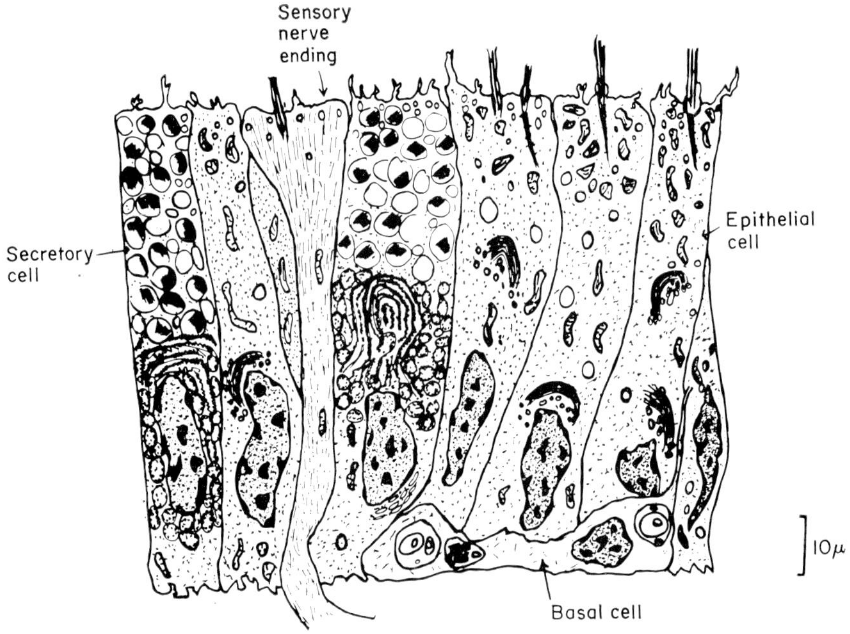

FIG. 1. Diagram of the structure of the osphradial epithelium of *Planorbarius corneus* showing three types of epithelial cell and the presumed sensory nerve ending, as seen with the electron microscope.

cortex of cell bodies. Within the ganglion monopolar, bipolar and multipolar neurones have all been observed. Some of the bipolar cells have been seen to send one axon to the neuropile and one to the epithelium; these are therefore most likely to be the sensory cells. Many of the other neurones are not directly concerned with perception. Electron microscopy has also revealed many interaxonal synapses in the neuropile and it would therefore appear that the osphradial ganglion has some integrative functions. The organisation of the sensory nerve endings passing through the osphradial epithelium thus seems similar in both *Planorbarius* and *Buccinum*. The main structural difference lies in the tubular nature of the sampling region in the former species.

The relationship between the osphradial ganglia and the central nervous complex of these two species may or may not be similar since the homologies of the molluscan central ganglia appear to be still a matter for some debate. The nervous connection to the central nervous system arises in both cases from the visceral-pleural-parietal complex although from opposite sides. The observation of Bullock and Horridge (1965) that the osphradial ganglion may represent "merely a cluster of primary cell bodies, without relay or independent reflex function" seems unjustified in the light of present work and the observations of Bernard (1890) of multipolar nerve cells linking the "neuro-epithelial" cells at various levels within the osphradial leaflets of a number of molluscs. It would appear therefore that betweeen the osphradium and the central nervous system in both *Buccinum* and *Planorbarius* there is interposed a nervous integrative centre of some complexity.

Possibly related to the complexity of innervation it has been found that the osphradia are capable of co-ordinated movements on local stimulation. Strong withdrawal contractions are elicited in the osphradium of *Buccinum* by noxious stimuli such as a strongly acidic solution. Light mechanical stimulation of the same organ causes the withdrawal of a small number of leaflets in the stimulated region. In *Planorbarius* a co-ordinated "clearing response" has been observed involving a strong contraction of the muscular wall of the canal forcing water out of the aperture. This response can be caused by noxious stimuli such as that above or when a particle of 5 μ diameter or greater enters the canal.

Attempts to record afferent activity in the osphradial nerves of *Buccinum* and *Planorbarius* have failed. This may, however, be related to the small diameter of the nerve fibres in these nerves. The largest axons in *Buccinum* osphradial nerves were found to be only about 1 μ in diameter and a similar picture emerges in respect of *Planorbarius* except for one fibre, consistently present, which was approximately 10 μ in diameter. The function of the latter remains unknown since no activity attributable to it has been observed. In *Buccinum* it has been possible to record the central expressions of chemoreceptors in the osphradium. Thus changes in the spiking activity of supra-intestinal ganglion cells were seen in response to a sea water extract of *Mytilus edulis*. This extract has previously been tested on *Buccinum* in storage tanks and found sufficiently attractive to them to elicit the proboscis extension response which is part of their normal feeding behaviour. The electrophysiological responses to the extract were thus taken to be a reasonable representation of the nervous activity in the normal animal. Therefore, in attempts to identify the active agents in the "*Mytilus* extract", the responses of the osphradial receptors to sea water solutions of various synthetic chemicals were compared with the response to the extract (Bailey and Laverack, 1963 and 1966). The chemicals causing a response most like that of the "*Mytilus* extract" were glutamic and aspartic acids, both of which are known to occur in free form in *Mytilus* muscle (Potts, 1958). Further tests with synthetic chemicals revealed that effective stimulation of the receptors was related to the dicarboxylic nature of the acids and the length of the carbon atom chains.

In *Planorbarius* it has proved possible to penetrate the osphradial ganglion cells with micro-electrodes and to record ongoing activity of 30–40 spikes/min from cells of 80–90 μ diameter, the frequency of discharge being similar in different cells of the same preparation (6–12 penetrations). So far, however, it has not been possible to record changes in nervous activity in response to either mechanical or chemical stimulation; the stimulus in the latter case being the algal homogenate found attractive to *Planorbarius* in the Y-maze tests. Intracellular recordings of activity within the left pallial ganglion of *Planorbarius* have also failed to reveal any central expression of chemical or tactile sensitivity of the osphradium.

Histological studies thus indicate similarity between the osphradia of *Buccinum* and *Planorbarius* in relation to the types of cell present and the organization of the innervation. They do not prove or disprove the homologous nature of these organs in the different species. If such structures are homologous, then the chemoreceptive nature of the osphradium in *Buccinum* is not a general feature and Professor Yonge's concept of secondary evolution of chemoreceptive function from a primitive mechanoreceptor is quite possible. The reactions of the osphradia in both *Buccinum* and *Planorbarius* in response to light mechanical stimuli indicate a sensitivity in this direction, but this is not reflected in the recorded nervous activity of the ganglia and may therefore be a purely local reflex. The absence of electrophysiological response in the osphradial ganglion of *Planorbarius* suggests yet another function for the osphradium. In the search for another receptive modality the effects of pH, temperature and inorganic ions have been tested, but without positive results. The same case applies in respect of extracts of reproductive snail peni, which reduces the possibility of an osphradial involvement in mating behaviour. Respiratory activity remains unaltered by the destruction of the osphradium and a connection with this process therefore seems unlikely. There remains, however, the possibility that the osphradium is an effector organ of some kind; a secretory function appearing the most likely. In relation to this it is perhaps relevant that in *Buccinum* osphradial nerves efferent nervous activity could be recorded which increased in frequency in response to noxious stimuli. This has previously been assumed to be related to the physical withdrawal of the osphradium but might also conceivably cause a secretion of some kind.

References

Bailey, D. F. and Laverack, M. S. (1963). Central nervous responses to chemical stimulation of a gastropod osphradium. *Nature, Lond.* **200**: 1122–1123.

Bailey, D. F. and Laverack, M. S. (1966). Aspects of the neurophysiology of *Buccinum undatum* L. (Gastropoda). I. Central responses to stimulation of the osphradium. *J. exp. Biol.* **44**: 131–148.

Bernard, F. (1890). Récherches sur les organes palleaux des gastéropodes prosobranches. *Annls Sci. nat. Zool.* (7) **9**: 88–404.

Brock, F. (1936). Suche, Aufnahme und enzymatische Spaltung der Nahrung durch die Wellhornschnecke, *Buccinum undatum* L. *Zoologica, Stuttg.* **34** (92): 1–136.

Brown, A. C. and Noble, R. G. (1960). Function of the osphradium in *Bullia* (Gastropoda). *Nature, Lond.* **188**: 1045.

Bullock, T. H. and Horridge, G. A. (1965). *Structure and function of the nervous system of invertebrates.* New York: Freeman.

Copeland, M. (1918). The olfactory reactions and organs of the marine snails *Alectrion obsoleta* (Say) and *Busycon canaliculatum* Linn. *J. exp. Zool.* **25**: 177–227.

Dakin, W. J. (1912). *Buccinum* (The Whelk). *L.M.B.C. Mem. typs Br. mar. Pl. Anim.* No. 20: 1–115.

Henschel, J. (1932). Untersuchungen uber den chemischer Sinn von *Nassa reticulata*. *Wiss. Meeresuntersuch.* **21**: 131–159.

Hulbert, G. C. E. B. and Yonge, C. M. (1937). A possible function of the osphradium in the Gastropoda. *Nature, Lond.* **139**: 840.

Potts, W. T. W. (1958). The inorganic and amino acid composition of some lamellibranch muscles. *J. exp. Biol.* **35**: 749–764.

Stork, H. A. (1935). Beiträge zur Histologie und Morphologie des Osphradiums. *Archs néerl. Zool.* **1**: 71–99.

Wölper, C. (1950). Das Osphradium der *Paludina vivipara*. *Z. vergl. Physiol.* **32**: 272–286.

Yonge, C. M. (1947). The pallial organs in the aspidobranch Gastropoda and their evolution throughout the Mollusca. *Phil. Trans. R. Soc.* (B) **232**: 443–518.

Symp. zool. Soc. Lond. (1968) No. 23, 269–277

TASTE RECEPTORS OF ARTHROPODS

EDWARD S. HODGSON

Columbia University, New York, New York, U.S.A.

SYNOPSIS

The best known taste receptors are the labellar chemoreceptors of flies. Specific receptors for cations, anions, sugars and water are known. There are no specific receptors for amino acids in flies, although these are found in crustacea. A high degree of integration, both peripheral and central, occurs in this sensory system.

The arthropods occupy a unique position in the study on taste receptors. Theirs were the first chemoreceptors (in any animals) which proved amenable to electrophysiological analysis, and they continue to yield important data concerning the relationships between sensory input and animal behaviour.

At the outset, it should be noted that the distinctions between taste and olfactory receptors are somewhat arbitrary. In arthropods, taste receptors are located on legs and many parts of the body surface, in addition to the mouthparts. Even calling them "contact chemoreceptors" is only a partial semantic aid, since olfactory receptors must also make contact with stimulating molecules. For convenience, we may say that the taste receptors are located predominantly on the mouthparts and are stimulated most often by relatively high concentrations of chemical stimuli acting in liquid phase. They are the sense organs which make the final chemical assays of potential food materials before these materials are either ingested or rejected. In this capacity, they have enormous economic, as well as theoretical, importance.

General reviews of this topic are numerous, and there seems little point in adding another at this time. [For background material on anatomy and electrophysiological techniques, see the recent reviews of Dethier (1963) and Hodgson (1965).] This is, however, a particularly appropriate time to examine the results from a decade of electrophysiological study on certain arthropod taste receptors—especially the labellar chemoreceptors of flies.

Within the past year, all the different receptor cell types associated with the labellar hairs of flies have been identified. In the same period, there has been a detailed study of the relationships between receptor activity and afferent impulses passing via the labial nerve to the brain.

This makes the receptors of the labellar hairs of flies probably the most thoroughly known of all taste receptors. The present coverage will concentrate on recent studies of them, plus certain chemoreceptors of crustacea which provide an interesting comparison.

ANATOMY OF LABELLAR RECEPTORS OF FLIES

The most commonly studied species is the blowfly, *Phormia regina*, although other flies (*Calliphora*, *Lucilia*, *Stomoxys*, etc.) have also been used. Prominent hairs on the labellum contain dendritic endings of two, three, or four bipolar neurons. The cell bodies of these taste receptors lie at the base of each labellar hair, within a membrane continuous with the hypodermis. The dendrites within the hair are enclosed in a chamber which is separate from the rest of the hair cavity. Movement of the hair can stimulate a mechanoreceptor cell attached to the base of the hair, but only the tip of a labellar hair is sensitive to chemical stimulation.

Perhaps the most interesting anatomical problems associated with these receptors concern the nature of the dendritic tip, contacted by the chemical stimuli, and the relationship between the receptor dendrites and the axons making up the labial nerve. With regard to the first problem, it is quite clear that there is an opening in the cuticle, and thus there is no rigid barrier between the sensory dendrite and the surrounding environment (Slifer, 1961). What is *not* clear is the nature of the fluid covering the receptor tip.

Stürckow (1967) noted the presence of a "sticky, slightly viscous" material, apparently extruded from the tips of the labellar hairs. In electronmicrographs of taste hairs of *Stomoxys*, this material appears to be coming from the pore of the channel which contains the dendrites. Electrical measurements indicate an electrolytic conductive path between the two chambers within the labellar hair, so this viscous fluid may get into the dendrite chamber at the base of the hair or else pass through the wall from the dendrite-free channel (Peters, 1965; Stürckow, 1967). In any case, the receptor surface is, in all probability, covered with a specialized secretion, the composition of which is still completely unknown.

Following up some of these observations in our laboratory, we have recently found that after injection of tritiated water into the hemolymph of *Phormia*, the radioactive material can be detected at the tip of a labellar hair in amounts proportional to the time which has elapsed after the injection. Moreover, if the tip of the labellar hair is removed, the recovery of radioactivity from the hair is sharply increased. This suggests that the size of the tiny opening in the cuticle really acts as the limiting factor in a continuous distal flow of fluid within the sensory hair. It also suggests that any freely diffusable substance in the hemolymph could have access to the sensory dendrites, and possibly to the entire sensory cell.

One possible functional significance of this is provided by the observation that hormones can affect the magnitude of response to a standardized chemical stimulus (E. S. Hodgson, Ishibashi and Wright, unpublished data). An example of this is shown in Table I, where

TABLE I

Hormone effects on labellar chemoreceptors

	No. of afferent impulses in successive seconds			
Second No.	Pringle's solution alone	Pringle's plus brain extract	Pr plus 5×10^{-5} M adrenalin	Pr plus 5×10^{-5} M dopamine
1	0	6	18	9
2	5	4	22	14
3	1	0	34	18
4	4	11	38	23
5	7	0	29	20

epinephrine and dopamine increase the frequency of afferent impulses from a cation receptor responding to Pringle's insect saline. The effect cannot be attributed to non-specific or osmotic properties of the stimulating solution, since brain extract is ineffective. Epinephrine closely mimics the effect of extracts of corpora cardiaca, to be considered as neurosecretory organs in this regard. It appears that, instead of a fixed threshold in these taste cells, there is some degree of lability according to the hormonal milieu.

The second anatomical problem has recently received detailed attention from Stürckow, Adams and Wilcox (1967). This is the old problem of the relation between the sensory cells and the axons going to the brain. The mere failure to find any synapses between receptors and afferent axons has been tentatively interpreted as evidence that nerve activity recorded at the periphery is an accurate reflection of what actually reaches the brain. When the axons in the nervus labialis were recently counted in electronmicrographs, the total numbers (within 95% confidence limits) corresponded to the numbers of sensory neurons in the labellum—about 800–850 in *Phormia*. Thus, there is no anatomical evidence of integrating internuncials. This is advantageous to the experimenter, in that the receptor activity recorded at the periphery does give a good picture of the "information" passing to the brain. However, it leaves problems in understanding how some variables in the chemical environment are integrated by the receptor cells (see below).

RECEPTOR CELL TYPES

The first two labellar taste receptors to be studied electrophysiologically were a receptor for sugars and one for salts (Hodgson and Roeder, 1956). Mellon and Evans (1961) subsequently detected a spike potential from a water receptor. The fourth chemoreceptor cell, present

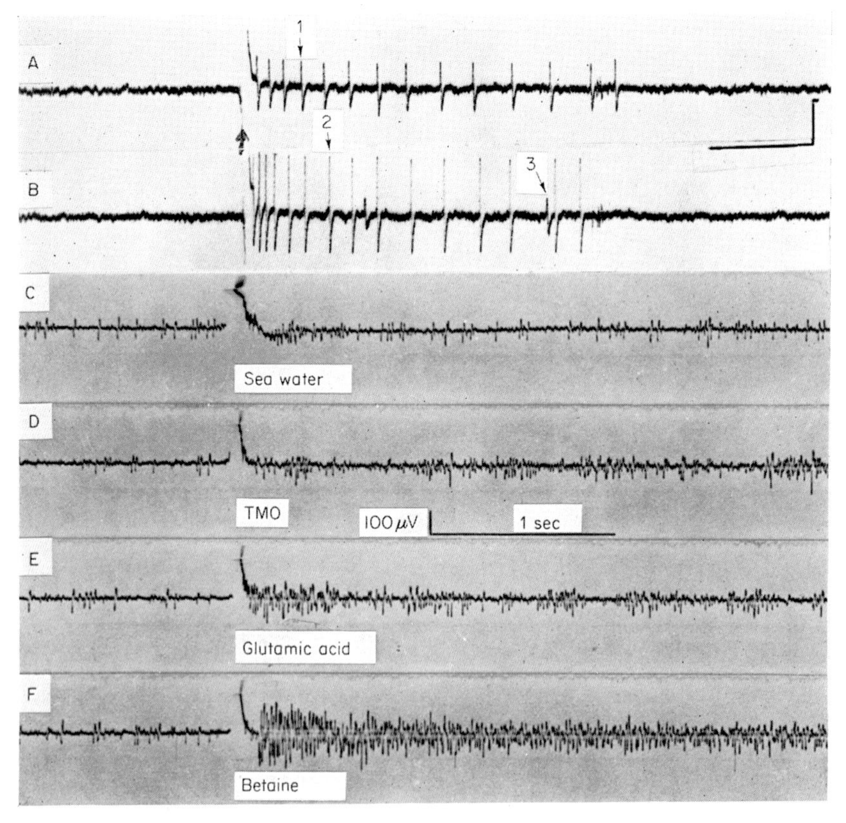

FIG. 1. Electrophysiological recordings of taste receptor activity. A and B: water, cation, and anion receptor impulses, recorded from a labellar taste hair of *Phormia* (courtesy Dr R. A. Steinhardt). C–F: dactyl responses to trimethyl amine oxide (TMO), glutamic acid, and betaine in *Panulirus*. For further details on all records, see text.

in some labellar hairs, has recently been described by Steinhardt (in press) as an anion receptor. Consequently, the old "salt" receptor, which was known to be largely sensitive to cations, should now be designated as the *cation* receptor.

Afferent impulses from the most recently studied labellar taste receptors are shown in Fig. 1. Stimulation with distilled water is shown in 1A, and the water receptor is designated 1. In trace 1B, 0·2 M NaCl is the stimulus, and the cation receptor spikes are designated 2. The much smaller anion receptor spikes are marked 3.

The obvious question is what can be the behavioural significance of having *two* receptors for electrolytes. Steinhardt (in press) has clarified this through an interesting experiment on water-deprived flies. Such

TABLE II

Labellar discrimination of salts

No. of flies	Test solution	% acceptance
First group with antennae removed		
10	distilled water	100
10	0·48 molal KCl	100
20	0·48 molal KNO_3	10
Retest of negative flies:		
20	0·48 molal KCl	70
Second group with antennae intact		
5	distilled water	100
13	0·48 molal KCl	84
14	0·48 molal KNO_3	0
Retest of negative flies:		
14	0·48 molal KCl	100

flies, of course, always drink distilled water. A more surprising observation is that they drink rather high concentrations of KCl, which predominantly stimulates the cation receptor (see Table II). At the same time, KNO_3 (which mainly stimulates the anion receptor) is not drunk. Retests of flies rejecting KNO_3 show high acceptance of KCl. It appears that only the *anion* receptor mediates rejection responses under all circumstances, even though both cation and anion receptors mediate rejection responses in water-satiated flies.

It should be noted that Stürckow (1960) tentatively identified an anion receptor in *Calliphora*, but the range of stimuli tested was quite limited and no behavioural significance was suggested. We cannot even state whether the anion receptor in *Phormia* corresponds to this, earlier described one in *Calliphora*. After testing a large series of cations and anions, Steinhardt (in press) concluded that the stimulating efficiencies of both cations and anions are related to their effective electrostatic field strengths, rather than simple ionic mobilities, as has been previously thought. This situation is much simpler physiologically from that in mammals, where the "water response" is mediated by "salt" fibres with both anionic and cationic groups in the membrane (Beidler, 1967).

AMINO ACID RESPONSES

Labellar receptors

Characterization of the sensitivity of the last of the labellar chemoreceptor cells leaves unexplained some of the apparent feeding responses of flies. In particular, behavioural studies indicated positive responses to protein infusions by *Phormia* (Dethier, 1961). Consequently, considerable effort has gone into attempts to determine the responses of the labellar taste receptors to amino acids or other protein components.

Inspection of spike potentials during amino acid stimulation, as carried out in our laboratory, has always given negative results. Wolbarsht and Hanson (1967), investigating the same possibilities, obtained negative results so far as spike potentials are concerned, but found that amino acid effects might be exerted in another way. The size of the receptor potential was slightly larger with amino acids than with pure water. Dicarboxylic amino acids and glutathione were found to inhibit the salt fibre, and a post-inhibitory rebound following the stimulus was also observed. This is not unlike the effects of long-chain hydrocarbons (Hodgson and Steinhardt, 1967). Hydrocarbons, when present in mixtures of compounds, inhibit the salt, water, and sugar receptors. The possibilities for coding chemosensory information must, therefore, include inhibitions of receptor activities as well as stimulations.

A possibility does remain that more specific receptors for amino acids or protein breakdown products will be found among the interpseudotracheal papillae of the labellum. These receptor organs, too, contain four receptors—one for sugars, another for "salts or certain carbohydrates such as L-arabinose", a mechanoreceptor, and one unknown type (Dethier and Hanson, 1965). In view of the small size and the difficulty in recognizing individual receptor spikes from these receptors, a good deal of caution will have to be used in interpreting such results. No water receptor has been identified, and one receptor cell remains uncharacterized.

Crustacean chemoreceptors

Since even in mammals, amino acids may act by affecting sugar receptor sites (Tateda, 1967), it seems appropriate to ask whether single-unit amino acid receptors can be found at all. Hodgson (1958) found chemoreceptors sensitive to glutamic acid and glycine on the antennules, chelae and protopodites of the first two walking legs of a large crayfish, *Cambarus bartonii*.

In marine Crustacea, the evidence is not entirely consistent. Case (1964) found glutamic acid a particularly effective stimulus for dactyl chemoreceptors of *Cancer*. Laverack (1963) found only trimethyl amine oxide (TMO) and betaine to be consistently stimulating in related species. However, in the spiny lobster, *Panulirus*, receptors for both glutamic acid and short-chain tertiary amines were detected electrophysiologically (Levandowsky and Hodgson, 1965). These receptors seem to be insensitive to a wide range of other stimuli, but it is not yet clear whether the same receptor cell can mediate responses to *both* amino acids and tertiary amines.

Sample data from the *Panulirus* experiments are shown in Fig. 1, C–F. These recordings are from nerves supplying the dactyl, and the stimuli are dropped onto the dactyl at the points where large deflections are found in the traces. Mechanoreceptor activity during the stimulus dropping is obscured, off the trace, by the large stimulus artifact. Glutamic acid appears to stimulate several fibres, although in other cases, single unit activity to glutamic acid stimulation has been identified. The betaine responses characteristically have a longer latency, sometimes several times that shown in this example, and this has led to the possible interpretation that injury effects may be involved. In view of the reversible nature of the response, it seems more likely that the betaine requires longer to reach some point (unknown) of actual stimulation. The lobsters give strong behavioural reactions to TMO, glutamic acid, and betaine—strengthening the argument that the responses noted electrophysiologically have some role in normal behaviour.

INTEGRATION OF AFFERENT INFORMATION

Completion of the electrophysiological inventory of labellar chemoreceptors, plus recent evidence of the one-to-one relationship between receptors and afferent nerve fibres, emphasizes the importance and extent of integration in the chemoreceptor system. Multiple stimuli do not simply give additive effects, in terms of afferent impulses. Nor does the integration take place entirely in the central nervous system.

Barton-Browne and Hodgson (1962) showed that part of the sugar suppression of activity could be accounted for in terms of lowered mobility of ions in a sugar solution. However, some of this seeming "interaction" of sugars and salts is still unaccounted for. Interactions of cations and anions, prior to stimulation or in the specific receptor activities, have not been investigated. Effects of certain stimuli on generator potentials only, such as Wolbarsht and Hanson (1967) postulate for amino acids, are another area in need of rigorous investigation.

Within the CNS, of course, very elaborate integrations must occur, and the problem is simply to find a convenient experimental approach to them. The possibility for studying behavioural results of stimulating a known few chemoreceptor cells offers interesting prospects in this regard. As an illustration, it now appears possible to get some precise measure of changes in central excitatory states (CES) by experimenting with the labellar chemoreceptor cells. Dethier, Solomon and Turner (1965) report that responsiveness to water can be reinduced in a water-satiated fly merely by touching another labellar hair with sucrose. The time course of the altered CES, its relations to previous feeding, etc. can all be investigated by this method. Obviously, resolution of the simpler problems with regard to these sensory receptor cells merely opens up a whole new area for physiological study.

Acknowledgements

Original investigations were aided by U.S. Public Health Research Grant AI-02271. Drs R. A. Steinhardt, T. Ishibashi, and H. Morita collaborated in some of the original experiments and discussions at Columbia University. It is a pleasure to acknowledge the debt to all of them.

References

Barton-Browne, L. and Hodgson, E. S. (1962). Electrophysiological studies of arthropod chemoreception. IV. Latency, independence, and specificity of labellar chemoreceptors of the blowfly, *Lucilia*. *J. cell. comp. Physiol.* **59**: 187–202.

Beidler, L. M. (1967). Anion influences on taste receptor response. In *Olfaction and taste* **2**: 509–535. Hayashi, T. (ed.) New York: Pergamon Press.

Case, J. (1964). Properties of the dactyl chemoreceptors of *Cancer antennarius* Stimpson and *C. productus* Randall. *Biol. Bull. mar. biol. Lab., Woods Hole* **127**: 428–446.

Dethier, V. G. (1961). Behavioral aspects of protein ingestion by the blowfly, *Phormia regina* Meigen. *Biol. Bull. mar. biol. Lab., Woods Hole* **121**: 456–470.

Dethier, V. G. (1963). *The physiology of insect senses.* London: Methuen.
Dethier, V. G. and Hanson, F. E. (1965). Taste papillae of the blowfly. *J. cell. comp. Physiol.* **65**: 93–99.
Dethier, V. G., Solomon, R. L. and Turner, L. H. (1965). Sensory input and central excitation and inhibition in the blowfly. *J. comp. physiol. Psychol.* **60**: 303–313.
Hodgson, E. S. (1958). Chemoreceptors of terrestrial and fresh-water arthropods. *Biol. Bull. mar. biol. Lab., Woods Hole* **115**: 114–125.
Hodgson, E. S. (1965). The chemical senses and changing viewpoints in sensory physiology. *Viewpoints in Biol.* **4**: 83–124.
Hodgson, E. S. and Roeder, K. D. (1956). Electrophysiological studies of arthropod chemoreception. I. General properties of the labellar chemoreceptors of diptera. *J. cell. comp. Physiol.* **48**: 51–75.
Hodgson, E. S. and Steinhardt, R. A. (1967). Hydrocarbon inhibition of primary chemoreceptor cells. In *Olfaction and taste* **2**: 734–749. Hayashi, T. (ed.) New York: Pergamon Press.
Laverack, M. S. (1963). Aspects of chemoreception in crustacea. *Comp. Biochem. Physiol.* **8**: 141–151.
Levandowsky, M. and Hodgson, E. S. (1965). Amino acid and amine receptors of lobsters. *Comp. Biochem. Physiol.* **16**: 159–161.
Mellon, D. and Evans, D. R. (1961). Electrophysiological evidence that water stimulates a fourth sensory cell in the blowfly taste receptor. *Am. Zool.* **1**: 372.
Peters, W. (1965). Die Sinnesorgane an den Labellen von *Calliphora erythrocephala* MG. (Diptera). *Z. Morph. Okol. Tiere* **55**: 259–320.
Slifer, E. H. (1961). The fine structure of insect sense organs. *Int. Rev. Cytol.* **11**: 125–159.
Steinhardt, R. A. (In press). Cation and anion receptors in the blowfly, *Phormia. J. gen. Physiol.*
Stürckow, B. (1960). Elektrophysiologische Untersuchungen am Chemorezeptor von *Calliphora erythrocephala. Z. vergl. Physiol.* **43**: 141–148.
Stürckow, B. (1967). Occurrence of a viscous substance at the tip of the labellar taste hair of the blowfly. In *Olfaction and taste* **2**: 707–720. Hayashi, T. (ed.) New York: Pergamon Press.
Stürckow, B., Adams, J. R. and Wilcox, T. A. (1967). The neurons in the labellar nerve of the blowfly. *Z. vergl. Physiol.* **54**: 268–289.
Tateda, H. (1967). Sugar receptor and α-amino acid in rat. In *Olfaction and taste* **2**: 383–398. Hayashi, T. (ed.) New York: Pergamon Press.
Wolbarsht, M. L. and Hanson, F. E. (1967). Electrical and behavioral responses to amino acid stimulation in the blowfly. In *Olfaction and taste* **2**: 749–760. Hayashi, T. (ed.) New York: Pergamon Press.

Symp. zool. Soc. Lond. (1968) No. 23, 279–297

CHECKLIST OF INSECT OLFACTORY SENSILLA

D. SCHNEIDER and R. A. STEINBRECHT

Max-Planck-Institut für Verhaltensphysiologie, Seewiesen über Starnberg, Germany

SYNOPSIS

Electrophysiological recordings with the use of microelectrodes permit the localization of olfactory sensilla in insects. In the antennal flagellum receptor cells belonging to the following four types of sensilla were found to respond to odours:

1. Sensilla trichodea (Lepidoptera);
2. S. basiconica (Lepidoptera, Diptera, Coleoptera, Orthoptera);
3. S. placodea (Hymenoptera);
4. S. coeloconica (Orthoptera, Hymenoptera).

In the first three types the cuticle of the sensillum is penetrated by tubular structures, each connecting an outer pore with the vacuolar space which surrounds the dendritic membrane of the receptor cell. The fourth type is a grooved cone without the typical pores and tubules.

The receptive unit being the sense cell, it is possible to state that cells belonging to a given antennal sensillum type react similarly in related genera. For instance, sensilla trichodea in male moths have one or several sexual attractant receptor cells (odour specialists). Other odour specialists reacting to pheromones, food, etc., are found in all four types of sensilla. Odour generalists with reaction spectra differing from cell to cell are observed only in basiconic and plate sensilla.

LOCALIZATION OF OLFACTORY ORGANS

The history of attempts to localize insect olfactory organs goes back to the 18th century, but decisive experiments were delayed until von Frisch (1919, 1921) demonstrated that honeybees are unable to find odorous food without antennae (see Schneider, 1957, 1964). Since then, students of insect physiology have agreed that the antennal flagellum is the main but not the only site of the olfactory power of insects.

Microscopic inspection of antennae immediately leads to the next problem: which of the many types of the antennal sensilla are responsible for olfaction? One can even go a step further and ask which cell of a group of sense cells innervating one morphological sensillum type, is olfactory. Identification of the olfactory function (modality) of one cell of a given sensillum is consequently followed by asking for the qualitative range of such a cell. Eventually we will have to enquire whether different peripheral dendritic loci of the sense cells serve different olfactory qualities. Such a sequence of questions will finally lead us to search for the location and chemical definition of the molecular acceptor structure, for presumably an acceptor molecule transduces the information transferred by the odour particle into the cellular response called excitation.

Methods for a sequential localization of olfactory function in insects begin with:

(1) Behaviour tests and progressive antennal amputation (see von Frisch, 1919, 1921) or nowadays also by recording summated electrical responses from the antenna during stimulation (Schneider, 1957).

(2) Identification of olfactory sensilla which can only successfully be done using local electrical recordings with fine electrodes (see the summary by Boeckh, Kaissling and Schneider, 1965) unless the number of possible olfactory sensilla on a given antenna is small enough to seal or amputate them one by one and to check the behaviour (Dethier, 1941, 1954; Wigglesworth, 1941; Bolwig, 1946).

(3) Progress towards an identification of the functional power of individual sensory cells, a step which depends upon the situation. This is easy if an olfactory sensillum has only one sense cell or several cells reacting identically. It is difficult when there is more than one sense cell per sensillum serving different qualities or even modalities.

(4) Topo- and biochemical identification of acceptor molecules is one of the future goals. The classical and logical biochemical approach must wait for methodological adaptations because of the small dimensions of the organ and the relatively small number of molecules involved in the reactions. Methods such as those used to identify the interaction of a receptor proteid of the cell nucleus with the estradiol molecule in the calf uterus might be helpful. This proteid is the first acceptor molecule ever identified in any chemoreceptor mechanism (Jungblut, Hätzel, de Sombre and Jensen, 1967).

The checklist in this paper gives a survey of all insect sensilla unequivocally identified as olfactory, using an electrophysiological micromethod. In this the cuticle is simply penetrated with a recording electrode near the sensillum to see whether there is a receptor (generator) potential and/or nerve impulse response to odorous stimuli. All those cases where amputation and behaviour responses were used to attribute olfactory powers to a sensillum are omitted here because of the difficulties as described before. The best known example for this is the bee antenna. Here, von Frisch predicted that the pore plate sensilla (sensilla placodea) would presumably be the organs of smell since their distribution corresponded to the results of amputation experiments. Later, Slifer and coworkers (Slifer, Prestage and Beams, 1959; Slifer and Sekhon, 1962, 1963, 1964a,b,c; Slifer, Sekhon and Lees, 1964) who found a porous cuticle in many peg-like sensilla which they thought to be olfactory considered the pore plates to be auditory organs (Slifer and Sekhon, 1960, 1961). The question was cleared by Lacher and Schneider (1963) and Lacher (1964) in favour of von Frisch's original prediction: the pore plates are olfactory organs.

MORPHOLOGY OF INSECT OLFACTORY SENSILLA

General morphology

Insect sensilla are not only morphologically but also morphogenetically defined units since all the cells belonging to one sensillum derive from one epidermal mother cell in a sequence of strictly determined mitotic steps (Henke and Rönsch, 1951).

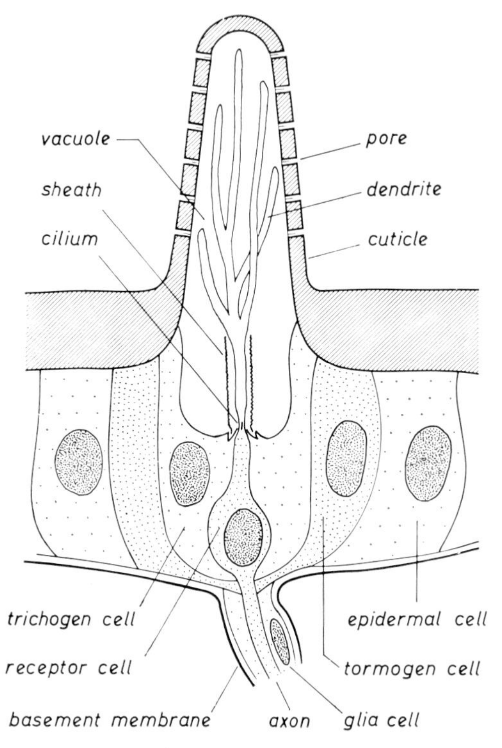

FIG. 1. Diagram of a peg-shaped insect olfactory sensillum in longitudinal section.

A typical sensillum consists of one or more bipolar receptor cells which are combined with a cuticular apparatus, for instance a peg (Fig. 1). This structure has been brought about by two formative cells, the trichogen and tormogen cell. The receptor cell is always a primary sense cell of which the axon is directly connected to the central nervous system. No synaptic contact between sense cells or axons has been

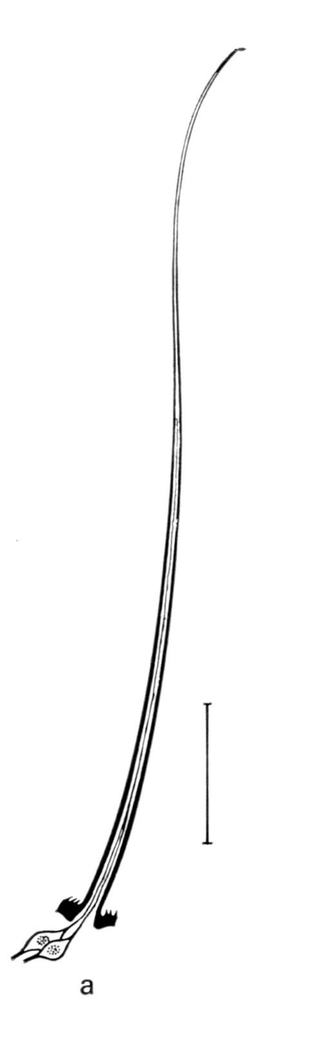
a

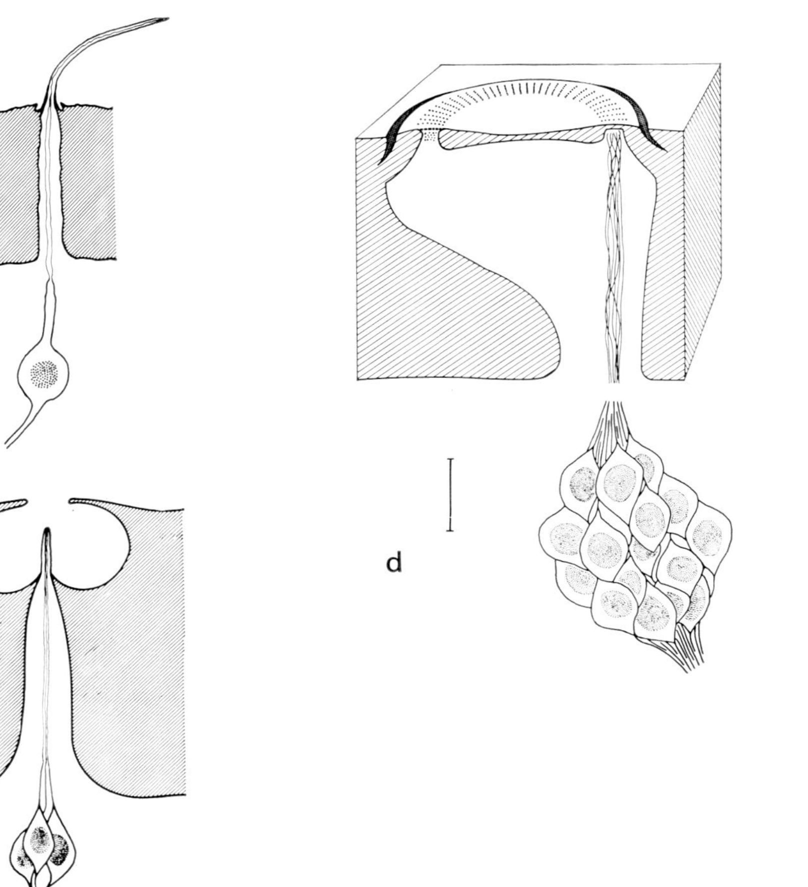
b
c
d

observed in the periphery. The dendritic process of the sense cell runs distally towards the cuticular apparatus of the sensillum. Its proximal part contains many mitochondria and other normal cell organelles and is surrounded by the trichogen and tormogen cell. A ciliary structure is observed in a short constriction of the dendrite, where the latter leaves the cellular envelope and invades the extracellular, fluid-filled hair vacuole. Distally from this point the dendritic cytoplasm shows only neurotubules and tiny vesicles. A sheath of unknown material, often called the "cuticular sheath", encloses the distal part of the dendrite at least up to the hair base. The olfactory dendrites, either branching or not, always enter the hair lumen. This is different from a mechanoreceptor, where the dendritic process ends at the hair base.

The most peculiar structural characteristic of insect olfactory receptors is the presence of numerous fine pores in the surface of the sensillum. These pores are complex systems and not simply holes penetrating the cuticle. Since their detection by Slifer *et al.* (1959), these pores have been observed in most olfactory sensilla. Contact chemoreceptors do not show these structures but have a single, comparatively wide hair-opening at the tip (Adams, 1961; Adams, Holbert and Forgash, 1965).

Cuticle pores of one kind or another have often been postulated to be a necessary condition of chemoreceptor sensilla. Before the use of ultrathin sectioning, Slifer (1954, 1955, 1956, 1960) found a number of basiconic sensilla to be permeable to dyes (e.g. crystal violet) in aqueous solutions, mostly following a normal fixation. A. G. Richards, D. Schneider and R. A. Steinbrecht (unpublished) found this to be true also with long antennal hairs (s. trichodea) of male moths. Sensilla chaetica, s. trichodea and s. styloconica with proved or suggested contact chemoreceptor function are stained in the tip region only (see e.g. Schneider and Kaissling, 1957; Boeckh *et al.*, 1960; Slifer, 1962). This observation is in agreement with electron micrographs showing relatively large openings in the tips of these sensilla (Adams, 1961).

FIG. 2. Examples of the four sensillum types hitherto proved to be involved in insect olfaction: (a) sensillum trichodeum (*Antheraea pernyi*); (b) sensillum basiconicum (*Necrophorus vespillio*); (c) sensillum coeloconicum (*Locusta migratoria*); (d) sensillum placodeum (*Apis mellifica*). In the figure of s. placodeum a surface view is combined with the longitudinal section to demonstrate the radial arrangement of the pores on top of a circular furrow. At the left the dendrites are seen in cross-section below a thin cuticular membrane, which is penetrated by the pores. At the right the section passes through a thickening which supports the thin membrane between two rows of pores; here the area is shown where the dendritic processes leave the furrow and travel towards the sense cell bodies. The scale indicates 50 μ in (a), and 5 μ in (b), (c), (d). Smaller details (e.g. pores, dendrites, axons, thin cuticular parts) are not drawn to scale.

The different olfactory sensillum types

While the typical organization of an olfactory sensillum, as given here, is rather constant, there is great variation in the shape of the cuticular processes. These cuticular parts of the sensillum are the basis of sensillum classification into hairs, cones, plates etc. Receptor cells reacting to odour stimuli have been found on four types of insect sensilla. Three of them are trichoid sensilla (sensilla trichodea = thick walled hairs—not to be confused with thick walled bristles; s. basiconica = thin walled pegs; s. coeloconica or/and s. ampullacea = pit-pegs), the fourth type is placoid (s. placodea) (Fig. 2).

Sensilla trichodea

The fine structure of sensilla trichodea has been studied in the male silk moths *Antheraea pernyi* and *Bombyx mori* (Steinbrecht unpublished, quoted by Schneider, Lacher and Kaissling, 1964 and Boeckh *et al.*, 1965). The dendrites do not branch and invade the hair lumen up to the tip. The cuticle of the hair wall measures one μ in thickness, but tapers progressively towards the tip. The whole surface is covered with minute pores, at the bottom of which several tubules insert, penetrating the rest of the cuticle independently (Fig. 3a). The dimensions of pores and tubules are given in Table I.

Sensilla basiconica

The thickness of the cuticle in the sensilla basiconica is relatively constant over the whole length of the peg and measures between 0·05 and 0·3 μ, depending on the species. In most cases (for exceptions see below) the dendrite(s) divide(s) into many fine branches at the base of the cuticular process. The pores may be of somewhat different size and shape (Fig. 3b, c, d). The tubules, however, have always a diameter between 100 and 200 Å, varying only in their length and number per pore (see Table I). These structures, here referred to as tubules, have been called "*Kanälchen*" by Richter (1962) and "pore filaments" by Slifer and Sekhon (1964a).

Morphological data similar to those of the species presented in Table I have been reported for basiconic sensilla of the honey bee *Apis mellifica* (Slifer and Sekhon, 1961), of the milkweed bug *Lygaeus kalmii* (Slifer and Sekhon, 1963) and of the beetle *Popilius disjunctus* (Slifer and Sekhon, 1964b), but here the proof of olfactory function is still lacking. In the honey bee none of the attempts to record electrophysiological responses to various odourous stimuli has so far been successful (Lacher, 1964). Possibly this sensillum showing a pore-tubulus-system is not of olfactory function *sensu stricto*.

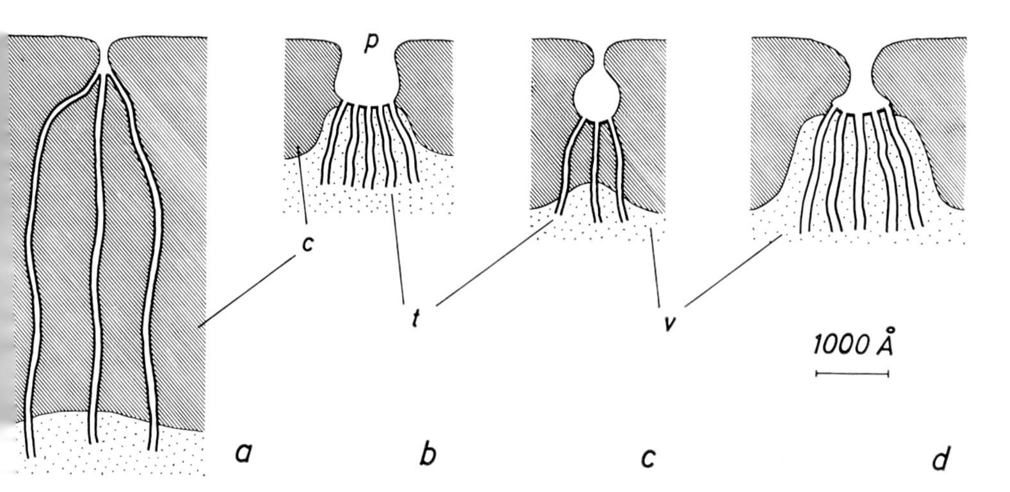

FIG. 3. Various forms of pore-tubulus-systems of olfactory sensilla in longitudinal section:
(a) *Antheraea pernyi* ♂, sensillum trichodeum;
(b) *Locusta migratoria*, s. basiconicum;
(c) *Necrophorus vespillio*, s. basiconicum;
(d) *Bombyx mori*, s. basiconicum.
The figures are drawn to scale after original electronmicrographs, but slightly schematized. The outside of the sensillum is up, the lumen of the hair down. *c* = cuticle; *p* = pore; *t* = tubule; *v* = fluid-filled hair vacuole.

There is, on the other hand, clear evidence of olfactory sensilla without such a pore-tubulus-system. This is the case in the "sharp-tipped, thin-walled hair" of the mosquito *Aedes aegypti* (Slifer and Sekhon, 1962; electrophysiology by Lacher, 1967). The fine structure of this sensillum which has been classified as a s. basiconicum by use of the light microscope, is similar to that of the s. coeloconica (see below). In the flies *Phormia regina* (Dethier, Larsen and Adams, 1963) and *Sarcophaga argyrostoma* (Slifer and Sekhon, 1964a) similar sensilla have been described, yet the exact physiological proof of olfactory function is difficult here, because these sensilla occur intermingled with "normal" s. basiconica, from which they cannot be distinguished by light microscopy.

Sensilla coeloconica

Only the pit-pegs of grasshoppers and the honey bee have been studied both electronmicroscopically and electrophysiologically. They all are longitudinally grooved, the flutes appearing as scallops in cross-sections, the number of which is 16 in *Romalea microptera* (Slifer *et al.*, 1959) or close to 8, 16, or 32 in *Apis* (Slifer and Sekhon, 1961). The

TABLE I

Fine structure of pore-tubule-systems in

Species and order	Sensillum type	Length of sensillum	No. of sense cells per sensillum	Pore diameter
Antheraea pernyi ♂ (Lepidoptera)	s. trichodeum	150–370 μ	1–3	~100 Å
Bombyx mori ♂ (Lepidoptera)	s. trichodeum	45–110 μ	1–2	100–150 Å
Melanoplus differentialis (Orthoptera)	s. basiconicum	12–20 μ	22–51	1000–2000 Å[4]
Locusta migratoria (Orthoptera)	s. basiconicum	~10 μ	~25–45	200–750 Å
Necrophorus vespilloides (Coleoptera)	s. basiconicum type I	30–40 μ	1	~100 Å
	s. basiconicum type II	15–20 μ	2	100–150 Å
Bombyx mori (Lepidoptera)	s. basiconicum	12–37 μ	1–3	150–300 Å
Protoparce sexta—larva (Lepidoptera)	s. basiconicum		4–7	100–200 Å
Aedes aegypti (Diptera)	s. basiconicum type A2[6]	20–40 μ	4–5[6]	~500 Å
Phormia regina (Diptera)	s. basiconicum (*a*) in pits	~13 μ	1–2	~350 Å
	s. basiconicum (*b*) on surface	~25 μ	"identical in nearly all respects"	
Sarcophaga argyrostoma (Diptera)	s. basiconicum (*a*) on surface	~25 μ	5–6	<300 Å
	s. basiconicum (*b*) in pits	~8 μ	2	similar to the
Calliphora erythrocephala larva—(Diptera)	s. basiconicum	~10 μ	21	~300 Å
Apis mellifica (Hymenoptera)	s. placodeum	ϕ~12μ	15–30 mean: 18	~150 Å

[1] tubule = "pore filament" (Slifer and Sekhon, 1964a) = "*Kanälchen*" (Richter, 1962);
[2] value based on rough estimates of pore density;
[3] preservation of tubules not good enough for measurements;
[4] these values, given 1959, have not been corrected 1964, but the electronmicrographs of the more recent paper indicate that they are incorrect; pore diameter and pore density look very similar to *Locusta*;

sensilla with proved olfactory function

No. of pores per sensillum	Tubule[1] diameter	No. of tubules per pore	Author Fine structure	Author Physiological proof
20000–50000[2]	~130 Å	3–5	K. D. Ernst and R. A. Steinbrecht (unpubl.)	Schneider *et al.* (1964)
1000–3000[2]	<200 Å		R. A. Steinbrecht (unpubl.)	E. Priesner (unpubl.)
~150[4]	~200 Å	23–24	Slifer *et al.* (1959); Slifer and Sekhon (1964c)	J. Boeckh (unpubl.)[5] in *Locusta*
~3000[2]	~125 Å		Steinbrecht (unpubl.)	J. Boeckh (unpubl.)
10000–15000	~100 Å	5–7	Schneider, Steinbrecht and Ernst (1966); K. D. Ernst (in preparation)	Boeckh (1962)
4500–6000	~100 Å	6–9		
	~150 Å	15–20	R. A. Steinbrecht (unpubl.)	E. Priesner (unpubl.)
	[3]		Schoonhoven and Dethier (1966)	
~1000	[3]		Slifer and Sekhon (1962)	Lacher (1967)
180–360[7]	[3]		Dethier *et al.* (1963)	J. Boeckh (unpubl.)[5] *in Sarcophaga*
with the pegs found in pits				
	~200 Å	20–30	Slifer and Sekhon (1964a)	J. Boeckh (unpubl.)[8]
surface peg				
~9000	~200 Å	3–5	Richter (1962)	Bolwig (1946)[5] larva of *Musca*
3000–4000	[3]	~4	Richards (1952); Slifer and Sekhon (1960, 1961) Krause (1960); R. A. Steinbrecht (unpubl.)	Lacher (1964) Boeckh *et al.* (1965)

[5] olfactory function suggested from work with homologous sensilla in *Locusta*, *Sarcophaga* and *Musca* respectively;
[6] Steward and Atwood (1963);
[7] counts of pore density in the published electronmicrographs suggest higher values;
[8] physiological proof difficult, see text.

dendrites do not branch and are enclosed in a dense sheath up to the tip of the peg. In serial ultrathin sections through several pit-pegs of *Locusta migratoria*, neither an opening in the tip, nor tubules combined with pores in the wall of the sensillum, could be observed (R. A. Steinbrecht, unpublished). But there are radially arranged cuticular bars, which are hollow and are open to the dendritic processes and to the outside as well. Some medium dense material seems to come out through these openings and one could tentatively speculate that these spots serve the same function as the pore-tubulus-systems of other olfactory sensilla.

Sensilla placodea

The pore plate of the honey bee *Apis mellifica* was the first insect sensillum studied with the electron microscope. In 1952, Richards had already demonstrated on free-hand tangential sections, that the very thin cuticular membrane around each plate (see Fig. 2d) is traversed by some 120–150 radially arranged thickenings. Between these rod-like structures he observed rows of holes in the thin membrane. The cytological organization has been described by Slifer and Sekhon (1960, 1961) and Krause (1960). Dendritic processes reach up to the plate and below the circular thin membrane. These authors confirmed most of Richard's findings on the cuticular architecture, but were unable to demonstrate the existence of the pores with certainty. This was accomplished recently with refined sectioning and staining methods (R. A. Steinbrecht, unpublished). The cuticle penetrated by these pores is less than 300 Å thick; at the inner end of the pores about four short tubulus-like structures arise (see Table I). A structurally different plate sensillum with unknown function has been described in aphids by Slifer *et al.* (1964).

Final remarks

The observations of Richards (1952) showed that lipid solvents may alter the appearance of cuticular material. One must not forget that the routine embedding procedures for thin sectioning involve such lipid solvents which might very well cause important changes, especially in the layers of the epicuticle. Future investigation of the pore-tubulus-systems should include the study of the nature and properties of the specialized sensillum cuticle.

Slifer and her coworkers (1959, 1961, 1964a,c) and Richter (1962) assume that the tubules (=pore filaments) are extremely fine microvillus-like protrusions of the dendritic cell membrane. Consequently in such a case the receptor cell would be in direct contact with the environment, its dendritic plasma membrane being apposed to it at the inner nd of the pores.

Contrary to this hypothesis there is now evidence that the tubules are extracellular material and not part of the dendritic cell membrane (K. D. Ernst, in preparation). In this context Locke's (1965) hypothesis that the tubules are "lipid-water liquid crystals in the middle phase" is of interest. Such liquid crystals might provide a trapping surface for odourous molecules and ensure their rapid diffusion to the receptor site as well.

The form of the connection of the tubules to the dendrites, however, remains still obscure. In most electron-micrographs they seem to end about 100 Å from the dendrite surface. In spite of this they might well have been attached to it *in vivo*.

FUNCTIONAL COMPLEXITY OF SENSILLA

Only the s. placodea seem—as a morphological type in hymenoptera—to be exclusively olfactory. S. basiconica with a corresponding porous fine structure are possibly also olfactory in general as predicted by Slifer *et al.* (1959), but sensilla of this general type could perhaps also be serving other modalities (see Loftus', 1966, and Schoonhoven's, 1967 experiments with thermoreceptors of the cockroach and moths). S. trichodea, relatively thick walled and rather long hairs, are olfactory in the Lepidoptera heterocera, but e.g. mechanoreceptors in the cockroach (Nicklaus, 1965). Here too, the presence or absence of pores in the hair cuticle may be the morphological distinction between functionally different hairs. There are cases where a distinction between bristles (sensilla chaetica) and hairs (s. trichodea) might be difficult. Some insect taste organs, notably in flies, look like the bristles of moths (for a definition see Schneider and Kaissling, 1957), some like hairs. The taste sensilla are an important example, because one of the four or five sense cells (Peters, 1961, 1963) is not a gustatory receptor but reacts to mechanical stimuli (Wolbarsht and Dethier, 1958). This demonstrates that receptor cells which serve different modalities may belong to one and the same sensillum. The individual taste cells of this sensillum, however, do react to different contact chemical stimuli, which means that they serve different taste qualities. This is also true for some of the olfactory receptors listed below. Individual cells of one sensillum may all react to the same odourous compounds though if, as frequently happens, they react to a series of different ones, the sensillum can be stimulated by different qualities of stimulus.

THE TWO PHYSIOLOGICAL TYPES OF OLFACTORY RECEPTOR CELLS

Experiences in our laboratory led us to distinguish two general types of olfactory receptor cells in insect sensilla. The basis for such a distinction is the qualitative reactivity range of the cells. Individual olfactory cells react to several or many odourous compounds. The range of compounds will—in analogy to visual phenomena—be called spectral range or reactivity spectrum. The use of the term spectrum, however, neither means that we understand the common stimulatory property of the compounds which belong to a cell spectrum, nor does it mean that we are able to arrange different, but overlapping spectra of olfactory cells linearly as it is possible to do with visual cell spectra (Schneider *et al.*, 1964; Boeckh *et al.*, 1965).

The olfactory receptor cells are the sensory units belonging to certain sensilla. There is no large scale correspondence of morphological characteristics of the cells or the sensilla with the physiological properties as there is, for example, with vertebrate cones reacting to different light spectra in the retina. However, in small systematic groups, some predictions can now be made. We are, for instance, able to say that all the sensilla trichodea of the moths very probably have at least one olfactory cell responding to the sexual attractant of the female.

The two cell types are:

(1) *Odour specialist.* These are sense cells responding to biologically important odours like sexual attractants, warning odours and specific food odours. Many such cells with identical spectra occur in either one or both sexes of one species. A *Bombyx* male has on both antennae approximately 40 000 sex-odour receptor cells belonging to the s. trichodea. All the specialized cells of this large population have the same reaction spectrum.

(2) *Odour generalists.* In a given species, each of the many cells of this type has a unique and stable odour spectrum. Spectra of individual cells are different, but overlap. Neighbouring olfactory receptor cells which belong to one sensillum may not only differ from one another as it was described for the generalists, but may also be generalists and specialists. While Lacher (1964) demonstrated that the pore plate cells of the honeybee are odour generalists, it was found later that some of these cells are specialists reacting to pheromones (Kaissling, see Boeckh *et al.*, 1965).

INSECT SENSILLA AS IDENTIFIED TO BE OF OLFACTORY FUNCTION

Sensilla trichodea

Bombyx has s. trichodea in both sexes (Schneider and Kaissling, 1957). Saturniids and most other Lepidoptera heterocera examined have s. trichodea only in the male sex (Boeckh *et al.*, 1960). These sensilla are innervated by two or three sensory nerve cells.

In the *male moth*, only one or more of these cells respond to the sexual attractant of the female of the same or of a related species (Schneider, 1962; Schneider *et al.*, 1964; Boeckh *et al.*, 1965) with extremely high sensitivity (Schneider, Block, Boeckh and Priesner, 1967). In *Bombyx* and some Saturniids, these cells have a stable side spectrum of less effective odorants like alcohols and terpenes (see Boeckh *et al.*, 1965, quoting unpublished experiments by E. Priesner). All the cells of these male sensilla which do not respond to the sexual attractant, respond to other compounds similar to the side spectrum of the bombykol receptor cell. Some compounds (oil of cloves in *Bombyx*; geraniol in *Antheraea*) were found to inhibit the pheromone receptor cell if applied in higher concentration.

Female *Bombyx* have s. trichodea like the males. The olfactory cells of these sensilla respond to alcohols, terpenes and some other compounds with relatively high threshold (depolarization plus nerve impulses). All the cells in the s. trichodea of females, then, have responses similar to those cells in the male hair sensillum which gave no response to the sexual attractant (E. Priesner, unpublished).

Sensilla basiconica

Lepidoptera heterocera

Antheraea pernyi ♂. The olfactory receptor cells are typical odour generalists, reacting to many compounds (alcohols, terpenes, essential oils etc.), either by de- or hyperpolarization, with relatively high threshold (Schneider *et al.*, 1964).

Hyalophora gloveri larva. Cell responses during recordings from s. basiconica of antenna and maxillary palpus are in principle identical to those of *Protoparce* larvae (see below, Schoonhoven and Dethier, 1966).

Bombyx ♂ ♀. Most of the receptor cells are presumably odour generalists and react to numerous compounds, including those in leaf extracts (green odours e.g. hexenal, hexenol etc.). In the female, some of the cells are specialized for the food plant odour (mulberry) and respond to minute quantities (E. Priesner, unpublished). Possibly, some of the male receptor cells of these sensilla are specialists, because they are hyperpolarized (inhibited) by the sexual attractant bombykol.

Bombyx larva, antennal s. basiconicum. Odour receptor cells respond with depolarization and impulse volleys to hexanol, hexenol, butyl-aldehyde, acetic acid, butyric acid and to mulberry leaves. Some of these compounds as well as ether and chloroform may either depolarize the cell and elicit the impulses or hyperpolarize and inhibit the impulses. Because of the mostly multicellular response, it is difficult to say whether the cells are specialists or generalists (Morita and Yamashita, 1961).

Sphinx pinastri ♂. Antennal cells of s. basiconica react to a number of alcohols, terpenes and essential oils. Inhibition was not observed during a preliminary study but is probably also present. The small number of recordings does not permit any further classification of the cells (Schneider and Boeckh, 1962).

Protoparce sexta larva. Odour receptor cells of the antenna and the maxillary palpus respond to a large number of leaf odours including the food plant and to some other compounds with either an increase or decrease of impulse frequency. The cells are presumably generalists (Schoonhoven and Dethier, 1966).

Coleoptera

Necrophorus, several species ♂ and ♀. Singly and doubly innervated sensilla in central fields of the antennal segments. The cells are odour specialists with a very low threshold for decomposing meat. They also respond to mercaptanes and amines (depolarization). Saturated homologous fatty acids either hyperpolarize (C_3) or depolarize (C_{4-10}) the cell. The threshold is minimal at C_6 = caproic acid (Boeckh, 1962).

Thanatophilus rugosus ♂ and ♀. Cell-situation and responses are in general similar to *Necrophorus*. Carrion odour specialists are inhibited by propionic and butyric acid. They are excited by the C_{5-8} fatty acids, by some aldehydes like hexenal, and some other compounds (Boeckh *et al.*, 1965; Boeckh, 1967a).

Diptera

Calliphora erythrocephala ♂ and ♀. Two types of antennal surface peg cells are odour specialists. One cell type responds to meat, carrion and cheese, presumably with very low threshold. This cell is inhibited by propionic, butyric and valeric acids and styrolyl acetate, but is excited by some alcohols, aldehydes, mercaptanes and other compounds. The other cell is excited by styrolylacetate but only weakly inhibited by carrion (Boeckh *et al.*, 1965; J. Boeckh, unpublished).

Sarcophaga sp., *Eristalomyia* sp., and *Lucilia* sp. have not yet been thoroughly investigated. The responses to decomposing meat are similar to those of *Calliphora* (J. Boeckh, unpublished).

Aedes aegypti ♀. Sense cells of surface pegs of the antenna respond to odour (Lacher, 1967). Three morphological types of sensilla are distinguished by their length (Steward and Atwood, 1963). Type A_1 cells (*ca* 60 μ in length) are inhibited by essential oils and excited by fatty acids. Type A_2 cells (*ca* 40 μ in length) are inhibited by C_{2-5} fatty acids and excited by C_{7-10} fatty acids. Some essential oils excite, others inhibit the cells. Type A_3 cells (*ca* 5 μ in length) are excited by fatty acids. Only a limited number of A_3-sensillum cells have been studied. All these sense cells seem to be odour specialists, possibly belonging to a number of types with slightly different response spectra.

Orthoptera

Locusta migratoria ♂ and ♀. Sense cells of short, thin-walled pegs with many pores respond to essential oils, terpenes and other flowery or fruity odours (J. Boeckh, unpublished).

Sensilla coeloconica

Apis mellifica ♂ and ♀. Some cells of the s. coeloconica (and/or s. ampullacea) respond to carbon dioxide. These cells are specialists not responding to any other "normal" odorant. Other cells of these sensilla respond to temperature and/or humidity (Lacher, 1964).

Locusta migratoria ♂ and ♀. In some of these sensilla, one sense cell responds to leaf-extracts (green-odours) such as hexenal and hexenol. More than 80 other compounds or derivatives have been tested qualitatively and some of them also quantitatively with this cell. (In other coeloconic sensilla of this animal we found cells responding to humidity and temperature.) Recordings from the olfactory cell showed that the simple C_6 compounds (paraffin, hexane, hexanol, hexanal) are without effect. Functional groups are essential for the stimulatory power of a compound (hexanol and hexanal are without effect, while hexenol, hexenal, hexanoic and hexenoic acids excite the cell). Of the homologous fatty acids, formic and acetic acids are without effect, while C_{3-8} acids are effective odorants exciting the cell. Maximal responses are elicited by caproic acid (C_6). No inhibition has been observed as yet with these cells (Boeckh *et al.*, 1965; J. Boeckh, 1967a and b).

Bombyx ♂ (♀). Sense cells of the pit-pegs respond to warm humid air (E. Priesner, unpublished).

Sensilla placodea

Apis mellifica ♂ and ♀. Several or many cells of the plate-sensilla are odour-generalists, responding to fruity, flowery odours and fatty acids by inhibition or excitation (Lacher, 1964). Other cells in drones, worker bees and queens were found to be odour specialist, one or more cells per sensillum are excited by the queen substance (9-oxo-decenoic acid) and by caproic acid in low concentration as well as by the Nasanoff gland odour (geranoic acid, etc.) in higher concentration. Another cell is specialized for the Nasanoff gland odour but also responds weakly to the queen substance (Boeckh *et al.*, 1965; K.-E. Kaissling, unpublished).

References

Adams, J. R. (1961). *The location and histology of the contact chemoreceptors of the stable fly*, Stomoxys calcitrans *L.* Ph.D. dissertation Rutgers—The State University of New Jersey, New Brunswick.

Adams, J. R., Holbert, P. E. and Forgash, A. J. (1965). Electron microscopy of the contact chemoreceptors of the stable fly, *Stomoxys calcitrans* (Diptera: Muscidae). *Ann. ent. Soc. Am.* **58**: 909–917.

Boeckh, J. (1962). Elektrophysiologische Untersuchungen an einzelnen Geruchsrezeptoren auf der Antenne des Totengräbers (*Necrophorus*, Coleoptera). *Z. vergl. Physiol.* **46**: 212–248.

Boeckh, J. (1967a). Inhibition and excitation of single insect olfactory receptors, and their role as a primary sensory code. *Proc. 2nd Int. Symp. Olfaction Taste*: 721–735. Oxford: Pergamon Press.

Boeckh, J. (1967b). Reaktionsschwelle, Arbeitsbereich und Spezifität eines Geruchsrezeptors auf der Heuschreckenantenne. *Z. vergl. Physiol.* **55**: 378–406.

Boeckh, J., Kaissling, K.-E. and Schneider, D. (1960). Sensillen und Bau der Antennengeissel von *Telea polyphemus* (Vergleiche mit weiteren Saturniden: *Antheraea, Platysamia* und *Philosamia*). *Zool. Jb.* (Anat.) **78**: 559–584.

Boeckh, J., Kaissling, K.-E. and Schneider, D. (1965). Insect olfactory receptors. *Cold Spring Harb. Symp. quant. Biol.* **30**: 263–280.

Bolwig, N. (1946). Senses and sense organs of the anterior end of the house fly larvae. *Vidensk. Medd. dansk naturh. Foren.* **109**: 80–217.

Dethier, V. G. (1941). The function of the antennal receptors in lepidopterous larvae. *Biol. Bull. mar. biol. Lab. Woods Hole* **80**: 403–414.

Dethier, V. G. (1954). The physiology of olfaction in insects. *Ann. N.Y. Acad. Sci.* **58**: 139–157.

Dethier, V. G., Larsen, J. R. and Adams, J. R. (1963). The fine structure of the olfactory receptors of the blowfly. *Proc. 1st Int. Symp. Olfaction Taste*: 105–110. Oxford: Pergamon Press.

Henke, K. and Rönsch, G. (1951). Über die Bildungsgleichheiten in der Entwicklung epidermaler Organe und die Entstehung des Nervensystems im Flügel der Insekten. *Naturwissenschaften* **38**: 335–336.

Jungblut, P. W., Hätzel, J., de Sombre, E. R. and Jensen, E. V. (1967). *Über Hormonrezeptoren.* Die Östrogen bindenden Prinzipien der Erfolgsorgane. Heidelberg: Springer.

Krause, B. (1960). Elektronenmikroskopische Untersuchungen an den Plattensensillen des Insektenfühlers. *Zool. Beitr.* NF. **6**: 161–205.

Lacher, V. (1964). Elektrophysiologische Untersuchungen an einzelnen Rezeptoren für Geruch, Kohlendioxyd, Luftfeuchtigkeit und Temperatur auf den Antennen der Arbeitsbiene und der Drohne (*Apis mellifica* L.). *Z. vergl. Physiol.* **48**: 587–623.

Lacher, V. (1967). Elektrophysiologische Untersuchungen an einzelnen Geruchsrezeptoren auf den Antennen weiblicher Moskitos (*Aedes aegypti* L.). *J. Insect Physiol.* **13**: 1461–1471.

Lacher, V. and Schneider, D. (1963). Elektrophysiologischer Nachweis der Riechfunktion von Porenplatten (Sensilla placodea) auf den Antennen der Drohne und Arbeitsbiene (*Apis mellifica* L.). *Z. vergl. Physiol.* **47**: 274–278.

Locke, M. (1965). Permeability of insect cuticle to water and lipids. *Science, N.Y.*, **147**: 295–298.

Loftus, R. (1966). Cold receptor on the antenna of *Periplaneta americana. Z. vergl. Physiol.* **52**: 380–385.

Morita, H. and Yamashita, S. (1961). Receptor potentials recorded from Sensilla basiconica on the antenna of the silkworm larvae, *Bombyx mori. J. exp. Biol.* **38**: 851–861.

Nicklaus, R. (1965). Die Erregung einzelner Fadenhaare von *Periplaneta americana* in Abhängigkeit von der Größe und Richtung der Auslenkung. *Z. vergl. Physiol.* **50**: 331–362.

Peters, W. (1961). Die Zahl der Sinneszellen von Marginalborsten und das Vorkommen multipolarer Nervenzellen in den Labellen von *Calliphora erythrocephala* Mg. (Diptera). *Naturwissenschaften* **48**: 412–413.

Peters, W. (1963). Die Sinnesorgane an den Labellen von *Calliphora erythrocephala* Mg. *Z. Morph. Ökol. Tiere* **55**: 259–320.

Richards, A. G. (1952). Studies on arthropod cuticle. VIII. The antennal cuticle of honeybees, with particular reference to the sense plates. *Biol. Bull. mar. biol. Lab. Woods Hole* **103**: 201–225.

Richter, S. (1962). Unmittelbarer Kontakt der Sinneszellen cuticularer Sinnesorgane mit der Außenwelt. Eine licht- und elektronenmikroskopische Untersuchung der chemorezeptorischen Antennensinnesorgane der *Calliphora*-Larven. *Z. Morph. Ökol. Tiere* **52**: 171–196.

Schneider, D. (1957). Elektrophysiologische Untersuchungen von Chemo- und Mechanorezeptoren der Antenne des Seidenspinners *Bombyx mori* L. *Z. vergl. Physiol.* **40**: 8–41.

Schneider, D. (1962). Electrophysiological investigation on the olfactory specificity of sexual attracting substances in different species of moths. *J. Insect Physiol.* **8**: 15–30.

Schneider, D. (1964). Insect antennae. *A. Rev. Ent.* **9**: 103–122.

Schneider, D. and Boeckh, J. (1962). Rezeptorpotential und Nervenimpulse einzelner olfaktorischer Sensillen der Insektenantenne. *Z. vergl. Physiol.* **45**: 405–412.

Schneider, D. and Kaissling, K.-E. (1957). Der Bau der Antenne des Seidenspinners *Bombyx mori* L. II. Sensillen, cuticulare Bildungen und innerer Bau. *Zool. Jb.* (Anat.) **76**: 223–250.

Schneider, D., Lacher, V. and Kaissling, K.-E. (1964). Die Reaktionsweise und das Reaktionsspektrum von Riechzellen bei *Antheraea pernyi* (Lepidoptera, Saturniidae). *Z. vergl. Physiol.* **48**: 632–662.

Schneider, D., Steinbrecht, R. A. and Ernst, K. (1966). Antennales Riechhaar des Aaskäfers. *Naturw. Rdsch., Stuttg.* **19**: H. 3, I u. III.

Schneider, D., Block, B. C., Boeckh, J. and Priesner, E. (1967). Die Reaktion der männlichen Seidenspinner auf Bombykol und seine Isomeren: Elektroantennogramm und Verhalten. *Z. vergl. Physiol.* **54**: 192–209.

Schoonhoven, L. M. (1967). Some cold receptors in larvae of three lepidoptera species. *J. Insect Physiol.* **13**: 821–826.

Schoonhoven, L. M. and Dethier, V. G. (1966). Sensory aspects of host-plant discrimination by lepidopterous larvae. *Archs néerl. Zool.* **16**: 497–530.

Slifer, E. H. (1954). The permeability of the sensory pegs on the antennae of the grasshopper (Orthoptera, Acrididae). *Biol. Bull. mar. biol. Lab., Woods Hole* **106**: 122–128.

Slifer, E. H. (1955). The distribution of permeable sensory pegs on the body of the grasshopper (Orthoptera, Acrididae). *Proc. R. ent. Soc. Lond.* **29**: 177–179.

Slifer, E. H. (1956). Permeable spots in the cuticle of the thin-walled pegs on the antenna of the grasshopper. *Science, N.Y.* **124**: 1203.

Slifer, E. H. (1960). A rapid and sensitive method for identifying permeable areas in the body wall of insects. *Ent. News* **71**: 179–182.

Slifer, E. H. (1961). The fine structure of insect sense organs. *Int. Rev. Cytol.* **11**: 125–159.

Slifer, E. H. (1962). Sensory hairs with permeable tips on the tarsi of the yellow-fever mosquito, *Aedes aegypti*. *Ann. ent. Soc. Am.* **55**: 531–535.

Slifer, E. H. and Sekhon, S. S. (1960). The fine structure of the plate organs on the antenna of the honey bee, *Apis mellifera* L. *Expl Cell Res.* **19**: 410–414.

Slifer, E. H. and Sekhon, S. S. (1961). Fine structure of the sense organs on the antennal flagellum of the honey bee, *Apis mellifera* L. *J. Morph.* **109**: 351–381.

Slifer, E. H. and Sekhon, S. S. (1962). The fine structure of the sense organs on the antennal flagellum of the yellow fever mosquito, *Aedes aegypti* L. *J. Morph.* **111**: 49–67.

Slifer, E. H. and Sekhon, S. S. (1963). Sense organs on the antennal flagellum of the small milkweed bug, *Lygaeus kalmii* Stal (Hemiptera, Lygaeidae). *J. Morph.* **112**: 165–193.

Slifer, E. H. and Sekhon, S. S. (1964a). Fine structure of the sense organs on the antennal flagellum of a flesh fly, *Sarcophaga argyrostoma* R.-D. (Diptera, Sarcophagidae). *J. Morph.* **114**: 185–207.

Slifer, E. H. and Sekhon, S. S. (1964b). Fine structure of the thin-walled sensory pegs on the antenna of a beetle, *Popilius disjunctus* (Coleoptera; Passalidae). *Ann. ent. Soc. Am.* **57**: 541–548.

Slifer, E. H. and Sekhon, S. S. (1964c). The dendrites of the thin-walled olfactory pegs of the grasshopper (Orthoptera, Acrididae). *J. Morph.* **114**: 393–410.

Slifer, E. H., Prestage, J. J. and Beams, H. W. (1959). The chemoreceptors and other sense organs on the antennal flagellum of the grasshopper (Orthoptera, Acrididae). *J. Morph.* **105**: 145–191.

Slifer, E. H., Sekhon, S. S. and Lees, A. D. (1964). The sense organs on the antennal flagellum of aphids (Homoptera) with special reference to the plate organs. *Q. Jl microsc. Sci.* **105**: 21–29.

Steward, C. C. and Atwood, C. E. (1963). The sensory organs of the mosquito antenna. *Can. J. Zool.* **41**: 577–594.
von Frisch, K. (1919). Zur alten Frage nach dem Sitz des Geruchssinnes bei Insekten. *Verh. zool.-bot. Ges. Wien* **69**: 17–26.
von Frisch, K. (1921). Über den Sitz des Geruchssinnes bei Insekten. *Zool. Jb.* (Zool. u. Physiol.) **38**: 449–516.
Wigglesworth, V. B. (1941). The sensory physiology of the human louse *Pediculus humanis corporis*, de Geer (Anoplura). *Parasitology* **33**: 67–109.
Wolbarsht, M. L. and Dethier, V. G. (1958). Electrical activity in the chemoreceptors of the blowfly. I. Responses to chemical and mechanical stimulation. *J. gen. Physiol.* **42**: 393–412.

Symp. zool. Soc. Lond. (1968) No. 23, 299–326

ON SUPERFICIAL RECEPTORS

M. S. LAVERACK

Gatty Marine Laboratory and Department of Natural History, University of St Andrews, Scotland

SYNOPSIS

Superficial receptors are those organs placed at or near a surface and monitoring the changes that take place in the ambient environment.

Two examples of such receptors are the sensory cells of the statocyst of the snail *Helix pomatia*, and of the pedicellariae of *Echinus esculentus*. In the first case a description is given of the fine structure of the statocyst, demonstrating that the classical description of Pfeil (1922) is erroneous. In particular the giant cell of the statocyst wall is statolith-producing, not sensory, whilst the so-called syncytium is composed of numerous small sensory cells. These sensory cells possess axonic processes that emerge into the statocyst nerve. The sensory surface is a much villified one, with many finely branched projections into the lumen of the statocyst. Hidden amongst the villi is a single small ciliary peg; a minute bump on the cell surface, that has the features of a basal body/centriole. These sensory cells are presumably mechanoreceptive, and a hypothesis is advanced as to the mode of function of the statocyst.

The pedicellariae of *Echinus* which have been investigated are of three types, the tridentate, the ophiocephalous and the gemmiform. Experimental analysis of the closing responses of these organs indicate that closure occurs in tridentate forms when the inner faces of the jaws are touched, but opening takes place when the outer surfaces are manipulated. Similar events take place in ophiocephalous forms. Chemical stimuli are without effect, and von Uexküll's autodermin response cannot be demonstrated. Gemmiform pedicellariae react to light mechanical stimuli by a further opening response. Very heavy mechanical stimulation may bring about closure. Chemical stimulation, by foreign material such as extracts of the tube feet of starfish, or pieces of *Homarus* muscle evoke closure which is rapid and usually irreversible. This response is mediated via the sensory hillock of cells with long cilia that occur on the inner faces of the gemmiform pedicellariae. This is demonstrated by the lack of response to chemical stimulation after cauterization of the sensory hillocks.

The cilia of the urchin sensory cells are long, and have internal structures that are typical of other types of cilia, and also have a very long ciliary root. In neither the sensory hillock, nor in the sensory cells of the *Helix* statocyst, can it be shown that the cilia are in any way involved in sensory transduction. The great extremes of structure shown, one very long, the other very short, and the apparent ability of statocyst cells to function as receptors with only a basal body present leads the author to suggest that ciliary structures are not necessary to sensory transduction; an observation in keeping with the findings of Tucker (1967) on olfactory cilia in the box turtle. The elongation of the cell surface provided by cilia may be of significance, but the part played by internal fibrils is only that of skeletal support.

INTRODUCTION

This paper would perhaps be better called "On surface receptors", except that this is ambiguous. An alternative version "The receptors of surfaces" is no more explicit. Superficial receptors are those placed at or close to a surface.

Surfaces are not necessarily those in direct contact with the external environment of an animal, although these probably spring to mind first. The internal side of tubes, cylinders and spheres also represent a surface, and in animals various fluid, and gas-filled bodies such as blood vessels or swim bladders, are also bounded by an epithelial layer that may be sensorily innervated. Thus superficial receptors are those sensory areas concerned with monitoring ambient influences. Such stimulants may be of several types; for example chemical, tactile, and thermal.

In the present paper I wish to report on two separate receptor systems. One has nothing to do with the outer surface of the animal, but is concerned with the conditions prevailing only within a circumscribed space contained within the animal body. This is the statocyst of the snail *Helix pomatia*, a roughly spherical entity that is filled with fluid in which plate shaped calcareous statoliths are suspended. The second example is that of an organ placed on the external surface of an appendage, which is linked directly with an effector organ apparently without the intervention of a "central nervous system" but which shows excellent reflex activities. This system is that of the receptors of the jaws of pedicellariae from the sea urchin *Echinus esculentus*.

Helix STATOCYST

An electron microscope investigation into the structure of the statocyst of *Helix* was carried out in order to ascertain whether or not it might provide a preparation for the study of primary receptor cells with intracellular microelectrode techniques. This was prompted by the observation of Pfeil (1922) and the review by Fretter and Graham (1962) which proposed a sensory function for cells termed "giant cells" that line a statocyst. If these cells were of the type and size indicated then a chance existed that the system would be amenable to electrophysiological analysis, as are the central ganglionic neurons of the same species. The short statocyst nerve that originates at the organ and passes to the cerebral ganglion does so through thick connective tissue sheaths, and hence is not approachable by usual wire recording electrodes, and studies in the cerebral ganglion suffer from being at best second order if not higher order responses.

General comments

The statocyst of the Roman snail *Helix pomatia* was described in detail by Pfeil (1922). Two of these organs are found lying laterally between the pleural and pedal ganglia in the circumoesophageal ganglionic complex of the nervous system. They are fluid filled spheres, containing plate-like calcified statoliths, and reach about 200 μ in diameter. The axons from the organ run directly to the cerebral ganglion.

The sensory epithelium that lines the organ is bounded by a thick collagenous capsule. This confers the spherical shape upon the statocyst, and the capsule merges imperceptibly with the sheath of the nervous system.

Two types of cell form the lining of the statocyst as described by previous authors.

Giant cells

The giant cells normally occupy the full depth of the epithelial lining of the statocyst capsule, but may in places be underlain by processes from other cells (Fig. 1). The cells are about 10 μ thick, or a little less, from the basement membrane on the outside to the lumen on the inside. The giant cell is star-shaped, extensions of the cell reaching out across the internal surface of the organ and in these areas the second type of cell obtrudes beneath. The giant cells reach 50–70 μ across in the greatest dimension. The nucleus is 5–10 μ in diameter and nearly fills the depth of the cell in one plane.

There are three main features of the giant cell. These are the ciliated surface, the mitochondrial inclusions and the associated lamellate structures.

The presence of cilia on the internal face of the statocyst was first described by Pfeil. The present studies show the cilia to be short and motile. Cilia can be seen in motion in statocysts freshly isolated in the snail's own blood and observed with a binocular dissecting microscope. Ciliary activity, however, ceases within 15 min of isolation. The activity is evident as a flickering motion, and causes movement of the smaller particles within the fluid filling the lumen. The large plate-like statoliths waver in the fluid currents, but there is no indication that they are moved around the statocyst by ciliary action. The only pronounced statolith movements occur upon passive alteration of the statocyst orientation. Under the influence of gravity the statoliths move as a collection towards the lower side of the organ.

The cilia show slight modification compared with other cilia already described in molluscs (*Anodonta*, Gibbons, 1961). Each is 5–6 μ in length and the diameter (*ca* 0·16–0·18 μ) is the same from base to tip apart from a short basal dilatation. Occasionally the enveloping membrane of the cilium is lifted to enclose a few vesicles approximately

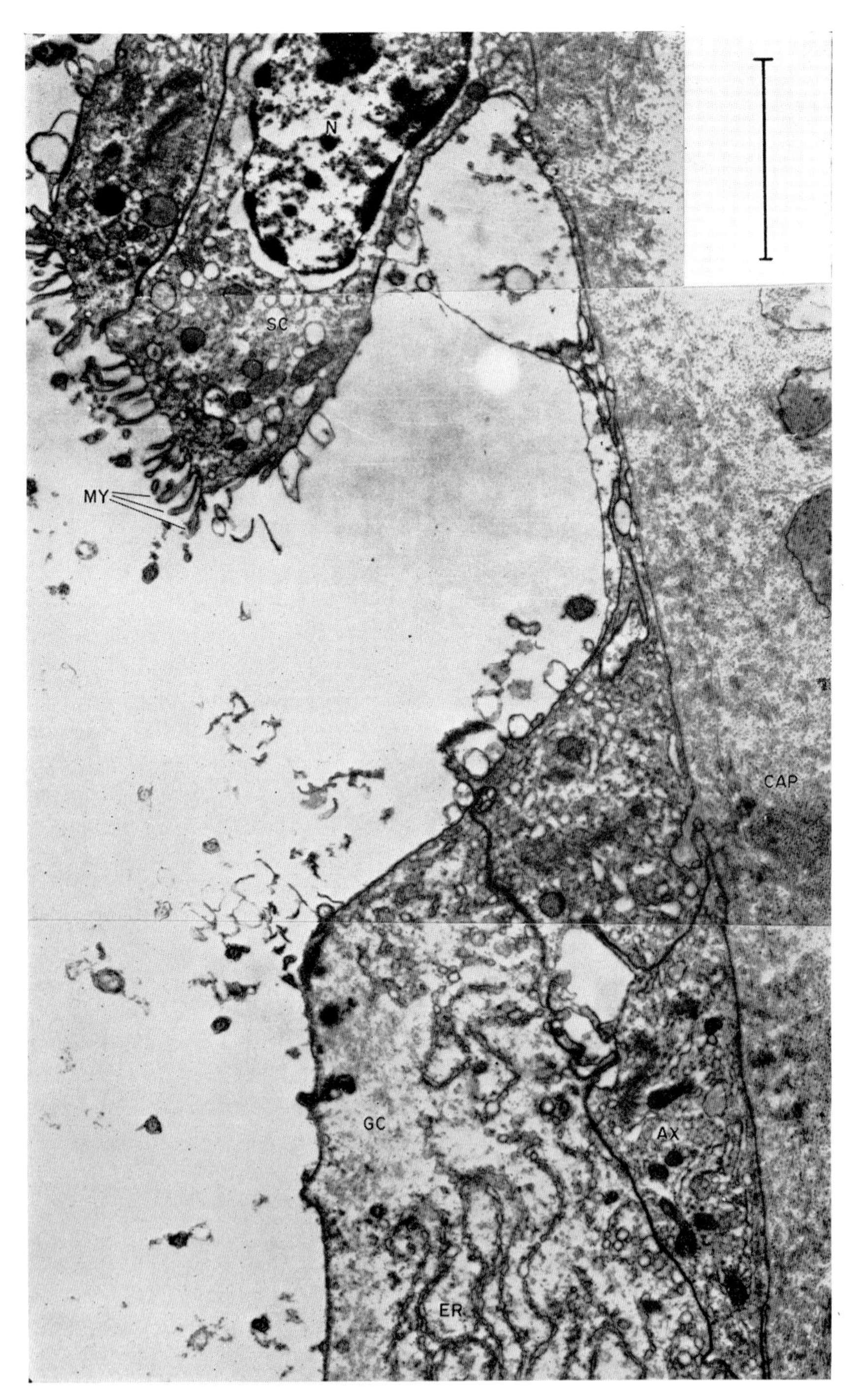

Fig. 1.

400Å in diameter within a small chamber alongside the axial structures of the cilium. The longitudinal structures of the cilium comprise nine outer pairs of outer fibrils arranged peripherally and a single pair of fibrils in the centre. A basal plate is found at the origin of the cilium just above the cell surface. At the base there is a basal body and a prolongation of the outer fibrils into the cytoplasm of the cell. A short projection extends into the cell below the basal body and represents all there is of the ciliary root. It is not possible to determine if this is banded or not. A prominent basal foot projects to one side of the ciliary base. The structure may be disposed to one side or the other in relation to the direction of the power stroke of the cilium, but there is no way of confirming this because of the difficulties of orientation of the tissue, and identification of individual cilia. The basal foot is not banded (Fig. 2).

The mitochondria of the giant cells are small, ranging from 0·5 to 1 μ in length and 0·2 to 0·3 μ in diameter, and possess many cristae. These organelles are arranged in a characteristic manner in the giant cell. They are most prominent at the periphery, occurring in large numbers along the surface of the cell, just below the ciliary region. A smaller number of mitochondria is found along the basal side of the giant cell where it is applied to the basement membrane and capsule.

The centre of the cell contains a considerable volume of cytoplasm with few mitochondria, but much endoplasmic reticulum and many lamellate structures. The endoplasmic reticulum is very extensive, consisting of fine channels about 300 Å in diameter that can be traced for distances of up to 20 μ. The surfaces of these channels are covered with ribosomes. Several variations on this basic simple tubular pattern are notable. Vesiculation, apparent as a beading of the tubules, appears to be followed by formation of blebs that are cut off from the main channel system. The vesicles reach 0·2–0·4 μ diameter. A second type of vesicle also is found but reaches only 600 Å diameter. Neither type of vesicle shows any consistent alignment with the boundaries of the cell.

Two major modifications of the cell cytoplasm should be mentioned. The first of these is the formation of large lamellate bodies (Fig. 3). The basic structure of these bodies, which are in close relation to the endoplasmic reticulum, is of fine tubular wrappings that are of similar

---

Fig. 1. The epithelium of the statocyst comprises giant cells (GC) and sensory cells (SC). The latter may lie between the giant cell and the capsule (CAP), and the axons (AX) run to the statocyst nerve. In the case in this figure part of the giant cell has detached from the capsule and left a gap in the continuity of the epithelial lining. Large channels of endoplasmic reticulum (ER) are notable in the cytoplasm of the giant cell, and microvilli (MY) extend from the lumen surface of the sensory cells. N = nucleus. Scale = 5 μ.

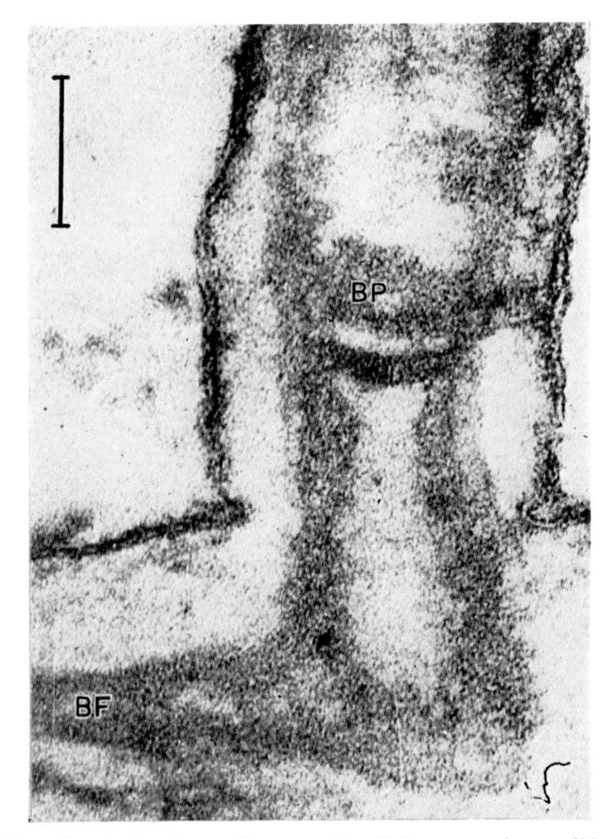

FIG. 2. The cilia of the giant cell are motile. Only one major modification on the usual cilium pattern is demonstrated, the basal foot (BF) extends to one side and seem to be unbanded. BP = basal plate. Scale = 0·1 μ.

dimensions to the reticulum, but it is not certain that the lamellae originate from the latter.

The large lamellate structures probably develop stage by stage from small beginnings. Small areas about 1 μ long and 0·5 μ across can be observed in the cytoplasm, and these are composed of two or three concentric membranes. Other more complex profiles occur that bridge the development of small to large lamellate bodies. These structures tend to occur in close proximity to one another, and may eventually take up a considerable proportion of the volume of the cell. Electron dense bodies also occur close to whorled structures but may be unrelated to them.

It has frequently been noticed that in regions of the giant cell where lamellate bodies are numerous the cell becomes detached from the

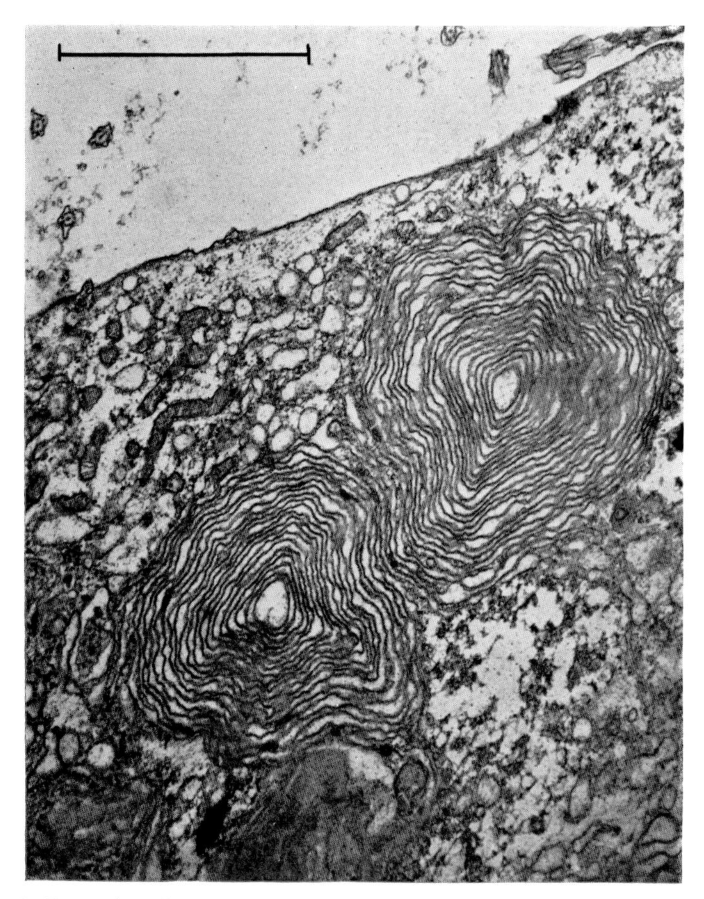

FIG. 3. Large lamellate structures appear in the cytoplasm of the giant cells. Apparently starting as small concentrically arranged membranous areas, they develop into the large structures of membranes arranged around a small clear core. These structures are three dimensional, not thin discs, and occur throughout the cell. Scale = 2·5 μ.

capsule, and in some cases completely separates from the surrounding tissue. The cell then disintegrates into the lumen of the statocyst. It remains uncertain whether the whole giant cell, or only parts of it, fall into the lumen. There is a reported constancy of giant cell numbers in the statocyst (Pfeil) and as active cell division has not been observed it may be that only portions of the cell are shed into the lumen. This portion may then be regenerated from the remainder of the cell.

The second type of cellular inclusion that is worthy of note is the statolith. Certain giant cells contain fully formed statoliths and in such cases the bulk of the calcareous material leads to herniation of the capsule (Fig. 4). The statoliths do not seem to be related in structure to the lamellated bodies mentioned previously, but when the cell contains a number of statoliths they are released into the lumen of the statocyst (Fig. 5), by disruption of the cell.

The fully formed statoliths in the lumen of the organ are plate-like calcified structures. Removal of the calcified material shows that the basic framework of the statoliths is a loosely connected fibrous network. In the longitudinal plane this fibrous material is seen to be arranged in

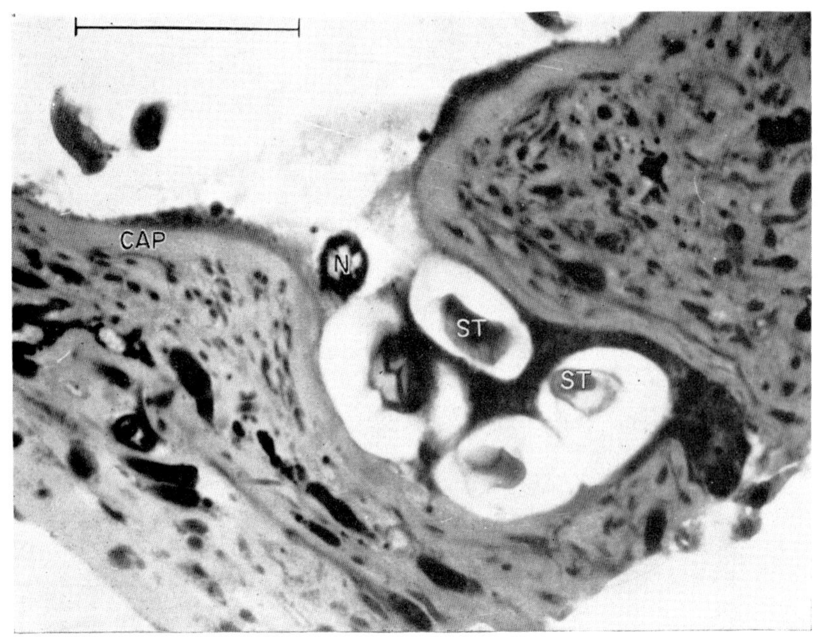

Fig. 4. Light microscope photograph to show that the giant cell may become loaded with statoliths that herniate the capsule. CAP = capsule, ST = statoliths, N = nucleus of giant cell. Scale = 100 μ.

concentric rings. In transverse section a system of interconnected fibres is apparent.

In the giant cell membranous structures which, when cut in the appropriate plane, have the characteristic appearance of statolith material are seen to differentiate in the cytoplasm. These areas may be modified cytoplasm which is not separated from the remainder of the cell contents by a membrane until a critical size, unknown at present, is reached. Fully formed statoliths (calcified, and hence difficult to cut in thin sections) may be found amongst disrupted cytoplasm. The early stages of statolith production have not been identified.

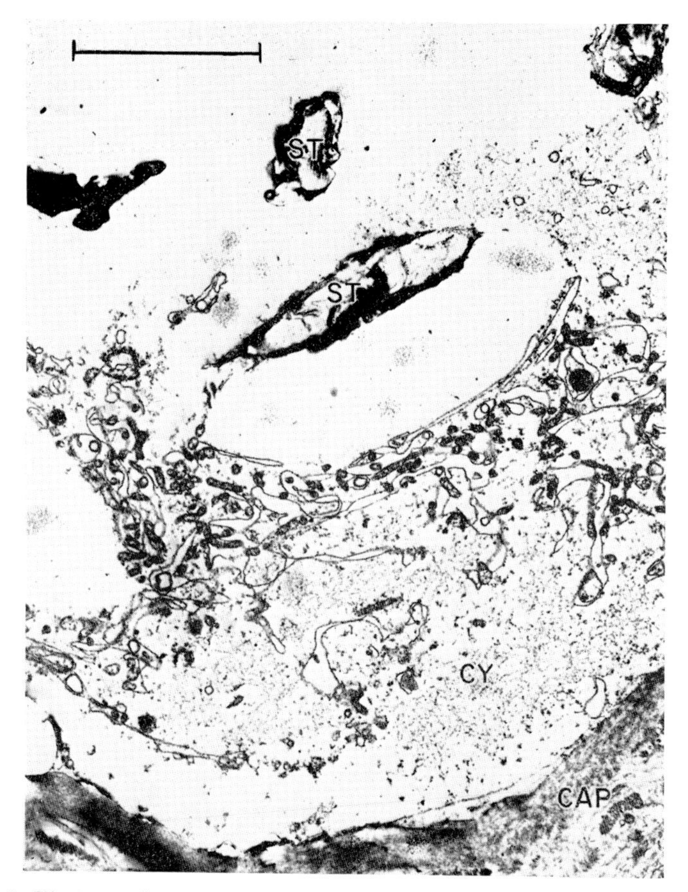

FIG. 5. Electron micrograph of the same cell as in Fig. 4. The cell has peeled away from the underlying capsule, (CAP) the cytoplasm (CY) has become much modified and the contained statoliths (ST) are released into the lumen. Scale = 50 μ.

Pfeil described a thin cuticular layer over some portions of the statocyst inner surface, and in particular over the giant cells. The cilia were figured as extending from the surface of the cuticle. No signs of such a structure have been observed in the present investigation.

Sensory cells

These cells are smaller than the giant cells, are distinct and discrete from one another and occur in patches on the inner surface of the organ, between, and, sometimes, underlying the arms of the giant cells. Pfeil shows a diagram in which the "syncytial" nuclei underly the giant cell, and this can now be reconciled with the observation above. At their largest, when they extend from the lumen to the capsule, the small cells are about 9–10 μ deep and approximately 1·5–1·75 μ in diameter. The nucleus is large in relation to the overall size of the cell, often reaching 6 μ in length and filling most of the basal part of the cell.

The cytoplasm of the cell contains many mitochondria which are rather larger than the corresponding structures of the giant cells, attaining 3 μ in length and 0·3 μ in diameter. There is no obvious localization of the mitochondria as in giant cells, and they are uniformly distributed through the cell. Vacuoles are widespread throughout the cell and may be formed from small lamellate and membranous systems which occur sparsely in each cell.

The surface of these small cells that borders the lumen of the organ shows interesting specializations. The most obvious surface modification of the cell is that of fine prolongations with simple internal structure. These are finger-like projections 0·5–0·75 μ long and 0·10 μ in diameter (Fig. 6). These are often single, blunt extensions of the cell surface, containing a few filaments, but they have no characteristic pattern. A second type of projection is sometimes seen in which the extensions are complex and probably branched. They are longer (up to 3 μ) and thinner (0·03 μ) than the first type. They form a fine mat of filamentous "fur" on the inner surface of the statocyst (Fig. 7). These filaments are different from the microvilli of the outer surfaces of *Helix* (Eakin and Westfall, 1964; Lane, 1963), but in distribution may resemble the network of ridges that Gibbons (1961) described on the surfaces of cells bearing frontal cilia in *Anodonta*.

Pfeil mentions cilia as being present on both giant and syncytial cells. In the present work cilia have been seen in a variety of planes of section throughout the lumen of the organ. Basal bodies can be readily seen at the surface of the giant cell, and the shafts of cilia extending above the cell surface. For a long while it was thought that cilia were found only on the giant cell, but on a few occasions basal bodies were noted at the periphery of the sensory cells (Fig. 8a, b). These centriole-like organelles form a small bump projecting into the lumen, and in one instance a short section of ciliary filaments were seen in such a bump (Fig. 9). This instance suffices to indicate that at least one short ciliary peg exists at the surface of the cell, though from the sparsity of records it seems likely that there is only one cilium per cell. A centriole is occasionally found elsewhere in the cell.

The giant cells sometimes overlie the basal regions of the sensory cells and the latter may in turn form very thin (50–160Å) outgrowths over the inner surface of the giant cells. This obviously greatly increases the effective surface area of these cells.

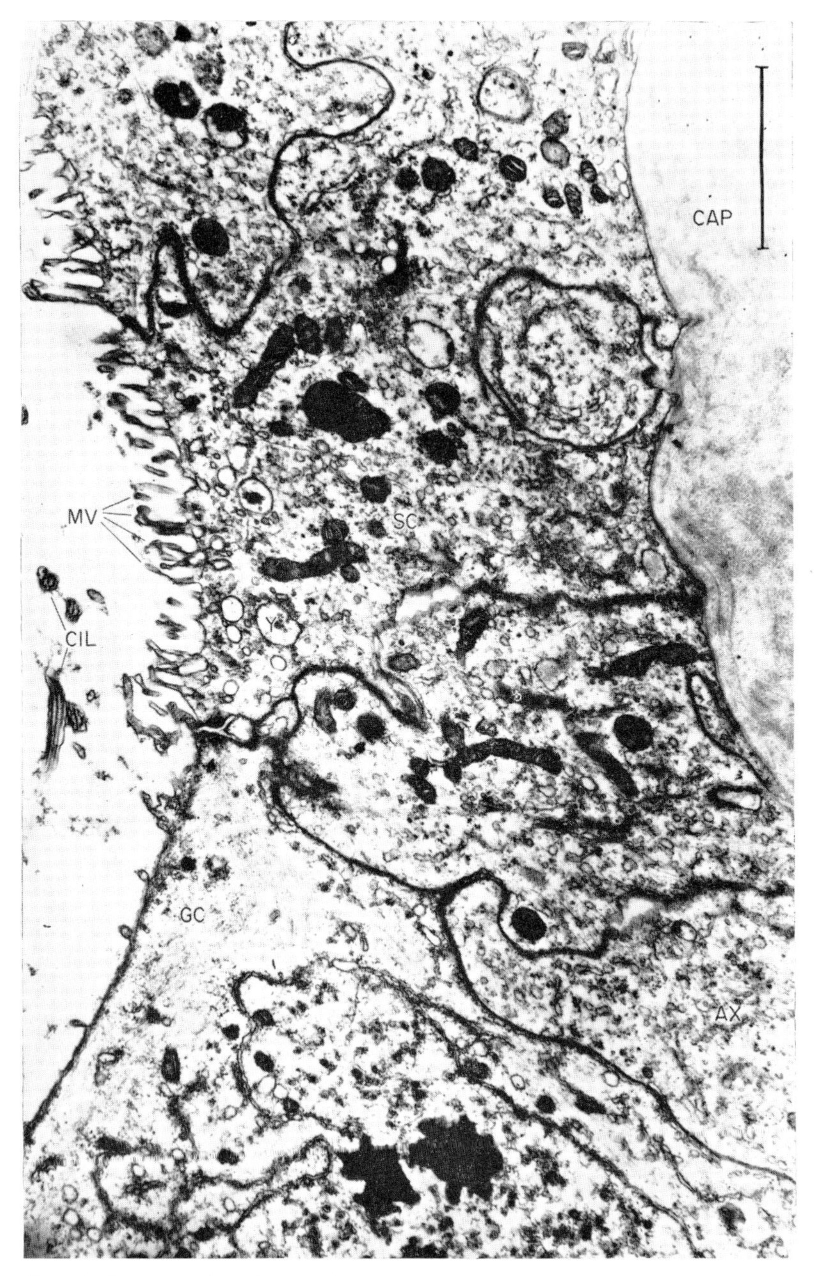

FIG. 6. Sensory cells (SC) alongside a giant cell (GC). Note the projections (MV) into the lumen, the extensions that form the axons (AX), the vesicles of the cytoplasm (V) and the capsule (CAP). CIL = cilia. Scale = 2 μ.

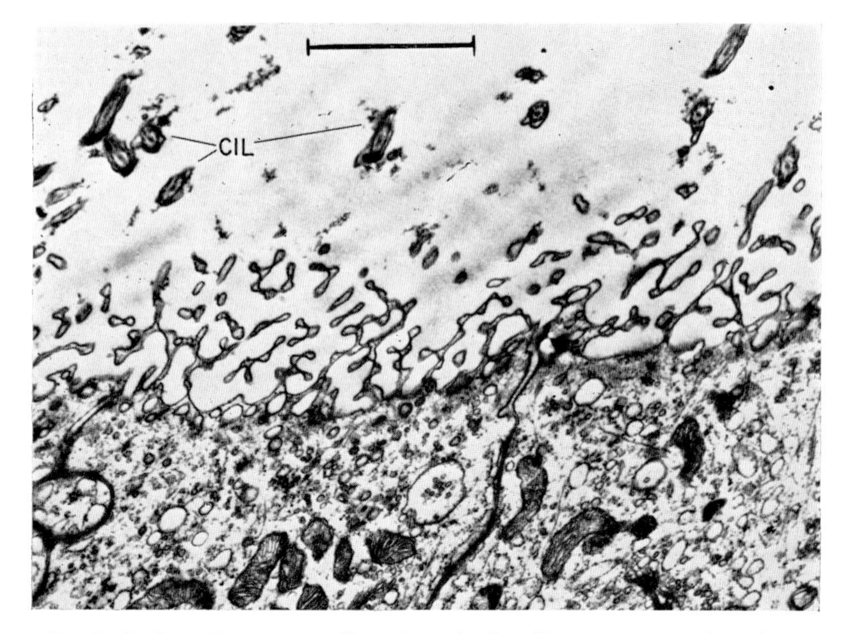

FIG. 7. Surface of a sensory cell, to show the fine filamentous extensions that probably form an interlinked series of ridges over the surface of the cell. These differ in dimensions and arrangement to those shown in Fig. 9. The cilia (CIL) in the lumen do not originate on the sensory cells. Scale = 2 μ.

The giant cells may also be separated from the bounding capsule by fine cellular extensions which represent the sensory axons of the primary receptor cells. These have in several cases been traced to their origin at the small cells. They have never been observed as originating at the giant cells.

At the position where the statocyst nerve leaves the statocyst capsule there is a plug of small cells with axons that run straight down into the nerve bundle.

Conclusions

Alas then for our hopes! The giant cells, of some $70 \times 10\ \mu$ in size, are not suitable subjects for the study of primary sensory processes since they are not receptor cells. The smaller receptor cells are not large enough for a direct approach with present methods. Mote (1967, personal communication) in California has attempted recording from higher order neurones with intracellular electrodes, but the stimulating conditions are stringent since the whole preparation needs to be rotated at the same time, and the difficulties of maintaining penetration whilst

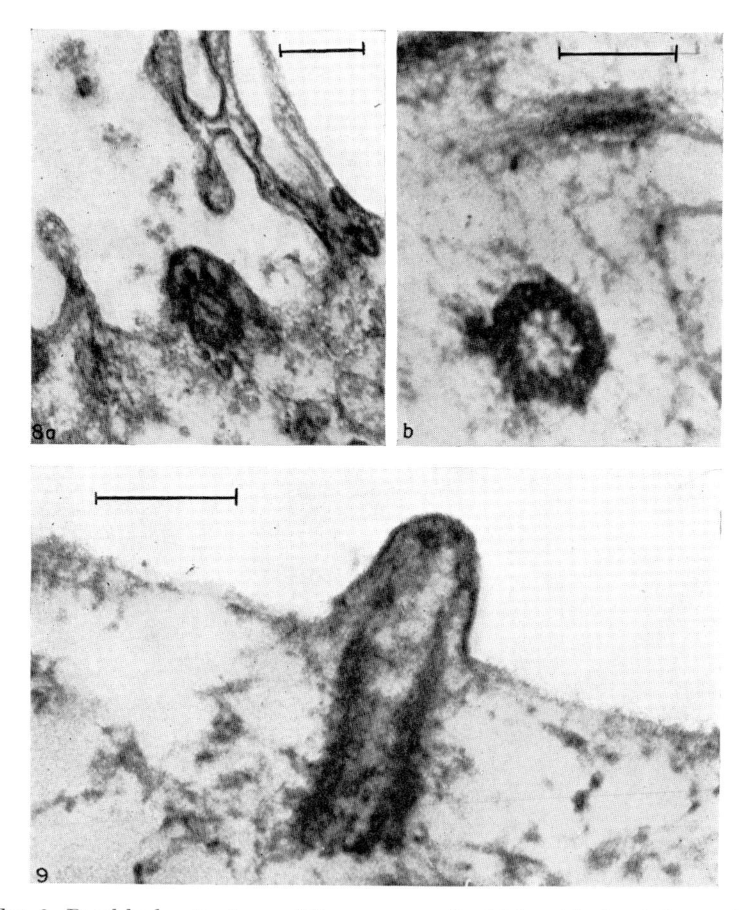

FIG. 8. Basal body structures of the sensory cells. (a) basal body at the surface of the cell, (b) section just below the cell surface. Scale = 0·25 μ.

FIG. 9. Longitudinal section to show formation of a small peg. The sensory cell ciliary peg has never been observed to protrude further than this. Scales = 0·25 μ.

moving the whole ganglionic mass can be imagined. Also there is no guarantee that the movement of the nerve mass does not stimulate impaled cells by some other means.

We may, however, advance a new hypothesis as to the mode of functioning of the statocyst.

If, as Pfeil suggested, the giant cells were the sensory cells, the field of sensitivity of these cells would be great by virtue of their extent over the surface of the statocyst. But the sensitivity would be limited by another consideration, since stimulation anywhere on the cell surface might be expected to initiate an impulse in a single sensory nerve. The sensitivity to displacement would then depend solely upon how many of the 11–13 giant cells were stimulated at any one time, surely not a very discriminating system. If each giant cell had many sensory axons passing to the cerebral ganglion it is conceivable that there would be a sufficient number of axons to account for the size of the statocyst nerve, and perhaps a greatly increased discrimination. But as we have seen the axons arise only from the smaller sensory cells.

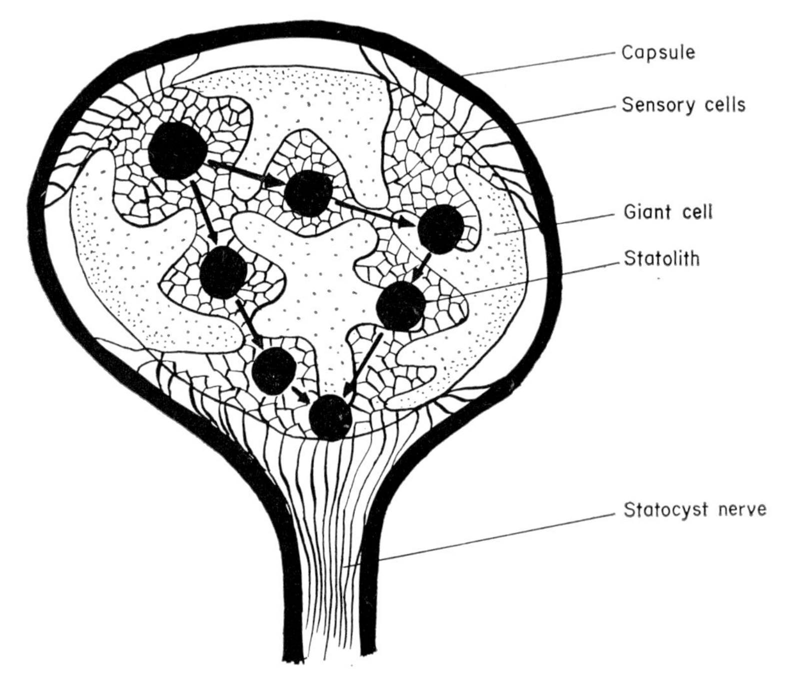

FIG. 10. Diagram to summarize the findings of this investigation in terms of the function of the organ. The sensory cells are the smaller entities of the epithelium, and the apparent sensitivity of the organ is increased manyfold due to the number of sensory elements. A statolith falling from one part of the organ to another thus stimulates numerous sensory cells.

In view of the undoubted sensory nature of the second type of cell, it can be seen that the sensitivity of the animal to its orientation in space must be greatly increased by reason of the number of end organs stimulated by falling statoliths. The secretory giant cells cover a large proportion of the inner surface of the statocyst, but between the radiating fingers of the cell bodies are the apical surfaces of the sensory cells. The shifting statoliths must fall upon different patches of sensory epithelium according to the orientation of the animal, and because of the small size of the sense cells a very small positional change could be monitored by a number of different units (Fig. 10).

PEDICELLARIAE OF *Echinus*

The second preparation is another in which individual components do not lend themselves too readily to experimentation. The nervous system of echinoderms is not well understood, but recently awakened interest as shown by the papers of Takahashi (1964), Sandeman (1965), Bullock (1965), Cobb and Laverack (1967), and the work summarized in Boolootian (1966) indicates that an attempt is now being made to determine at least some of the primary properties of echinoderm nerves. The structure of the nervous system is by and large amenable only to electron microscope study, though this does not as yet give details of tracts of nerves, and brings as many difficulties as it solves. Functionally the small size of the nerve fibres has precluded much electrophysiological analysis, but determined efforts are now being made by several groups to elucidate details in this respect.

The pedicellariae of the echinoids and the asteroidea are both obvious and well-known, but no original work on the function and movements of these structures seems to have been done since von Uexküll's classic descriptions of 1899. During this gap of seventy years our ideas on the function of nerves have changed radically, of course, and instrumentation now enables us to demonstrate responses in a more reliable and repeatable way than the written word transcribed from a worker's notebook.

Our work has been done on the three largest types of pedicellariae from the test of *Echinus esculentus*. These are the tridentate, the ophiocephalous and the gemmiform pedicellariae (a fourth type, the trifolicates has not so far been investigated). A full description of the apparatus used will appear elsewhere in due course, but in short the method is as follows.

A small light source (a 1·5 V torch bulb) was placed outside a small glass aquarium. On the far side, and opposite the source was situated a photoelectric cell (sometimes masked by a shield with a pin hole through it, otherwise unmasked). Pedicellariae were removed from a fresh urchin in good condition. They were placed in the aquarium, and held by gentle suction in the open end of a fine bore plastic tube. The bore of the tube must be very slightly larger in diameter than the diameter of the stalk that bears the pedicellaria. This prevents the organ from slipping too far down the tube. In this way the pedicellariae remain upright and the jaw movements are not impaired by the upper edge of the holding device. It is now a comparatively easy matter to arrange for the shadow of the jaws to be projected onto the photoelectric cell behind the aquarium. In the case of the tridentate pedicellariae the shadow is large enough to enable recording without further ado, but the gemmiform and ophiocephalous organs require interpolation of a microscope eyepiece ($\times 15$) in the light path before projecting a shadow of sufficient dimension to be readily recorded. Movements of the jaws now directly affect the amount of light falling on the photocell, and the varying output from this cell is used to drive a Devices pen recorder. The records shown here illustrate the typical results obtained.

The questions we posed were as follows; first what stimulus is appropriate to each type of pedicellariae, second what is the response of each type, third, can evidence be obtained to support von Uexküll's report of chemical priming, fourth can evidence be acquired to support von Uexküll in the matter of auto-recognition, lastly what information can be obtained into the matter of integration of the stimulus and the muscle response? Subsidiary problems also presented themselves, important among which were the details of morphology of the system (see Cobb, 1968).

All the varieties of pedicellariae of *Echinus* show certain characteristic movements, namely they close the jaws, and then they open them. The smallest type, the trifoliate, carry out these actions continually, scouring the immediate surrounding portion of the test. The largest form, the tridentate, close rapidly when mechanically stimulated, reopen and close again upon further stimulation; the ophiocephalous variety show similar responses, but the gemmiform type normally show an opening response under low intensity mechanical stimulation, to close finally and irreversibly upon high intensity stimulation followed by discharge of the poison glands with which this type is endowed.

It is immediately obvious that our measure of receptor activity is an indirect one. The response we record is of a whole independent organ perched on the top of a stalk, and which seems to have very little to do with events occurring closer to the test. It is not known if the pedicellariae heads send information centrally, or whether the radial cords (or epidermal plexus) affect the jaws (though they certainly affect the stalks). Nonetheless a lot of observations on each type lead us to make certain generalizations.

Mechanical stimulation

In response to tactile stimulation (presented by a fine probe mounted on the end of a relay) the tridentate pedicellariae close rapidly (Fig. 11a,

c) as also do the ophiocephalous variety (Fig. 12). The gemmiform organs show what we have called a "further opening response" (FOR) that exposes more of the epithelium that covers the internal faces of the jaws. The recording situation does not allow us to discriminate between

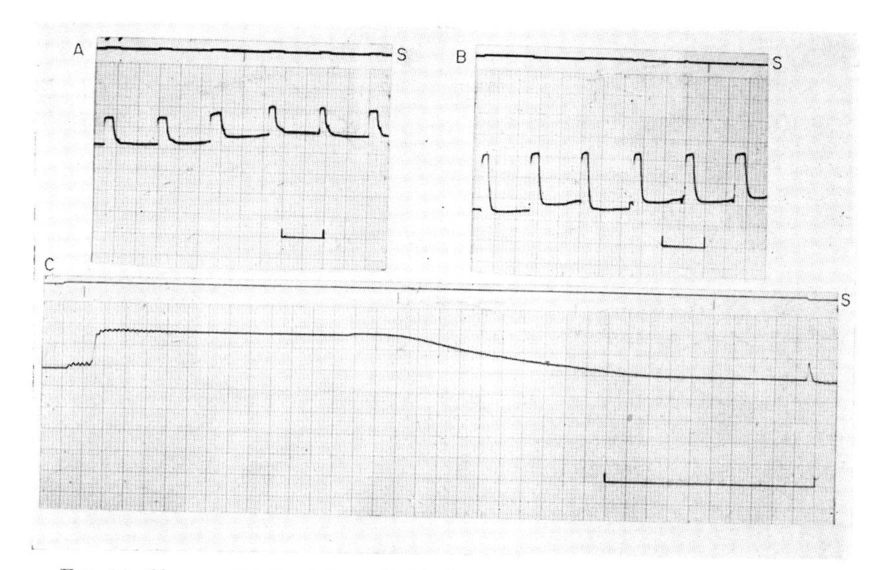

FIG. 11. Closure of tridentate pedicellariae. (a) Mechanical stimulation by a probe, several times repeated. (b) Stimulation by a spine from the urchin's own body. (c) Fast record of closure. Note rapid closure, with a distinct latency, and slow relaxation. Closure represented by upward movement of trace. Time scale (a) and (b) 10 sec, (c) 1 sec. S = stimulus marker.

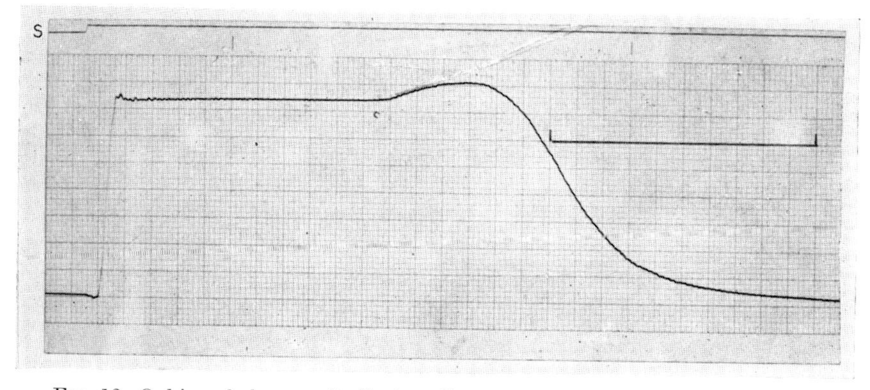

FIG. 12. Ophiocephalous pedicellariae. Closure to mechanical stimulation is very rapid and relaxation is slow. The bump in the trace prior to relaxation is due to recording conditions. Time scale 1 sec. S = stimulus marker.

these two directions of movement which is still observed visually. The responses described here occur if stimulation is carried out on the inside of the jaws.

If the outer sides of the jaws are touched the tridactyl type open when touched at the base of the jaws, the ophiocephalous pedicellariae do not open, whilst gemmiform organs may sometimes show this response, but the latter are open wide normally in any case.

The site of stimulation may be important in determining the response. Thus in tridentate pedicellariae tactile stimulation of the upper third of the jaw, or the apex, is followed by closure. Ophiocephalous type show no localization of receptor sites and close after stimulation anywhere on the surface. Gemmiform organs have a well-described sensory hillock on the inner face of the jaws and stimulation of this area leads to a somewhat different response when judged by the criteria of latency and FOR duration.

Von Uexküll (1899) differentiated between mechanical stimulation with an inert object and with a chemically significant one, particularly with a portion of the host urchin's own body. He stated that pedicellariae will not close on any isolated part of the urchin (tube foot, spine etc.) because of the presence of a specific substance. This he called the autodermin response.

No difference was found in the present investigation between different types of applied tactile stimulation. If a spine or other part of the urchin was used the response (latency, duration) remained as for stimulation with an inert object (Fig. 11b). If a sea water "extract" of urchin material was prepared and this added to the water in the test chamber whilst a probe was used to stimulate there was still no effect on the response. We conclude that for tridentate and ophiocephalus pedicellariae there is no autodermin response or recognition. Gemmiform organs showed FOR.

Chemical stimulation

It is generally believed that pedicellariae are involved in preventing fouling of the urchin test by removing small settling organisms that may otherwise cover the surface of the animal. Whether or not the conditions of the urchin skin, in fact, are suitable for settlement anyway is a moot point, but the pincer-like action of the pedicellariae would obviously crush the smaller planktonic animals. In this case one might suspect that chemical stimulation may play an important part in triggering the snap action of the organ.

In experiments on chemical stimulation most of the stimulants used were perhaps representative of those encountered in normal environmental conditions by a sea urchin. The following materials were tested; *Asterias* coelomic fluid, podia, podia extract in sea water; *Homarus* muscle, blood; filter paper, filter paper soaked in *Homarus* blood; *Carcinus* muscle, blood; *Cancer* muscle, blood; *Xenopus*, crushed tadpole; ox, liver; alga, *Laminaria sp.*, fronds; plankton, whole and crushed.

Tridentate and ophiocephalous pedicellariae showed no response to any of these substances that could be attributed to chemical stimulation. Extracts of these various tissues invoked no response. Solid objects, such as tube feet or algal fronds, invoke a response when offered but this is certainly due to a tactile stimulus. No potentiation or facilitation

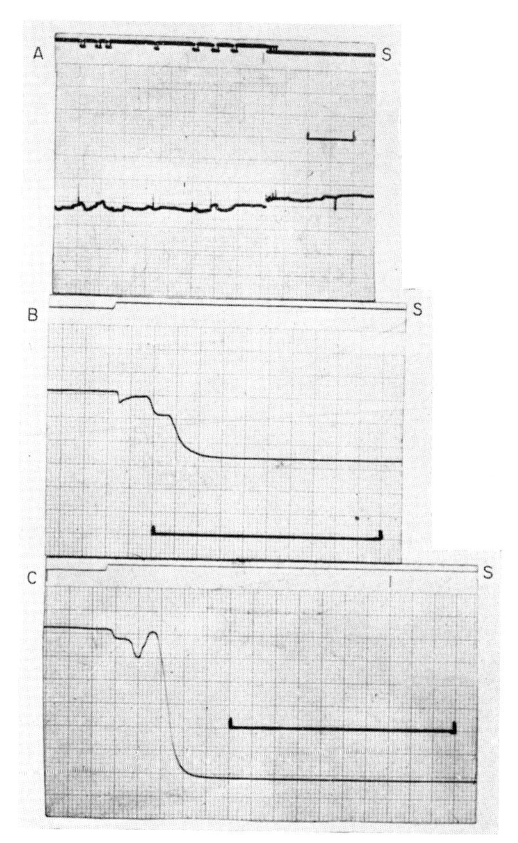

FIG. 13. Gemmiform pedicellariae. (a) Small FOR (Further opening responses) to mechanical stimulation of the head. (b) Closure on stimulation with a tube foot from *Asterias* (downward movements of record shows closure). (c) Closure on stimulation with a piece of muscle from *Homarus* (closure indicated as in (b) above). Time scale (a) 10 sec, (b), (c) 1 sec. S = stimulus marker.

of the normal closure response was seen when the pedicellaria was first immersed in an extract. Plankton, live and mobile, or dead and aromatic, had no effect.

Gemmiform pedicellariae showed a distinct increase in activity, in the form of rocking of the head, when immersed in an extract of *Asterias* podia (Fig. 13a). If now a solid object is brought into contact

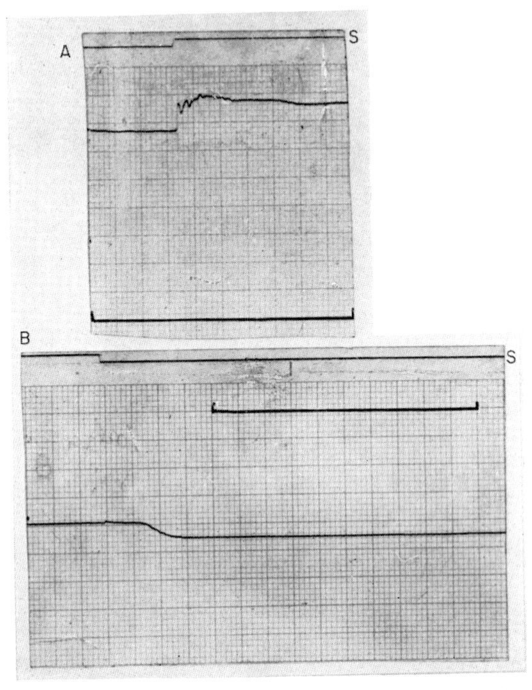

Fig. 14. Gemmiform pedicellariae. (a) FOR in response to mechanical stimulation of the sensory hillock with a piece of filter paper. (b) Closure of the same gemmiform after soaking the filter paper in *Homarus* extract. Time scale 1 sec. S = stimulus marker.

with a gemmiform pedicellaria it opens with more certainty than usual when stimulated by an inert object (this is the opposite effect to that described by von Uexküll).

The gemmiform closes rapidly on any solid foreign tissue presented to the gape of the jaws (Fig. 13b, c). After this closure the pedicellaria did not usually reopen. If the poison, contained in a gland in each jaw, was released after closure, re-opening occurred.

Closure took place only if stimulation occurred in the region of the sensory hillock of this type. If a foreign body contacted the sensory hillock, or came very close to it closure took place (Fig. 14b), the latency of the response increasing with increase of distance from the sensory

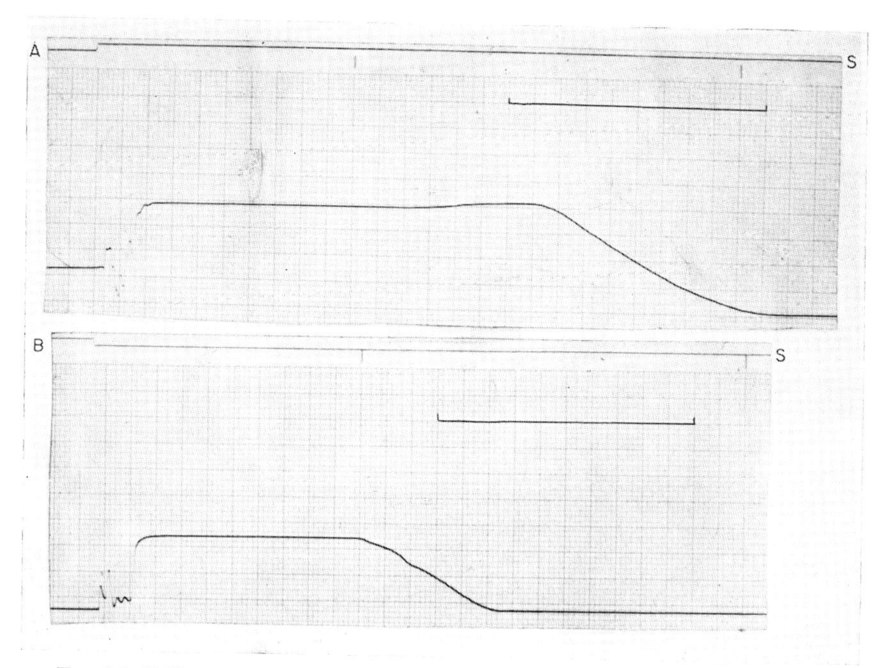

FIG. 15. Tridentate pedicellariae. Closure in response to mechanical stimulation (a) organ intact. (b) After cutting smooth muscle tracts. Note decrease in duration of closure. Time scale 1 sec. S = stimulus marker.

hillock. Distal stimulation was followed by a further opening response (Fig. 14a).

If stimulation was carried out with a chemically inert material, such as filter paper, on the sensory hillock FOR resulted. If now the filter paper was soaked in *Homarus* blood and offered again to the same area a closure took place. When poison is released the organ seems incapable of opening again, and in the normal animal the gemmiform is shed at this stage.

Interference with response mechanisms

It is possible to disrupt the typical responses outlined above by destruction of certain parts of the pedicellariae. Tridentate forms possess both striated and unstriated adductor muscles. The distribution of these muscles makes it possible to cut the smooth muscle with varying degrees of success. The apparent results of such dissection is a decrease in the duration of the closure reponse, suggesting that the fast closure is due to the striated muscle, but the long maintained phase of the closure is the function of the smooth muscle, the striated muscle meanwhile relaxing (Fig. 15).

Gemmiform pedicellariae have a sensory hillock on each jaw. It is possible to cauterize this area by electrical means. After cautery of the three hillocks mechanical stimulation of the pedicellariae is followed by the normal FOR, but chemical stimulation with foreign tissue is not followed by closure.

Structure of the receptors

J. L. S. Cobb (1968, and unpublished) has recently carried out a survey of the fine structure of the epithelium of *Echinus* pedicellariae. He concludes that all the cells of the epithelium should be considered as sensory since each appears to bear an axonic process (see also Kawaguti and Kamishima, 1964). As we have seen above tactile stimulation almost anywhere on the surfaces of tridentate and ophiocephalous forms is followed by either opening (outer face) or closure (inner face). The structure of the general epithelial covering shows that small cells bear cilia encircled by a ring of microvilli, and a prolongation of the cell passes towards the muscle of the jaws.

The only area that has been described which may have a special function is the sensory hillock of gemmiform pedicellariae. In this case a chemosensory function is indicated, and it might be anticipated that some difference in structure exists between the cilia here and the cilia elsewhere on the jaws. No structural difference at all has been discerned in electron microscope studies. The behaviour of the cilia is not that typical of motile ones, since they only beat occasionally by bending from the base. Internally the interesting structures are the ciliary root which extends deep into the cell, past the nucleus, and which J. L. S. Cobb (personal communication, Fig. 16) believes may eventually form the neurotubules of the axon; and the occurrence of a second basal body alongside the root of the cilium close to the cell surface. Neither of the features immediately suggest what part they may play, if any, in transduction of a chemical stimulus.

DISCUSSION

The evidence presented in this paper is concerned with superficial receptors. These receptors are involved in monitoring events occurring in the environment external to the cell. In the examples given these events are either mechanical (contact, deflection and vibration) or chemical (alien animal and plant material).

The statocyst of *Helix pomatia* is fluid filled, and the fluid is in constant motion due to the activity of long motile cilia on the giant cells. This motion is insufficient to cause more than slight wavering motion of the plate-like statoliths which fill the lumen of the organ. These move

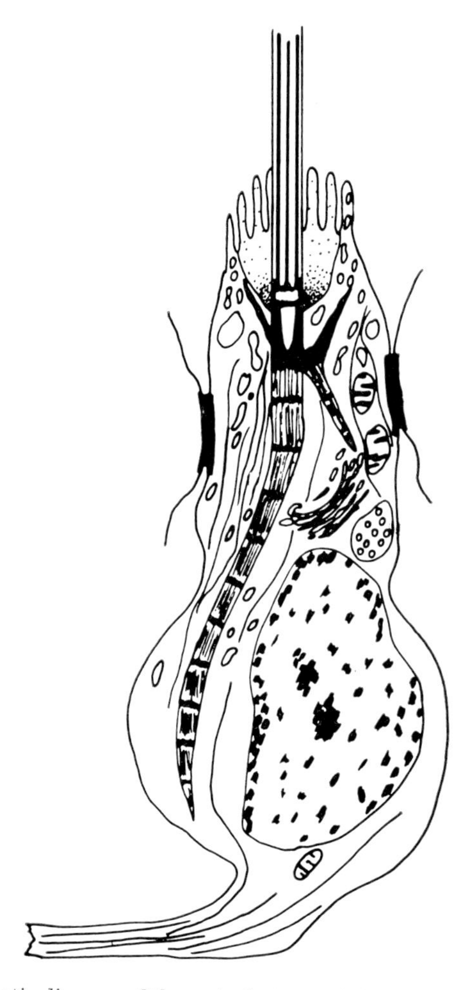

FIG. 16. Schematic diagram of the main features of the component cells of the sensory hillock of gemmiform pedicellariae. There is a single long cilium, surrounded by a number of smaller microvilli. The internal structures of the cilium are similar to those of normal motile cilia, with 9 + 2 filaments. There is a doubled basal body just internal to the cell, and a very long ciliary root that extends deep into the cell, past the region of the nucleus. These cells are believed to be chemoreceptive (from J. L. S. Cobb 1968, personal communication).

only under the influence of gravity when the snail changes position relative to the horizontal plane. Lever and Geuze (1965) have shown that removal of the statocyst of *Lymnaea* affects the behaviour of this pulmonate and I assume similar events could be shown in *Helix*.

The important features of this investigation are that Pfeil's original description suggested the wrong functions for the cell types; the giant cell not being sensory but concerned in statolith production and movement of fluid, whilst the so-called syncytium is composed of small cells with axons, and these are the primary receptor cells. These receptors have a microvillified surface that forms a fine mat lining the statocyst, and within this mat lies a single ciliary basal body just in contact with the lumen of the organ.

It is believed that acceleration of the fluid contents of the statocyst will be ineffective in stimulating receptors for two reasons; one is that

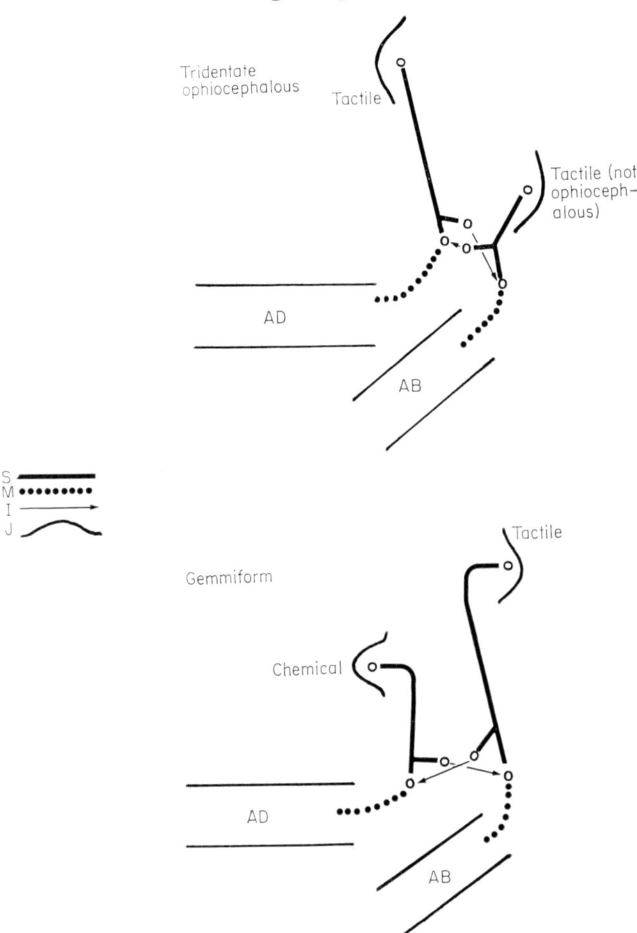

FIG. 17. Hypothetical linkages to account for the responses of pedicellariae to mechanical and chemical stimulation. (a) for tridentate and ophiocephalous, (b) for gemmiform types. Note that adductor (AD) and abductor (AB) muscles are present, and that the receptors make connection with these muscles, perhaps directly, or indirectly via the complex neuropile that exists at the base of each jaw. M = motor, I = inhibitor, S = sensory neurones, J = Jaw surface.

the fluid contents are in movement the whole time, though this is probably not in any definite direction, and it seems unlikely that such capricious fluid movements would be of significance to the animal. This mollusc thus differs from cephalopods (Maturana and Sperling, 1963), Crustacea (Cohen, 1960) and vertebrates (e.g. Lowenstein and Roberts, 1950) in all of which fluid movement is important in stimulating certain of the receptor cells, and in which solid bodies such as statoliths play another role. The second reason for suggesting fluid movements are unimportant rests on the lack of a long extension into the lumen of the organ. In the other invertebrates mentioned above the sensory cells bear long ciliary processes (or hairs, in Crustacea) that project into the statocyst cavity, and fluid movements move these together with any accessory structures (stereocilia) that may be present. *Helix* has only a short ciliary peg, unlikely to be moved by fluid alone, surrounded by a fine fur of surface projections that would interfere with fluid flow.

It is thus suggested that in this case the statoliths play an important role and must press upon the sensory areas directly to stimulate the receptors. What constitutes a sensory area is uncertain. There is no evidence to indicate that the cilium peg is of importance in transduction, or that the stimulating energy is channelled towards this structure, as might be supposed by the anatomy of other equilibrium organs. The entire outer surface of the cell may be important in reception. Other information bearing on this matter will be presented elsewhere (*Proceedings of the International Conference on Gravity and the Organism*, Tuxedo, New York, September 1967).

The pedicellariae of sea urchins also possess sensory cells with axons that run towards a small area of neuropile situated at the base of each jaw of the organ. These sensory cells all bear cilia that structurally do not differ from motile cilia, save perhaps in the possession of long roots as in the gemmiform type. Again it is impossible to be certain that the cilia are important in reception or transduction. The cilia may only provide an extension of the cell into the surrounding environment. At present it is not possible to mount a preparation in such a way that stimulation of a single cilium alone provides the only input, and hence it is not possible to demonstrate the importance of a lone ciliated cell.

It is, however, possible to make a few comments on the sensory apparatus of pedicellariae in general. It seems that all the epithelial surface is a sensory one. The outer areas are linked with abductor muscles that bring about opening of the jaws; the inner areas are linked with adductor muscles that cause closing. Each is also probably linked by an inhibitory system to the antagonist muscle. A hypothetical scheme for the integration of this system is proposed in Fig. 17.

The classical proposal that pedicellariae are concerned in maintaining the test free of settling organisms may still be true, but the pedicellariae evidently differ in their abilities to handle the stimulating organism. The tridentate group are the largest, close only when the inner face and the tips of the jaws are touched, and are not at all sensitive to vibration. They therefore will close only on organisms directly touching the jaws, and do not appear to close on planktonic larvae that swim between the jaws. Ophiocephalous forms are almost identical in function and differ only in size. Gemmiform pedicellariae gape wider when touched, presenting the sensory hillocks to the environment. If these are stimulated by an appropriate chemical the organ closes. This group may therefore be selective in closing on animal or plant material only and certainly inject poison into the prey.

Pedicellariae therefore are either crushers or poisoners, stimulated either by touch or chemicals, and may be retained (tridentate, ophiocephalous) or cast off (gemmiform) after use. No information is available on trifoliates. An interesting problem is raised by the presence of certain commensals such as *Astacillus* (isopod) and certain polychaetes that roam at will on urchins, and yet do not stimulate these organs, reminding one of the commensals of coelenterates.

It is noteworthy that structures with the morphology of cilia are involved in these two very different sensory systems. The theory of the involvement of cilia in the transduction of a stimulus is one that needs confirmation. It has never yet been shown that it is the *ciliary* nature of the structure that is important. The range of modifications of ciliary shaft, $9+2$, $9+0$, $9+1$, paracilia etc. only serves to indicate that receptor cells do not require a rigid conformation of internal structures apart, perhaps, from an elongation of the surface, with some skeletal support. In polarized systems, e.g. vertebrate equilibrium organs, polarity is a function of the whole structure (stereocilia, basal foot, central fibrils), and not for any one specific part. The present cases of *Helix* statocyst sensory cells, and *Echinus* pedicellariae may indicate that the basal body region is most important, if any ciliary component is involved at all (see pp. 308, 320). In *Helix* only basal bodies are present; in *Echinus* only the basal body-rootlet system shows any evident modification. Specializations of basal body regions have been shown several times in receptors but have still not been linked to the transducer process. Tucker (1967) has recently shown that 99% of the cilia of olfactory cells of the box turtle can be removed by treatment with detergents, and yet the organ still responds to stimulation with amyl acetate. Electron microscope studies of the epithelium after detergent treatment showed that the basal bodies were retained in the cells, but again there is no reason to suspect their implication in the primary events of reception. Matsusaka (1967) has recently shown that adenosine triphosphatase activity is associated with ciliary rootlets of mammalian retinal rods. He suggests that when the outer retinal segments do not move the requirements for metabolic energy indicated by this enzymic activity may be a reflection of conduction in the rootlets. In the present examples the pedicellariae possess very long rootlets, but the sense cells of *Helix* have none. It is therefore presumably possible for the initial generator potential to be set up without the presence of ciliary rootlets.

Acknowledgements

I am indebted to Dr J. L. S. Cobb and Mr A. C. Campbell for permission to quote from their, as yet, unpublished results. Part of the work described here was supported by a research grant from the Science Research Council.

References

Boolootian, R. A. (ed.) (1966). *Physiology of Echinodermata.* New York: Interscience.

Bullock, T. H. (1965). Comparative aspects of superficial conduction systems in Echinoids and Asteroids. *Am. Zool.* **5**: 545–562.

Cobb, J. L. S. (1968). The pedicellariae of *Echinus esculentus.* II. Sensory system. *Jl Rl microrc. Soc.* **88**: 223–233.

Cobb, J. L. S. and Laverack, M. S. (1967). Neuromuscular systems in Echinoderms. *Symp. zool. Soc. Lond.* No. 20: 26–50.

Cohen, M. J. (1960). The response patterns of single receptors in the crustacean statocyst. *Proc. R. Soc.* (B) **152**: 30–49.

Eakin, R. M. and Westfall, J. (1964). Further observations on the fine structure of some invertebrate eyes. *Z. Zellforsch. mikrosk. Anat.* **62**: 310–332.

Fretter, V. and Graham, A. (1962). *British prosobranch molluscs.* London: Ray Society.

Gibbons, I. R. (1961). The relationship between the fine structure and direction of beat in gill cilia of a lamellibranch mollusc. *J. biophys. biochem. Cytol.* **11**: 179–205.

Kawaguti, S. and Kamishima, Y. (1964). Electron microscopic study on the integument of the echinoid, *Diadema setorum. Annotnes zool. jap.* **37**: 147–152.

Lane, N. J. (1963). Microvilli on the external surfaces of gastropod tentacles and body walls. *Q. Jl microsc. Sci.* **104**: 495–504.

Lever, J. and Geuze, J. J. (1965). Some effects of statocyst extirpation in *Lymnaea stagnalis. Malocologia* **2**: 275–280.

Lowenstein, O. and Roberts, T. D. M. (1950). The equilibrium function of the otolith organs of the Thornback ray (*Raja clavata*). *J. Physiol., Lond.* **110**: 392–415.

Matsusaka, T. (1967). ATPase activity in the ciliary rootlet of human retinal rods. *J. Cell Biol.* **33**: 205–208.

Maturana, H. R. and Sperling, S. (1963). Unidirectional response to angular acceleration recorded from the middle cristal nerve in the statocyst of *Octopus vulgaris*. *Nature, Lond.* **197**: 815–816.

Pfeil, E. (1922). Die statocyste von *Helix pomatia*. *Z. wiss. Zool.* **119**: 79–113.

Sandeman, D. C. (1965). Electrical activity in the radial nerve cord and ampullae of sea urchins. *J. exp. Biol.* **43**: 247–256.

Takahashi, K. (1964). Electrical responses to light stimuli in the isolated radial nerve cord of the sea urchin *Diadema setosum* (Leske). *Nature, Lond.* **201**: 1313–1314.

Tucker, D. (1967). Olfactory cilia are not required for receptor function. *Fedn Proc. Fedn Am. Socs exp. Biol.* **26**: abs. no. 1609.

Von Uexküll, V. (1899). Die physiologie der Pedicellarien. *Z. Biol.* **37**: 334–403.

NOTE ADDED IN PROOF

For a somewhat shorter and different interpretation of the structure of the *Helix* statocyst, the reader is referred to the following paper which was published subsequent to the delivery of the present manuscript to the editor.

Quattrini, D. (1967). Osservazioni preliminari sulla ultrastruttura della statocisti dei molluschi gasteropodi polmonati. *Boll. Soc. ital. Biol. sper.* **43**: 785–786.

On the physiology and responses of pedicellariae reference should also be made to:

Jensen, M. (1966). The response of two sea urchins to the sea star *Marthasterias glacialis* (L.) and other stimuli. *Ophelia* **3**: 209–220.

Campbell, A. C. and Laverack, M. S. (In press). The responses of pedicellariae from *Echinus esculentus*. *J. exp. mar. biol. ecol.* **2**.

AUTHOR INDEX

Numbers in italics refer to pages in the reference lists at the end of each article.

I

J

K

L

M

SUBJECT INDEX

C

D TE